This book was written chiefly to help those physicists, physical chemists, metallurgists, and engineers who need to carry out investigations at low temperatures. It deals with the production and measurement of low temperatures, the handling of liquefied gases on the laboratory scale, the principles and some details of the design of experimental cryostats, including the problems of heat transfer and temperature control. Physical data on heat capacities, expansion coefficients, and the electrical and thermal conductivities of materials used in making low-temperature equipment are given. While there are many references to technical details, enough of fundamental principles is included to make the book useful to the advanced university student or graduate student.

Additional material in the second edition includes the use of helium-3 for cooling and superconducting magnets.

MONOGRAPHS ON THE PHYSICS AND CHEMISTRY OF MATERIALS

General Editors

WILLIS JACKSON H. FRÖHLICH N. F. MOTT

MONOGRAPHS ON THE
PHYSICS AND CHEMISTRY OF MATERIALS

General Editors

WILLIS JACKSON H. FRÖHLICH N. F. MOTT

MULTIPLE-BEAM INTERFEROMETRY OF SURFACES AND FILMS. By S. TOLANSKY

THEORY OF DIELECTRICS, DIELECTRIC CONSTANT AND DIELECTRIC LOSS. By H. FRÖHLICH

PHYSICS OF RUBBER ELASTICITY. By L. R. G. TRELOAR

PHYSICAL PROPERTIES OF GLASS. By J. E. STANWORTH

NEUTRON DIFFRACTION. By G. E. BACON. *Second edition*

THE INTERFERENCE SYSTEMS OF CROSSED DIFFRACTION GRATINGS. By J. GUILD

DIELECTRIC BREAKDOWN OF SOLIDS. By S. WHITEHEAD

THE PHYSICS OF ELECTRICAL CONTACTS. By F. LLEWELLYN JONES

THE PHOTOGRAPHIC STUDY OF RAPID EVENTS. By W. D. CHESTERMAN

DETONATION IN CONDENSED EXPLOSIVES. By J. TAYLOR

GRAIN BOUNDARIES IN METALS. By D. MCLEAN

THE DETECTION AND MEASUREMENT OF INFRARED RADIATION. By R. A. SMITH, F. E. JONES, and R. P. CHASMAR

MOLECULAR DISTILLATION. By G. BURROWS

DIRECT METHODS IN CRYSTALLOGRAPHY. By M. M. WOOLFSON

OXIDE MAGNETIC MATERIALS. By K. J. STANLEY

PHYSICAL PRINCIPLES AND APPLICATIONS OF JUNCTION TRANSISTORS. By J. H. SIMPSON and R. S. RICHARDS

THE THEORY OF DIELECTRIC BREAKDOWN OF SOLIDS. By J. J. O'DWYER

ULTRASONIC ABSORPTION. By A. B. BHATIA

EXPERIMENTAL TECHNIQUES IN LOW-TEMPERATURE PHYSICS

BY
GUY KENDALL WHITE

SECOND EDITION

OXFORD
AT THE CLARENDON PRESS
1968

Oxford University Press, Ely House, London W.1
GLASGOW NEW YORK TORONTO MELBOURNE WELLINGTON
CAPE TOWN SALISBURY IBADAN NAIROBI LUSAKA ADDIS ABABA
BOMBAY CALCUTTA MADRAS KARACHI LAHORE DACCA
KUALA LUMPUR HONG KONG TOKYO

FIRST EDITION 1959
SECOND EDITION 1968

PRINTED IN GREAT BRITAIN BY
JOHN WRIGHT AND SONS LTD., AT THE STONEBRIDGE PRESS, BRISTOL

PREFACE TO THE SECOND EDITION

SINCE the publication of the first edition of this monograph, there has been a continued growth of the study of low temperatures, largely because of the steady increase in the number of people and laboratories concerned with the properties of matter. This is partly a reflection of the need for cryogenic information among technically based industries concerned with communications, aerospace, gases, etc.

A number of new books have been written to meet the requirements of the times and it is pertinent to mention briefly their coverage *vis-à-vis* that of the present book. Among smaller texts which are of general interest to university students, both graduate and undergraduate, are the latest edition of *Low temperature physics* by L. C. Jackson (Methuen, London, 1962), *Cryophysics* by K. Mendelssohn (Interscience, New York, 1960), *The quest for absolute zero* by the same author (Wiedenfeld and Nicolson, London, 1966), and *Cryogenics* by M. McClintock (Reinhold, New York, 1964). These all give interesting accounts of the scope and applications of physics at low temperatures without any particular attention to the techniques or methods used.

The properties of solids (therefore not including liquid helium) at low temperatures have been surveyed by H. M. Rosenberg in *Low temperature solid state physics* (Clarendon Press, Oxford, 1963): this contains no treatment of either laboratory techniques or the principles of thermometry and refrigeration.

A larger compendium called *Cryogenic technology* (Wiley, New York, 1963) produced under the editorship of R. W. Vance includes chapters by some experienced contributors on the properties of solids, the principles of refrigeration, superconductivity, thermometry, as well as applications of cryogenics in biology, space research, cryopumping, etc.

Among the books devoted primarily to the techniques, methods, and properties of technical materials used at low temperatures is *Cryogenic engineering* by R. B. Scott (Van

Nostrand, Princeton, 1959); as its name implies, this emphasizes large-scale engineering techniques and information rather than those used on the laboratory scale. On the other hand, *Low temperature techniques* by A. C. Rose Innes (Van Nostrand, Princeton, 1964) is a useful practical guide for the laboratory worker who needs to handle refrigerant liquids but is not concerned with making them or with cooling by demagnetization methods. A more comprehensive treatment is in *Experimental cryophysics* edited by F. E. Hoare, L. C. Jackson, and N. Kurti (Butterworths, London, 1961): various experienced physicists have contributed the different chapters which deal with all the aspects of producing and handling low temperatures.

The present monograph remains a compromise in size between the last two. The revision is intended to add new material, chiefly concerning the use of helium-3 as a refrigerant and superconducting magnets, and to bring chapters on liquefaction, temperature measurement, and the properties of materials up to date. Minor changes have been made to other chapters to include useful new information or to rectify omissions in the earlier edition. In the process some of the tables which are referred to most frequently have been gathered together in the Appendix but tables of limited or specific interest remain through the text. A short list of firms which presently supply low-temperature equipment is included in the Appendix.

Some preparatory work for this revised edition was done while I was enjoying the hospitality of the Bell Telephone Laboratories at Murray Hill and I am very grateful to a colleague there, Dr. Eric Fawcett, for his help. Here at the C.S.I.R.O. National Standards Laboratory I have discussed many cryogenic problems with my colleagues and I am most grateful to them, particularly W. R. G. Kemp, for their patience; Mrs. P. Riley has aided greatly with her secretarial help.

G. K. W.

C.S.I.R.O. Division of Physics
Sydney, Australia
January 1967

PREFACE TO THE FIRST EDITION

IT is not very many years since most low-temperature physicists—those doing research on physical properties at the temperatures of liquid helium or liquid hydrogen—were trained in the techniques of this particular field at one of the comparatively few centres of low-temperature research such as Leiden, Berlin, Berkeley, Oxford, Cambridge, Toronto. Today, with the advent of increased research grants, defence contracts, and Collins helium liquefiers, many physicists in many laboratories around the world wish to carry out physical investigations in the low-temperature range and are faced with problems of designing cryostats, filling them with liquid helium, maintaining and measuring various temperatures. Often these problems are not very difficult, but nevertheless the technical information and published experience which may help to solve them are spread over a wide range of years and journals. There appears to be a need for a book which gives details of this information, including physical data for the technical materials used in cryostat design, methods of measuring and controlling temperatures, and associated problems. This book is an attempt to meet this need.

In a first flush of enthusiasm I hoped to include the full technical details of many operations, for example the winding of a countercurrent heat exchanger, but soon realized that lack of space made this impossible. Some chapters, notably those dealing with gas liquefaction and magnetic cooling, are merely brief discussions of the principles involved with examples and references to more detailed work. They are intended not only to give continuity to the subject-matter and to introduce those not familiar with the subject to the principles and the literature but also to act as a guide for anyone wishing to design a liquefier or adiabatic demagnetization cryostat; these are both subjects to which complete books could be devoted and recent reviews by Daunt, Collins, Ambler and Hudson, and de Klerk, and earlier books by Ruhemann, Keesom, Casimir, and Garrett cover these subjects far more competently than I could hope to do. Any reader wishing for a more complete survey of low-temperature physics, its scope and its achievements rather than its techniques, should prefer the texts of E. F. Burton, H. Grayson Smith, and J. O. Wilhelm (*Phenomena at the temperature of liquid helium*, Reinhold, 1940), L. C. Jackson (*Low temperature physics*, Methuen, 2nd edn, 1948), or C. F. Squire (*Low temperature physics*, McGraw-Hill, 1953).

I would like to acknowledge my debt of gratitude to the late Sir Francis Simon for his kindness and patience when I started the study of low-temperature physics; he asked me, as one of his students, to design and build a Linde helium liquefier. I hope this book is rather more successful than my first effort at solving the problems of helium liquefaction.

In preparing this book I have been helped considerably by various friends who have read and criticized individual chapters: these include my colleagues Drs. T. H. K. Barron, J. S. Dugdale, D. K. C. MacDonald, F. D. Manchester, and S. B. Woods of the Division of Pure Physics (National Research Council), Dr. J. A. Morrison of the Division of Pure Chemistry (National Research Council), Dr. H. Preston-Thomas of the Division of Applied Physics (National Research Council), and Dr. R. P. Hudson of the National Bureau of Standards in Washington.

As will appear throughout the text many publishers and learned societies have kindly granted permission for the reproduction of figures which originally appeared in their books and periodicals. The facilities and the co-operation of such sections of the National Research Council as the Central Drafting Office, Duplication, and the Typing Pool have been of great assistance.

Finally, I am happy to thank Dr. M. T. Elford for his help in the proof-reading of this book.

G. K. W.

Ottawa, Canada
July 1957

CONTENTS

PART I. GENERAL

PART II. THE RESEARCH CRYOSTAT

PART I

GENERAL

CHAPTER I

PRODUCTION OF LOW TEMPERATURES

1. Isentropic cooling

Introduction

SINCE the entropy or degree of disorder of a system at constant volume or constant pressure is a monotonically increasing function of temperature, any process of cooling may be regarded as one of ordering or entropy reduction. In the words of Simon, a refrigerator is a form of 'entropy-squeezer'. This 'squeezing' is possible since entropy S is a function of other variable parameters as well as temperature, e.g. $S = S(T, X)$ where the parameter X is a physical property of the system which can be varied within limits so as to change the entropy.

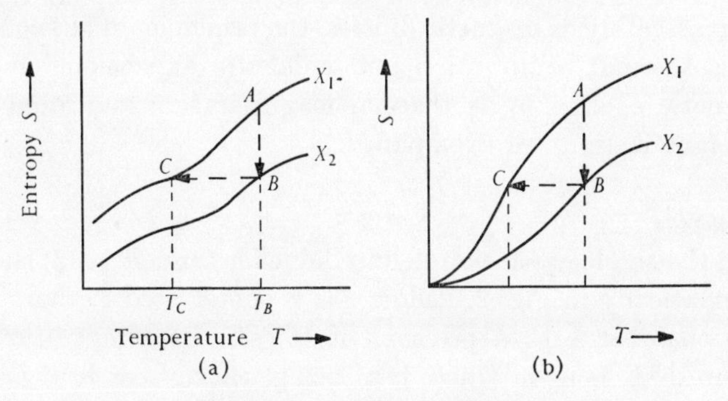

FIG. 1. A cooling process.

Fig. 1 (a) shows that when X is altered isothermally from X_1 to X_2 the entropy is reduced. By further varying X from X_2 to X_1 under isentropic conditions, a lowering in temperature from a temperature T_B to temperature T_C is achieved; an isentropic change is adiabatic since $\Delta S = \int dQ/T$. The process is a reversible one and therefore by the second law of thermodynamics is the most efficient means of cooling, in terms of the external work required. From Fig. 1 (a) it would seem possible, in principle at least, to cool the system to the absolute zero of temperature by a limited number of such steps as $A \rightarrow B \rightarrow C$. However, if the situation is as depicted in Fig. 1 (b), this is no longer the case; that 1 (b) is correct, rather than 1 (a), is implied by the third law of thermodynamics, which in the form due to Simon (1930, 1956) states: 'at absolute zero the entropy differences disappear between all those states of a system which are in internal thermodynamic equilibrium'.

The equivalence of this statement to the alternative statement that 'it is impossible by any procedure, no matter how idealized to reduce the temperature of any system to the absolute zero in a finite number of operations', has been demonstrated by Guggenheim and is strongly suggested by Fig. 1 (b). However, we still have a means of lowering the temperature—even if not to absolute zero—at our disposal and in practice such methods have been widely used. Associating the parameter X with the pressure p applied to a gas or with the magnetic field H applied to an assembly of magnetic dipoles, the principles of gas cooling by isothermal compression and adiabatic expansion, and of magnetic cooling by isothermal magnetization and adiabatic demagnetization are exemplified.

Examples

In the single-expansion helium liquefier (Simon 1932) shown schematically in Fig. 2, helium gas is compressed isothermally into chamber 1 to a pressure of 100 atm; a temperature of about 15° K is maintained, heat being transferred through the medium of helium exchange gas in the space 2 to a bath of liquid hydrogen boiling under reduced pressure in the dewar vessel 3. Thus the initial temperature $T_B \simeq 15°$ K; in practice T_B may be lowered to 10° K by reducing the pressure above the evaporating hydrogen to well below its triple-point pressure. Then the exchange gas is removed from 2 and the compressed helium in chamber 1 is expanded through the valve V to a pressure of 1 atm, so that the gas remaining in the chamber is adiabatically cooled to the final temperature T_C (equal to the liquefaction temperature); a substantial fraction of the chamber is left filled with liquid helium. The importance of the metal chamber 1 having a heat capacity small in relation to the gas is paramount, and is easily realized with helium gas at about 10° K, but would not be the case for compressed air at 200° K, for the heat capacity of the containing pressure vessel at this higher temperature would be considerable.

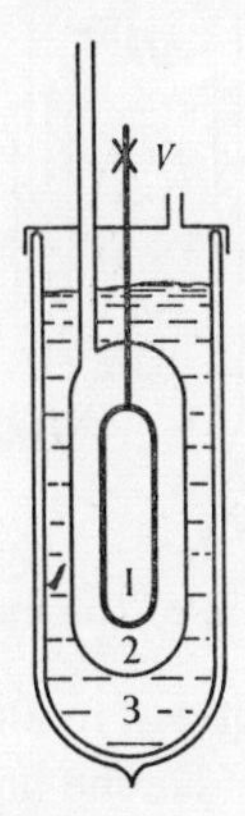

FIG. 2. Schematic diagram of the Simon expansion liquefier.

The desorption helium liquefier first discovered by Simon in 1926 and developed further by Mendelssohn (1931) is very similar. In this the inner container is partially filled with activated charcoal and the process $A \to B$ of Fig. 1 is simply the isothermal adsorption of helium gas to a pressure of 1–5 atm on the charcoal, the heat of adsorption being removed via exchange gas to the liquid-hydrogen bath. This is followed by an adiabatic desorption or pumping away of the adsorbed helium gas during which the temperature of the remaining gas, charcoal, and the container fall toward or to the liquefaction temperature.

The suggestion that a process of ordering by a magnetic field could be applied to an assembly of weakly interacting magnetic dipoles, as in a paramagnetic salt, and that a subsequent adiabatic demagnetization would cause cooling, was made independently in 1926 by Debye and Giauque. Within a few years Giauque and MacDougall (1933), de Haas, Wiersma, and Kramers (1933), and Kurti and Simon (1935) verified this experimentally; Fig. 3 illustrates the process schematically. A pill of a paramagnetic salt (gadolinium sulphate in Giauque's

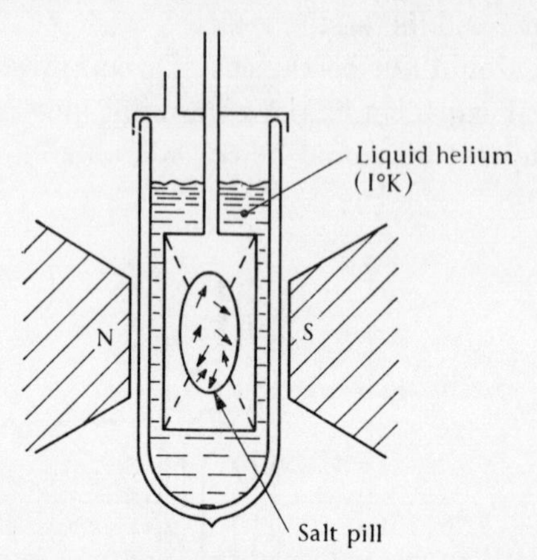

FIG. 3. Schematic diagram of the magnetic cooling process.

early experiments) is magnetized in a field of a few kilo-oersteds at a temperature of about 1° K, the heat of magnetization being transferred through helium exchange gas to the pumped liquid helium in the surrounding dewar vessel. After removal of the exchange gas, the magnetic field is reduced and the temperature of the salt pill falls. There is a further analogy with the expansion liquefier in that for appreciable cooling to occur, the lattice vibrational specific heat at 1° K of the salt pill must be small in comparison with the 'magnetic heat capacity', i.e. with the thermal energy of the disoriented magnetic ions; if this were not so, the reduction in the entropy on magnetization would be small in comparison with the total entropy of the crystal lattice.

2. Isenthalpic cooling

Introduction

The discovery, nearly a century ago, by Joule and Thomson that a gas undergoes a temperature change when it expands slowly through a porous plug, has been widely applied to gas refrigeration. The cooling on adiabatic expansion discussed in § 1 is a property of the perfect gas—in which attractive or repulsive forces are zero—and occurs for real gases at all temperatures by virtue of their performing 'external' work. However, the Joule–Thomson effect for any real gas depends both in magnitude and sign on the temperature and is zero for a perfect gas at all temperatures. This effect is sometimes called an 'internal work' process because the temperature change is determined by the change in energy of gas when the average separation between the gas molecules is increased.

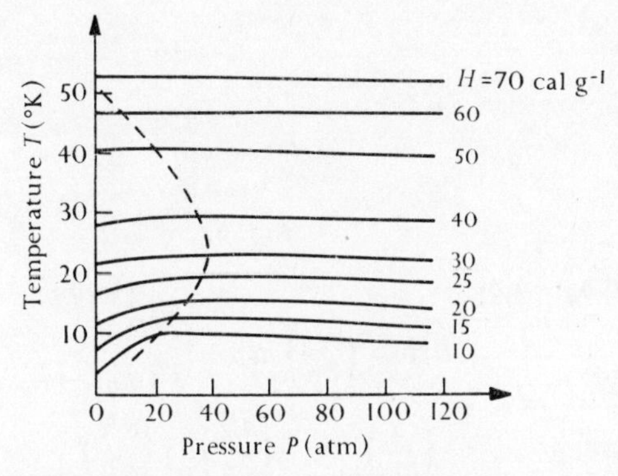

FIG. 4. Isenthalpic curves of helium.

In the Joule–Thomson process a gas undergoes a continuous throttling or expansion as it is driven by a constant pressure p_1 on one side of the expansion valve (or porous plug) and expands to a lower pressure p_2 on the other. Considering a fixed mass of gas passing the valve, it can easily be shown that the total heat or enthalpy $H = U+pV$ is unchanged in passing from state 1 (pressure p_1) to state 2 (pressure p_2); U is the internal energy per unit mass and V is the volume.

Since $dH = 0$ such a process is called isenthalpic.

Performing such a throttling experiment for helium we should obtain a set of values of T_2—the temperature after expansion—lying on a smooth curve. If T_1—the temperature before expansion—is below the so-called inversion temperature, such curves have a maximum as seen in Fig. 4. The locus of the maxima encloses a region within which the differential Joule–Thomson coefficient

$$\mu = \left(\frac{\partial T}{\partial p}\right)_H$$

is positive and hence a cooling results on expansion.

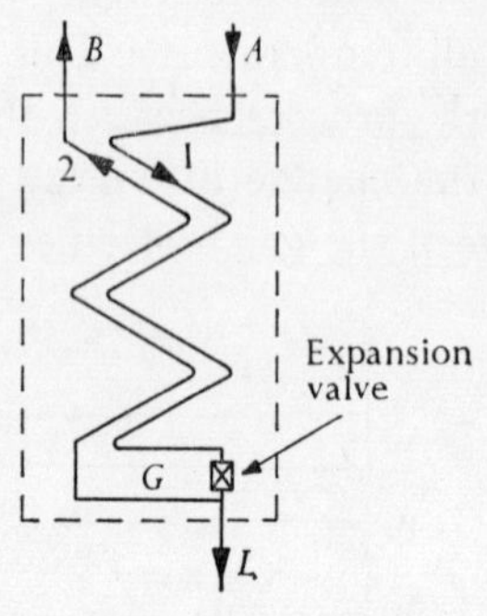

FIG. 5.

We may show:

$$dH = \left(\frac{\partial H}{\partial p}\right)_T dp + \left(\frac{\partial H}{\partial T}\right)_p dT,$$

therefore

$$\left(\frac{\partial T}{\partial p}\right)_H = -\left(\frac{\partial H}{\partial p}\right)_T \Big/ \left(\frac{\partial H}{\partial T}\right)_p$$

$$= -\frac{1}{C_p}\left(\frac{\partial H}{\partial p}\right)_T$$

or, since $dH = dU + p\,dV + V\,dp = T\,dS + V\,dp,$

$$\left(\frac{\partial T}{\partial p}\right)_H = -\frac{1}{C_p}\left\{T\left(\frac{\partial S}{\partial p}\right)_T + V\right\}$$

$$= \frac{1}{C_p}\left\{T\left(\frac{\partial V}{\partial T}\right)_p - V\right\}$$

and this vanishes for the perfect gas since $pV = RT$.

This isenthalpic process is important as it forms the final stage for nearly all 'circulation' liquefiers ('circulation' liquefiers exclude the Simon expansion and desorption liquefiers). Such a final stage, including expansion valve and heat interchanger, is shown schematically in Fig. 5, where compressed gas at pressure p_A, temperature T_A, and enthalpy H_A per gram enters at A. After expansion the gas (state p_B, T_B, H_B) leaves through the interchanger at B and liquid L (state p_L, T_L, H_L) collects at L.

If the liquefaction efficiency is denoted by $\mathscr{E}$, then since the process is isenthalpic

$$H_A = \mathscr{E}H_L + (1-\mathscr{E})H_B,$$

therefore the efficiency $\mathscr{E} = (H_B - H_A)/(H_B - H_L)$.

Heat interchanger

So far the heat interchanger has not been mentioned, but it plays a vital role in determining the liquefaction efficiency as it determines T_B and therefore the enthalpy H_B. The exchanger efficiency, η, is usually defined as the ratio of the actual heat transferred from stream 1 to stream 2, to the total heat available for transfer.

As $\eta \to 1$, $T_B \to T_A$.

If $\eta = 0$, i.e. if there is no interchanger, then $T_B = T_G$ (where T_G is the temperature of the gas immediately after expansion), and so $H_B = H_G$.

For liquefaction to occur we require $T_G = T_L$, therefore $H_G = H_L + \lambda$, where λ is the latent heat, and so, for $\mathscr{E} \geqslant 0$, we should require $H_A \leqslant H_L + \lambda$ if $H_B = H_G$ $(\eta = 0)$. For example, in the case of helium, $H_L + \lambda \simeq 7$ cal/g. The necessity that H_A be less than 7 cal/g would require that T_A should be less than about 7·5° K (from the enthalpy diagram for helium.) Since this is clearly too low a temperature to reach by the use of any other liquid refrigerants, and difficult but not impossible by the use of a helium expansion engine, it is desirable to have an efficient interchanger.

In the case where the interchanger is 100 per cent efficient, i.e. $\eta = 1$, we may write $H_B = H_B^0$ and $\mathscr{E} = \mathscr{E}_0$.

Then $$\mathscr{E}_0 = \frac{H_B^0 - H_A}{H_B^0 - H_L} \quad \text{for } \eta = 1.$$

It is easily shown in a practical case where $\eta < 1$ that

$$\mathscr{E} = \mathscr{E}_0 - (1-\eta)(1-\mathscr{E}_0)\frac{H_B^0 - \lambda - H_L}{H_B^0 - H_L}.$$

Since generally $0{\cdot}8 < \eta < 1$, this can be reduced to a simpler form, $\mathscr{E} = \mathscr{E}_0 - \alpha(1-\eta)$, where the factor $\alpha < 1$. Under the normal range of conditions met with in the final stage of a helium liquefier, α has a value in the vicinity of 0·6 so that it is necessary in designing an interchanger that its inefficiency $(1-\eta)$ shall not be comparable with or greater than the maximum liquefaction efficiency $\mathscr{E}_0$ of the last stage, often called the Linde stage of the liquefier.

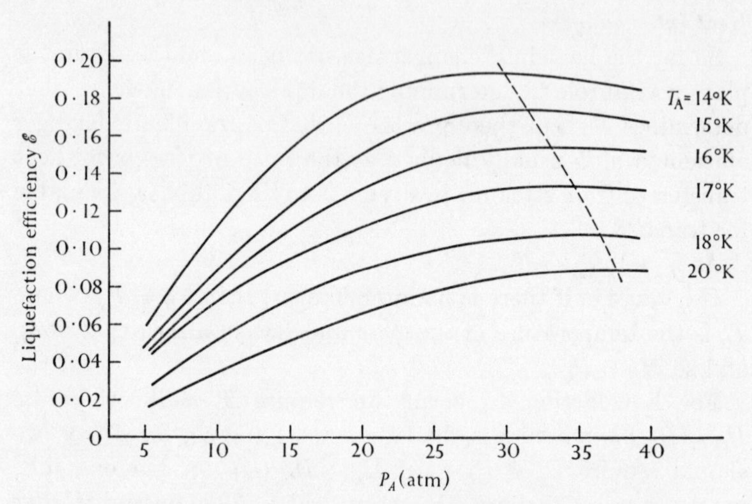

FIG. 6. Theoretical liquefaction efficiency of the Linde stage of a helium liquefier.

Fig. 6 shows the liquefaction efficiency $\mathscr{E}_0$ for helium as a function of the pressure p_A and temperature T_A, assuming $\eta \simeq 1$; the curves have been calculated from enthalpy data for helium given by Keesom (1942) and Van Lammeren (1941).

Since $\mathscr{E}$ is a maximum when H_A is a minimum, the conditions for maximum efficiency of this liquefaction stage include that

$$\left(\frac{\partial H_A}{\partial p_A}\right)_{T=T_A} = 0.$$

It follows from $$C_p\mu = \left(\frac{\partial H}{\partial T}\right)_p \left(\frac{\partial T}{\partial p}\right)_H$$

that $\mu = 0$ satisfies this condition.

Thus the optimum entry pressure is a value of p_A lying on the inversion curve (see Figs. 4 and 6).

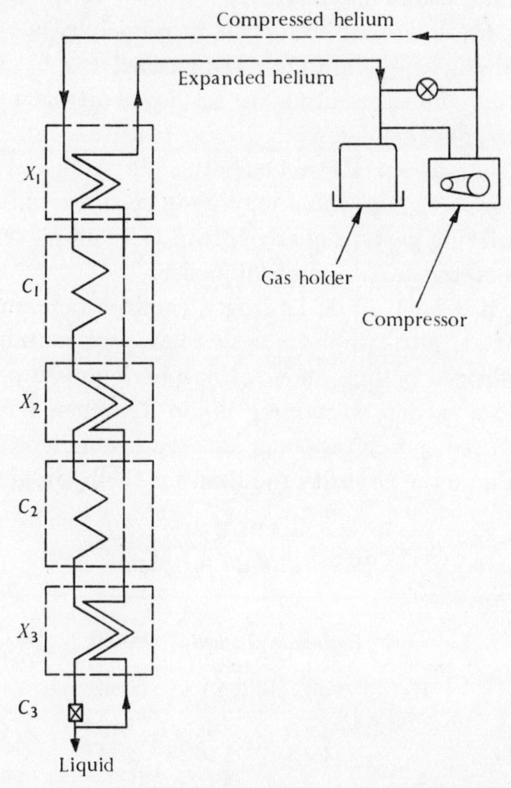

FIG. 7. A generalized flow circuit for a helium liquefier.

Principles of a Joule–Thomson liquefier

Consider the schematic diagram (Fig. 7) of a helium liquefier which relies on Joule–Thomson cooling for its final or liquefaction stage. The hydrogen liquefier is similar in principle but is more simple as one cooling stage can be omitted.

C_1, C_2, C_3 are cooling stages, the final stage C_3 being the Joule–Thomson valve; X_1, X_2, X_3 are heat interchangers. The chief points that emerge are that:

(i) The cooling stage C_2, which may be liquid hydrogen or an expansion engine device, should be capable of producing an exit temperature for the gas stream of well below the inversion temperature and preferably $\leqslant 15°$ K for high liquefaction efficiency.

(ii) X_1, C_1, X_2 are all inessential in priniciple, but their form and efficiency determine the character of C_2; in practice C_1 may be either liquid air (or liquid nitrogen or oxygen) or an expansion engine.

(iii) C_1 should generally act as both a purifying and precooling stage, although these may be in separate units with the purifying part—a charcoal trap at liquid-air temperature —external to the main liquefier.

(iv) C_2, if it be a liquid-hydrogen cooling stage, may also be divided into two parts with a first part containing liquid hydrogen boiling under atmospheric pressure at 20·4° K and a second containing liquid hydrogen boiling under reduced pressure; such an arrangement considerably reduces the capacity required for the hydrogen pump.

TABLE I

Physical data for gases

Gas	*Inversion temp.* (° K)	*Boiling point* (° K)	*Critical temp.* (° K)	*Critical pressure* (atm)	*Triple point* (° K)	*Triple-point pressure* (mm Hg)
Helium (^{3}He)	—	3·2	3·34	1·15	—	—
Helium (^{4}He)	51	4·2	5·19	2·26	—	—
Hydrogen	205	20·4	33·2	13·0	14·0	54
Neon	—	27·2	44·4	25·9	24·6	324
Nitrogen	621	77·3	126	33·5	63·1	94
Argon	723	87·4	151	48	83·9	512
Oxygen	893	90·2	154	50	54·4	1·2
Krypton	—	121·3	210	54	104	—

A more detailed discussion of these considerations and the general methods of constructing such a liquefier will be given below in § 5, with the aid of the examples offered by some of the many liquefiers described in recent years.

In Table I are the Joule–Thomson inversion temperatures for some common gases and other data governing their suitability as refrigerant liquids.

3. Air liquefiers

Linde and Hampson

To the low-temperature physicist who either buys liquid nitrogen or oxygen commercially for 15, 30, or perhaps 50 pence per litre, or who draws freely on the output of a large central laboratory liquefier, a description of the principles involved in air liquefiers may seem out of place in this text. However, these principles serve to illustrate and introduce the various types of helium and hydrogen liquefiers, and to explain historically the terms such as Linde, Hampson, Claude, which are frequently applied to particular designs of liquefiers. Because of the higher temperatures involved they introduce the methods of liquefaction without the complexity of a pre-cooling stage.

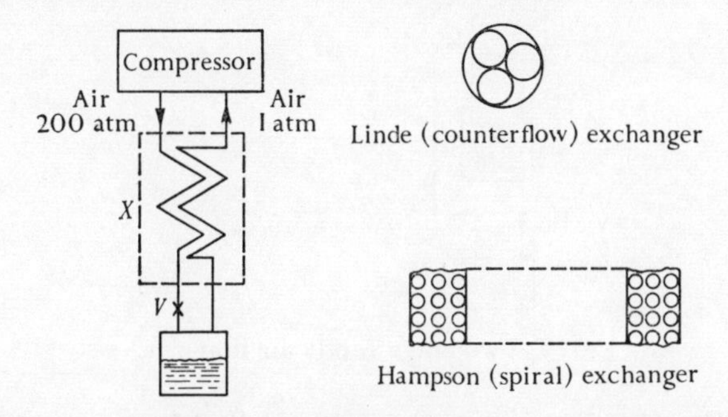

Fig. 8. Simple Linde air liquefier and section views of Linde and Hampson heat exchangers.

In the simple Linde air liquefier of Fig. 8, a compressor, heat exchanger X, and expansion throttle V are the important features. In 1895, the same year as Linde first used this simple pattern for air liquefaction, Hampson developed his air liquefier, which used a similar flow circuit but a different design of heat exchanger. Linde employed a countercurrent heat exchanger (see Fig. 8 and Chapter III) in which the two gas streams passed through concentric tubes. In the Hampson exchanger the high-pressure gas flows through a tube wound in a spiral pattern with the spiral enclosed in the annular space between two cylinders, and the returning low-pressure gas passes up this annular space past the spiral of tubing. Linde later improved the efficiency of his air liquefier by letting the gas expand in two stages (Fig. 9), first through a valve V_1 from 200 to 40 atm after which about 80 per cent of the gas returns through exchanger X to be recompressed; the remaining fraction is expanded through V_2 to 1 atm. The resultant saving in work of compression improved the practical efficiency by about 100 per cent.

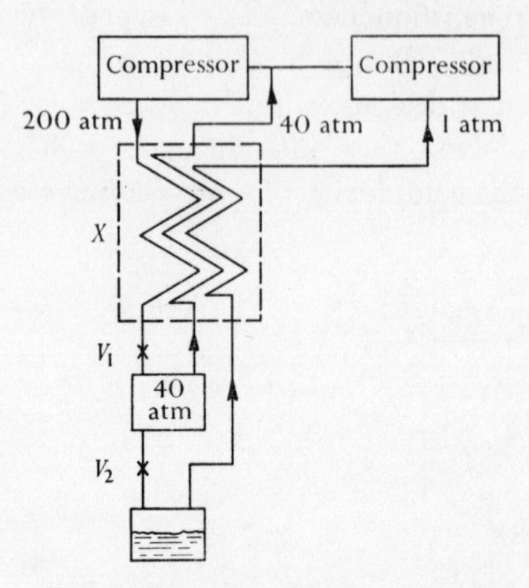

Fig. 9. Two-stage Linde air liquefier.

A further improvement occurred when Linde introduced a liquid ammonia pre-cooling stage so that the compressed gas entered X at about $-48°$ C.

Claude

As the name Linde is invariably associated with the Joule–Thomson process of cooling, so Claude is associated with expansion-engine liquefiers, although in the later form used by Claude and also developed by Heylandt early in this century both an expansion engine and a Joule–Thomson valve were incorporated (Fig. 10). Claude's first machines used simply isentropic expansion with liquid air being produced in the engine. However, the advantages of dividing the high-pressure stream and using Joule–Thomson expansion for the final liquefaction stage were considerable and this form has been used successfully over the years. Claude used a pressure of about 40 atm and allowed 80 per cent of the gas to pass through the engine; Heylandt used 200 atm, passed about 60 per cent through the engine, and dispensed with exchanger X_1. Thus in the Heylandt case gas enters the engine at room temperature which gives a somewhat higher operating efficiency.

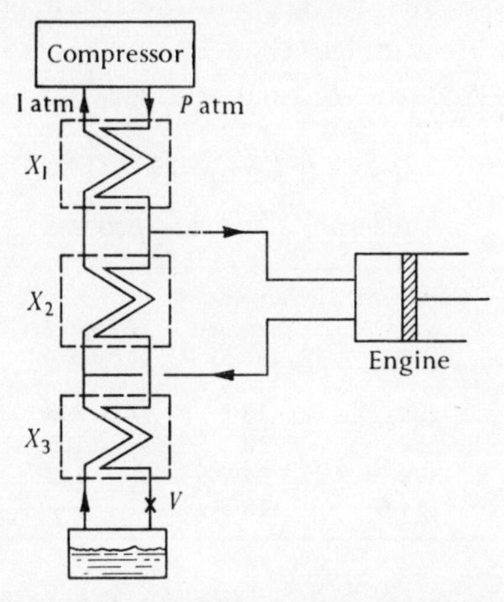

FIG. 10. Claude–Heylandt system of air liquefaction.

The overall operating efficiencies of the Linde and Claude systems are not very different and both are still widely used together with the Philips–Stirling cycle and turbine-expanders.

Cascade liquefiers

Pictet's first partially successful attempt to liquefy oxygen in 1877 used the cascade process, although the 'mist' of liquid which he produced was probably a result of adiabatic expansion plus cascade cooling rather than purely the cascade process itself. In this method a suitable gas A is liquefied at room temperature under pressure and then expanded to evaporate under atmospheric or under reduced pressure at a lower temperature; provided that this lower temperature is below the critical temperature of a second gas B, then this gas B may be liquefied likewise by isothermal compression; thus in stages the critical temperature of air may be reached. In Pictet's case gas *A* was sulphur dioxide and gas *B* was carbon dioxide. It is interesting to note that Keesom (1933) analysed the efficiency of a four-stage cascade process for air liquefaction using NH_3 (as gas *A*), C_2H_4 (*B*), CH_4 (*C*), and using N_2 (*D*) as the final stage; he found the efficiency to be higher than in the conventional Claude or Linde processes.

Keesom chose his refrigerants so that no liquid had to evaporate at a pressure less than 1 atm (Table II) in order to get appreciably below the critical temperature of the next gas.

TABLE II

Keesom's cascade process

	Condenser temp. (° K)	*Condenser pressure* (atm)	*Evaporation temp.*	*Evaporation pressure*
$A = NH_3$	298	10·2	240	1
$B = C_2H_4$	242	19	169	1
$C = CH_4$	172	24·7	112	1
$D = N_2$	114·6	18·6	—	—

However, the use of such a process for helium or hydrogen liquefaction seems unlikely in view of the critical temperatures shown in Table I and the limited choice of refrigerants available for cooling in the range below 50° K.

Philips–Stirling cycle

Kohler and Jonkers (1954) at the Philips Company developed a small air liquefier, based on the principles of the Stirling hot-air engine, which is suitable for laboratory use. The closed cycle shown in Fig. 11 (Kohler 1960) consists of compression of gas in a compression space followed by a transfer of the gas through a regenerative heat exchanger to an expansion space where the gas is expanded and cooled before it is returned through the exchanger. This action is achieved by the out-of-phase motion of the two pistons B and C moving in the cylinder A. C may be called the displacer. The four distinct phases of the cycle are:

(I) Compression in space D by piston B.

(II) Transfer of gas through the regenerator G to space E by the movement of the displacer.

(III) Expansion in the cold space E by the movements of C and B.

(IV) Transfer of gas back through G to space D.

The heat of the compression is removed by the water-cooling coil H and the expansion is used to cool the cylinder head or the coil J. With helium or hydrogen as the working substance in the closed circuit, air is condensed from the atmosphere onto the cooler cylinder head.

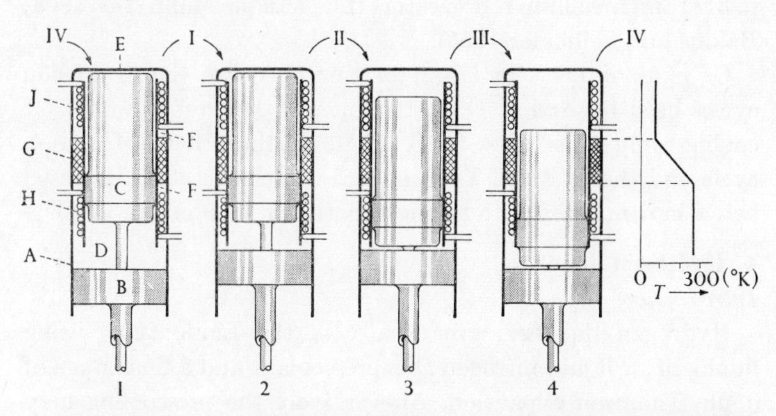

Fig. 11. The successive phases of the Philips–Stirling cycle, e.g. Phase I represents the compression of gas in space D by the piston B with heat of compression being removed in the cooler H (after Kohler 1960).

The smaller unit produced by Philips liquefies about 6 litres of air per hour with a starting-up time of about 10 min. The larger four-cylinder model liquefies about 30 l/h and includes a refrigeration-drier which allows over 100 h of continuous operation. If desired, fractioning columns may be attached to deliver liquid nitrogen.

A further development of this cycle has been to expand the gas in two stages (see Prast 1963, 1965) and has resulted in a refrigerator called the Philips Type A-20 'Cryogenerator' which uses helium as the working gas and can reach temperatures as low as 12° K or liquefy hydrogen at about 5 l/h.

Others

Various other forms of expansion cooling have been used in air liquefaction. Since the pioneering efforts of Kapitza and others, turbine-expanders have become widely used in air liquefiers, but only in plants of large capacity; these may yield 100 tons of liquid per day or more, not exactly of laboratory size. Their role in air liquefaction is discussed by Collins and Cannaday (1958), Scott (1959), Hoare (see Hoare, Jackson, and Kurti, 1961, p. 54). More recently turbines have been used also in large hydrogen liquefiers (Scott, Denton, and Nicholls 1964, p. 318) and in helium refrigerators (for example Mann *et al.* 1963; Baldus and Sellmaier 1965).

Of more interest in laboratory practice are the expansion cycles used by Arthur D. Little, Inc., in their refrigerators for cooling helium gas below 20° K. These are the Gifford–McMahon cycle and the modified Taconis cycle which are both discussed below in connexion with the liquefaction of helium (see § 1.5).

4. Hydrogen liquefiers

Introduction

Hydrogen liquefiers are usually of the Linde type, using liquid air or liquid nitrogen as a pre-coolant and a final stage of Joule–Thomson expansion. Alternatively the pre-cooling may be done by expansion devices such as piston engines, turbines, Philips–Stirling cycle, Gifford–McMahon cycle, etc. But most liquefiers, representing all sizes, have used liquid air or nitrogen for the pre-cooling (see for example Scott *et al.* 1964). Some important factors which affect the design of hydrogen liquefiers are the possible need for *ortho-para* conversion, safety considerations, and the relatively high pressures of 150 atm required for optimum Joule–Thomson cooling. This latter factor has led to the construction of a number of successful small liquefiers which are supplied by cylinders, obviating the need for high-pressure compressors.

Small hydrogen liquefiers

The design principles and the expected performance of cylinder-operated hydrogen liquefiers have been discussed in detail by Starr (1941) who gave the figures shown in Table III for the percentage liquefaction under various conditions of gas pressure and pre-cooling temperature. A schematic diagram of Starr's design is shown in Fig. 12.

TABLE III

Hydrogen liquefaction percentage (after Starr 1941)

Pre-cooling temp. (° K)	*Pressure* (atm)	*Efficiency of X* 100%	90%	80%
77	150	18	14	9
	100	16	11	6
	50	10	5	−1
	25	4	−2	—
63	150	29	26	22
	100	26	22	18
	50	19	14	10
	25	8	3	−3
55	150	40	38	35
	100	37	35	32
	50	26	24	20
	25	12	8	3

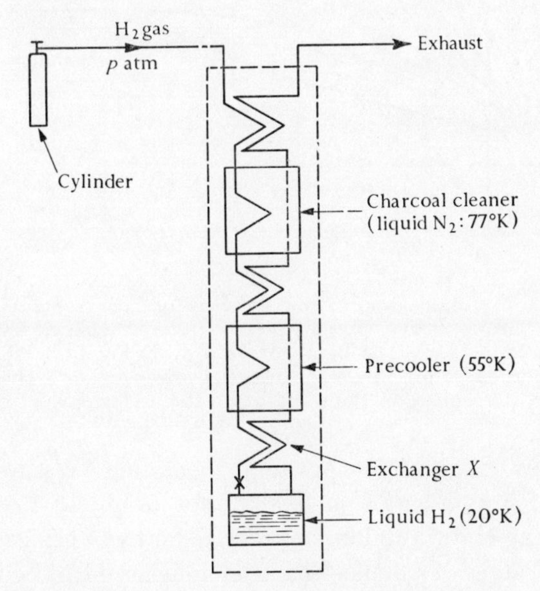

FIG. 12. Flow circuit of a small hydrogen liquefier.

The isenthalps and the inversion curve for normal hydrogen are shown in Fig. 13 from Woolley, Scott, and Brickwedde (1948). By comparison with data for helium (Figs. 4 and 6) it is

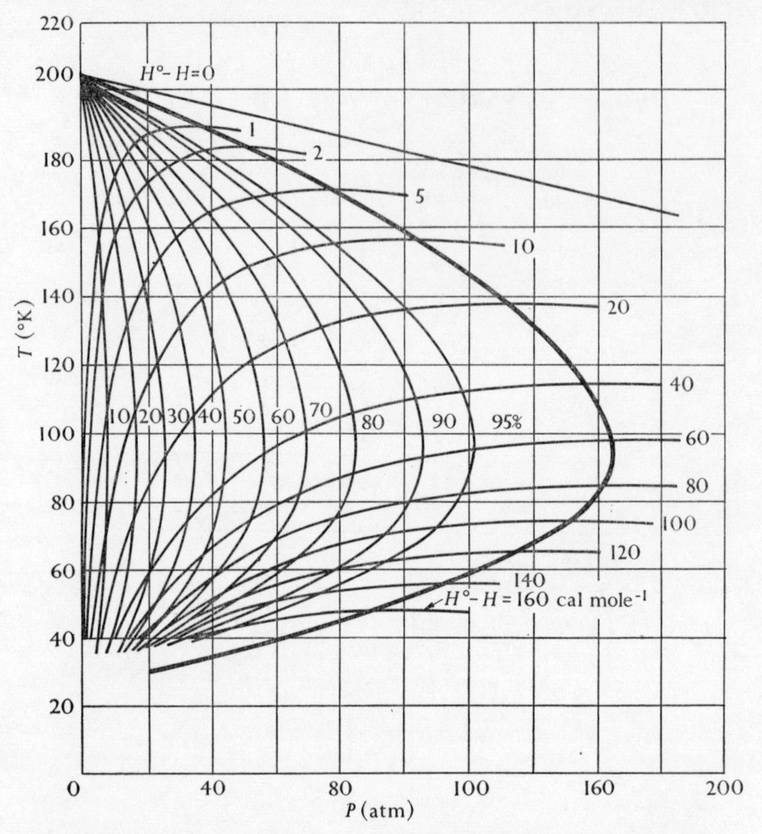

FIG. 13. Isenthalps (approximately horizontal curves) and inversion curve (heavy curve) for hydrogen (after Woolley, Scott, and Brickwedde 1948). The light curves drawn roughly parallel to the inversion curve are lines at which the refrigeration is the marked percentage of 100 per cent on the inversion curve.

seen that much higher optimum operating pressures are required for hydrogen liquefaction; for example, a convenient input temperature and input pressure for hydrogen gas entering the final stage of a Joule–Thomson liquefier might be 60° K and 130 atm.

Among the small hydrogen liquefiers which may be usefully studied by anyone who is concerned with design and construction are those described by Ahlberg, Estermann, and Lundberg (1937); Keyes, Gerry, and Hicks (1937); Fairbank (1946); and DeSorbo, Milton, and Andrews (1946).

With the advent of commercial helium liquefiers and general availability of liquid helium, small hydrogen liquefiers are becoming rather a rarity although medium-sized and large liquefiers are not uncommon. The reasons are many and various. Large liquefiers are needed to satisfy the requirements of bubble chambers and rocketry. Medium-sized liquefiers supply pre-cooling for helium liquefiers of the Linde pattern. Most low-temperature research can be performed quite adequately with liquid helium and liquid oxygen or nitrogen provided that suitable temperature-control methods are used to bridge the temperature region in between 4 and 54° K; there is, however, an occasional need for hydrogen in some laboratories to provide a thermometric fixed point.

Medium-sized hydrogen liquefiers

Linde liquefiers with capacities ranging from 7 to 50 l/h have been used in Berlin (Meissner 1928), Leiden (see, for example, Hoare, Jackson, and Kurti p. 81), Oxford (Jones, Larsen, and Simon 1948; Croft and Simon 1955), Berkeley, Zurich (Clusius 1953), Bellevue (Spoendlin 1954), the Royal Radar Establishment at Malvern (Hoare *et al.* p. 84), Moscow (Zeldovitch and Pilipenko 1960), and the Pennsylvania State University. Most of these have operated with a hydrogen input pressure of approximately 150 atm, pre-cooling temperature (that is an entry temperature to the final Linde exchanger) of 63–66° K; they have liquefaction efficiencies in the range from 22 to 30 per cent and consume 1–1½ litres of liquid air or liquid nitrogen per litre of liquid hydrogen.

The later Oxford liquefier (Croft 1964), shown schematically in Fig. 14, produces over 25 l/h. It operates at a pressure of only 60 atm because of economic considerations in obtaining a compressor. The hydrogen gas is produced electrolytically, is compressed, and passes successively through an oil separating column, an activated alumina drier, 'Deoxo' purifier, a further alumina drier, a refrigerator-drier, and then into the charcoal purifier at 77° K which is shown in the diagram. The gas flow is then split by the valves marked NV, 3 and 4, so that the major fraction goes through valve 3 to the exchanger Q1. Heat exchangers are of the Collins–Hampson pattern made from thin

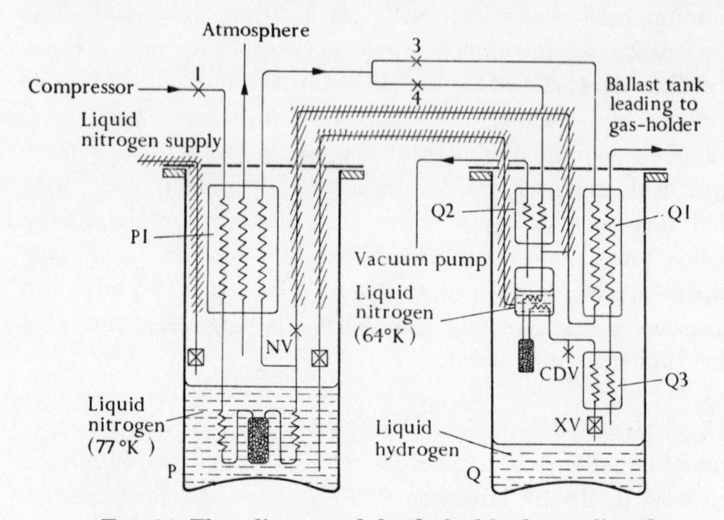

Fig. 14. Flow diagram of the Oxford hydrogen liquefier (after Croft 1964).

copper tubing (Integron tubing from Imperial Chemical Industries) wound in a helix between two cylindrical sheets. Thermal insulation is provided by metal dewars for these stages. No effort is made to promote *ortho-para* conversion since the major part of the liquid hydrogen is used soon after production.

By contrast the large liquefier at the National Bureau of Standards in Boulder, Colorado (Johnson and Wilson 1955; Scott 1959, p. 43), can produce over 350 l/h of normal liquid hydrogen or 240 l/h of *para*-hydrogen. The hydrogen gas passes successively through a 'Deoxo' purifier (palladium catalyst supported on silica gel), compressor, oil and water separator, refrigeration drier (cooled to below −100° C), and silica gel purifier (to adsorb N_2 and any remaining O_2); it then pre-cools the refrigeration drier, and passes on into a Joule–Thomson liquefaction stage.

Details of construction of these various liquefiers are given in the original papers and their merits are compared in reviews by Daunt (1956) and Croft (Scott, Denton, and Nicholls p. 356). Larger liquefiers have been constructed since 1955 in the U.S.A. to supply fuel for rockets. Most of these include expansion engines or turbines in the cooling cycle and are generally of greater efficiency and therefore of greater complexity than required for laboratory use. Van der Arend and Chelton discuss such large-scale facilities in the monograph by Scott *et al.* (1964, p. 79).

5. Helium liquefiers

Introduction

In many places liquid helium can now be bought if it is needed. Prices vary from $5 to $15 (or say, £2 to £6) per litre, depending on such factors as geographical location, return or non-return of gas, average consumption, local competition, etc. This may seem expensive and yet for an occasional user of small quantities it is cheaper than installing and maintaining a helium liquefier. The economic considerations are complicated as they depend on how a particular research laboratory receives its annual finances; whether it can find major sums for capital equipment or continuing finance for annual maintenance, also reliable technicians and available space and time in which to operate the liquefier. There is no universal answer to this problem and therefore it seems desirable that physicists using liquid helium should have at least a passing acquaintance with the various types of liquefiers, including the commercially available models and those which can be constructed most easily in a research laboratory.

Apart from the ability of a liquefier to supply a demand for a liquefied gas, two major features which determine its value are ease of construction and ease of maintenance. Except in the case of an expansion-engine liquefier where engines and crosshead impose their own constructional difficulties, the major constructional problem is usually the heat exchanger and the primary maintenance problem that of keeping the heat leakage at a sufficiently low value. If the heat exchanger allows of relatively easy construction and assembly, and the high vaccum surrounding the liquefier is maintained without much labour, the liquefier should be a successful design. In small-capacity liquefiers this first requirement can be met with Linde-pattern exchangers and the second can often be satisfied by exposing as few as possible of the working parts to the high vacuum space, as is done in the Collins machine.

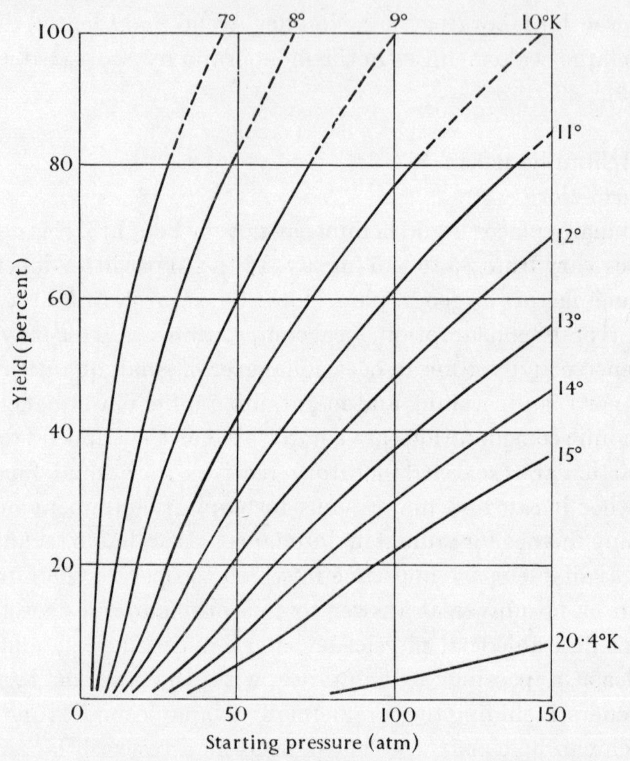

FIG. 15. Liquid-helium yield in the Simon expansion process (after Pickard and Simon 1948).

Simon expansion helium liquefiers

The principle of operation of single-expansion liquefiers was discussed in § 1 above, and Fig. 15 (Pickard and Simon 1948) illustrates the dependence of percentage yield (i.e. percentage filling of the expansion vessel with liquid helium) on the pressure and temperature from which the expansion begins. While requiring a high-pressure helium supply together with liquid hydrogen, such a liquefier is itself of comparatively simple

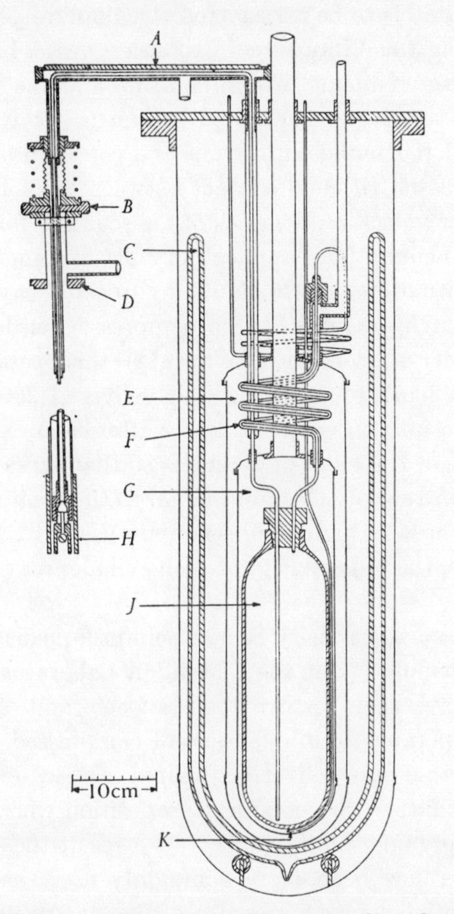

FIG. 16. Section of a Simon helium expansion liquefier (after Croft 1952). *E*, vacuum space; *F*, heat exchanger; *G*, liquid-hydrogen space; *H*, outlet valve; *J*, helium bottle; *K*, extraction tube for the liquid-air pre-coolant.

construction. Fig. 16 gives a sectional view of the liquefier constructed by Croft (1952) which yields 1·2 litres per expansion from 95 atm and 11° K and consumes about 5 litres of liquid hydrogen per expansion. As in the earlier liquefiers described by Cooke, Rollin, and Simon (1939), and Scott and Cook (1948), expansion takes place through a valve at the end of the transfer siphon and assists in pre-cooling the siphon and the dewar into which the liquid is to be transferred. The liquid-hydrogen space G surrounding the Vibrac steel expansion vessel has a volume such that after reducing its temperature and the temperature of the steel vessel and contained helium to 11° K, very little hydrogen is left. Radial copper fins and copper braiding on the steel bottle assist thermal contact between the solid hydrogen and the inner vessel, reducing the time required for cooling the compressed helium. The applicability of this single-expansion system to situations where liquid hydrogen is available, but there is no high-pressure helium compressor, is demonstrated by the liquefier of Scott and Cook (1948); this operated successfully from a bank of cylinders and delivered 300–400 cm^3 of liquid helium into an external dewar after each expansion. By judicious use of the bank of cylinders so that those cylinders at lowest pressure supply the first helium to the cooling expansion vessel, and those at highest pressure are used at the end of the cooling stage, the full capacity of each cylinder may be gainfully employed.

While the small-capacity Simon helium liquefier has had an undoubted popularity at the Clarendon Laboratory, comparatively few have been constructed elsewhere and of these most have been operated from cylinders of compressed gas. A compressor suitable for delivering helium gas at over 100 atm pressure is a rather uneconomical proposition when only called upon to supply one small liquefier. As a result the small Linde-type liquefier has been more commonly used, as it operates efficiently with an input pressure of 20–30 atm and therefore can be supplied by a suitably modified commercial air compressor or from cylinders fitted with a pressure-reducing valve.

Linde helium liquefiers

Typical of the simplest and the earliest of these small Joule–Thomson–Linde liquefiers is that of Ruhemann (1930) (Fig. 17).

It has a small heat exchanger—a narrow copper tube R_1 inside a slightly larger German silver tube R_2—wound into a spiral and enclosed by the vacuum jacket A. The expansion valve D in Ruhemann's liquefier is adjusted by a small brass screw before

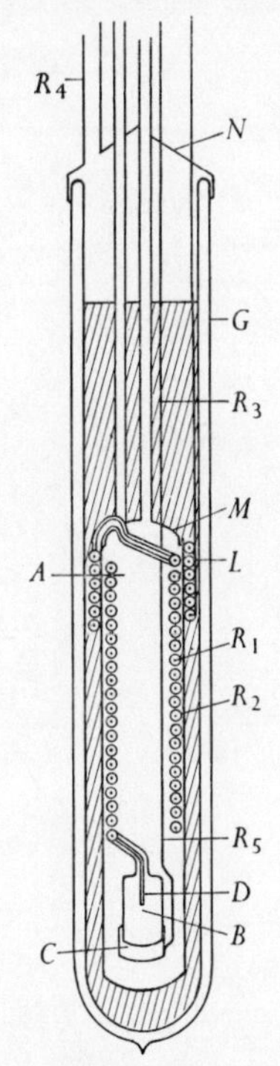

FIG. 17. Miniature Linde liquefier (after Ruhemann 1930).

attaching the vacuum jacket and the whole assembly is surrounded by a dewar of liquid hydrogen or liquid air. This miniature liquefier yields about 40 cm³/h of liquid helium or hydrogen and may be supplied from cylinders.

An example of the more usual type of liquefier, in which the expansion valve can be continuously controlled from outside and in which a liquid-air cooling stage is used, is that shown in Fig. 18 (after Daunt and Mendelssohn 1948). This liquefier may be easily constructed and assembled in three or four weeks by an experienced technician; it has a glass tail, attached by a copper–glass Housekeeper seal and Wood's metal joint to the brass vacuum jacket. Giving about 150 cm³ of liquid helium per hour with less than one hour starting time, this liquefier is very suitable for the visual observations on liquid helium for which it was intended. It has also served as a prototype for other small

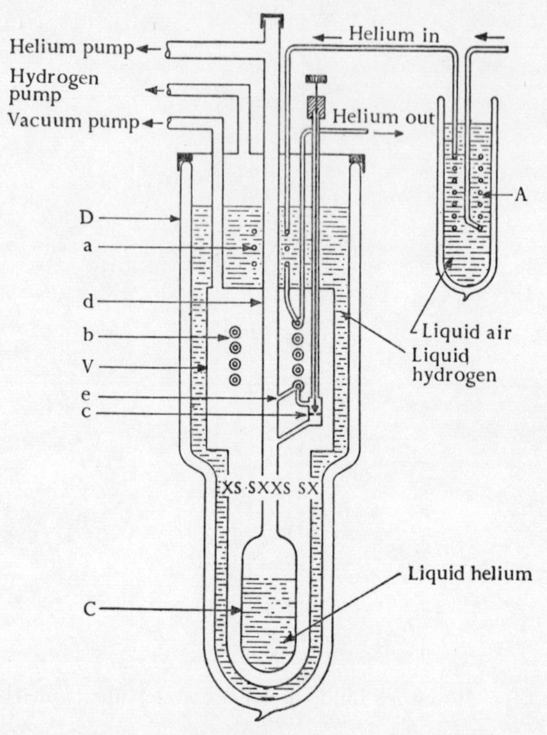

Fig. 18. Diagram of helium liquefier (after Daunt and Mendelssohn 1948).

and larger liquefiers, e.g. the 1 l/h liquefiers described by Dash *et al.* (1950), Hercus and White (1951), Parkinson (1954) and Brewer and Edwards (1956). Ambler (1951) has also given a very informative description of a small liquefier producing about 300 cm³/h. The Joule–Thomson liquefier at Ohio State University (Daunt and Johnston, 1949) with a production of 7·5 litres of liquid helium per hour and a liquid hydrogen consumption of 1·3 litres per litre of helium is a further example of the development. In this liquefier a Hampson-type heat exchanger is used to transfer heat from the incoming compressed gas to the low-pressure helium stream and to the evaporated hydrogen gas; this exchanger is designed to give a low pressure drop with the high flow rate used.

The helium liquefier at the Clarendon Laboratory (Croft, 1961) is a good example of a larger Linde helium liquefier which can be used very successfully when adequate supplies of liquid hydrogen are available. As shown in Fig. 19 there are three

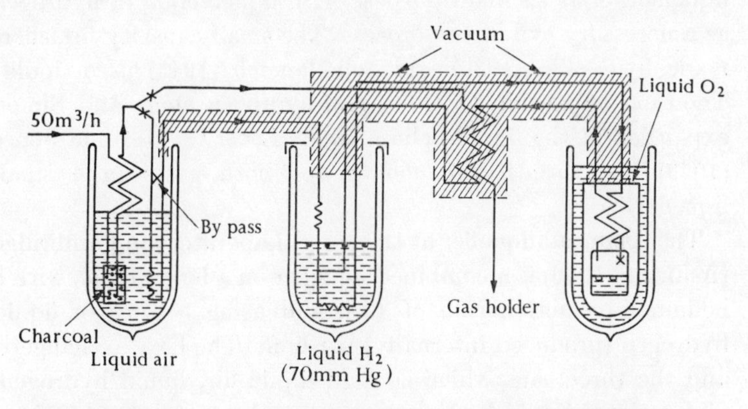

Fig. 19. The large Linde helium liquefier at the Clarendon Laboratory (after A. J. Croft, private communication, 1957).

distinct stages, only one of which, comprising the final exchanger and liquid-helium chamber, is enclosed in a vacuum space. In the first liquid-air dewar, the helium gas is purified by passing over charcoal and part of the helium stream is pre-cooled. The other part of the stream is pre-cooled by the returning cold helium gas; the incoming compressed helium is then further cooled by liquid hydrogen (reduced to a pressure of about 70 mm Hg, say 14° K) in the second dewar vessel before entering the final Linde stage in the third dewar. With a two-stage Reavell compressor 50 m³ of helium per hour are compressed to 27 atm and the liquefier yields about 12 litres of liquid helium per hour; the liquid hydrogen consumption is approximately 1·5 litres of hydrogen per litre of helium.

Parkinson (1960) has described a rather similar liquefier used at the Royal Radar Establishment, Malvern. This has finned-tubing Hampson exchangers and makes about 8 litres of liquid helium per hour with a consumption of 1·3 litres of hydrogen per litre of helium. The final exchanger is a Linde exchanger comprising a bundle of seven tubes.

Linde hydrogen–helium liquefiers

The combined hydrogen–helium liquefier usually has a helium liquefaction as its main purpose with liquefaction of hydrogen as a necessary evil of the process. The small-capacity liquefiers made by Rollin (1936) and Schallamach (1943) used Joule–Thomson expansion at the liquid-hydrogen stage and Simon expansion at the liquid-helium stage. Later Chester and Jones (1953) used isenthalpic cooling for both gases in a small liquefier.

The Ashmead liquefier at the Mond Laboratory in Cambridge (Ashmead 1950) is a combined liquefier on a larger scale, with a helium liquefaction rate of 4 l/h and using 5 litres of liquid hydrogen (produced internally) per hour. The heat exchangers, and the three cans which contain liquid air, liquid hydrogen, and pumped liquid hydrogen respectively, are supported from the main top plate and are surrounded by the main vacuum jacket. Problems of vacuum maintenance have been reduced by brazing the slightly domed brass faces onto each can and soldering the cooling coils, which carry the compressed gas, onto the outside of these cans. Perhaps the only more successful way of avoiding trouble over years of operation is that of the Collins liquefier in which an atmosphere of the gas being circulated surrounds all the working parts and is reasonably free from convection; however, in a large liquefier this method does require a rather large surrounding dewar vessel to isolate the interior.

Another combined liquefier was designed by Lacaze and Weil (1955); it has an output of about 7 l/h of liquid helium and 13 l/h of liquid hydrogen, and either liquid can be transferred to an external vessel if needed. The gas streams are cleaned with activated alumina at room temperature and with charcoal at 64° K. Heat exchangers and the liquid chambers are all surrounded by a high-vacuum enclosure unlike the Collins and

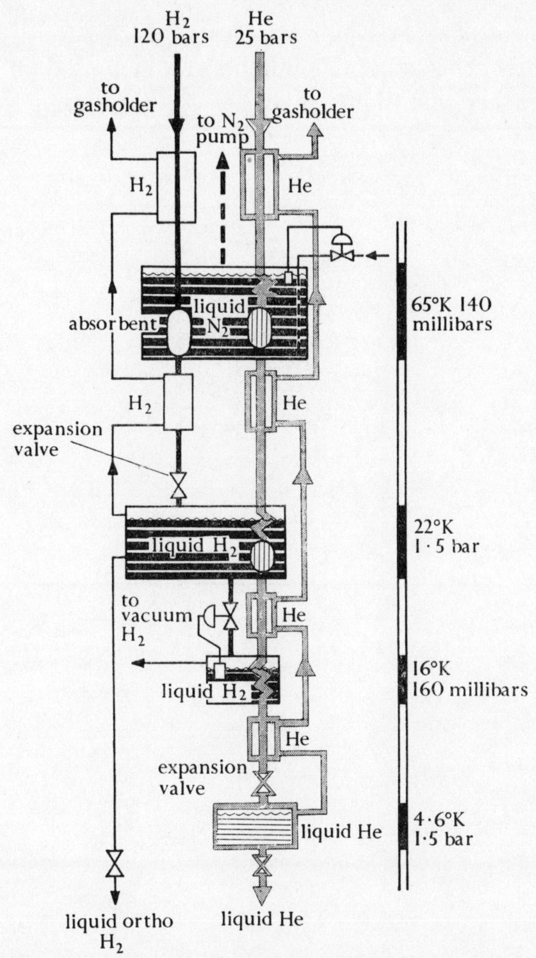

Fig. 20. Flow diagram of the combined hydrogen–helium liquefier made by T.B.T. of Grenoble, similar to the original model of Lacaze and Weil (1955).

Meissner liquefiers in which a stagnant atmosphere of helium surrounds most of the working parts. The commercial version of this helium liquefier (Fig. 20) is produced by T.B.T. o-

Grenoble. A number of these liquefiers have been used successfully in Europe now for some years despite the potential vacuum problems.

Kapitza's expansion-engine liquefier

The first successful helium liquefier which used an expansion engine rather than liquid hydrogen for pre-cooling was that of

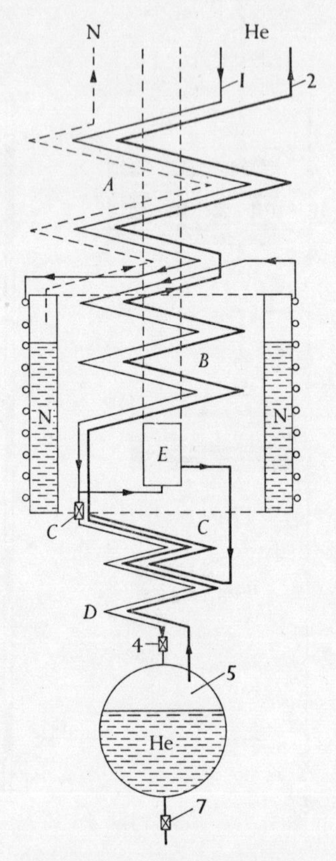

FIG. 21. Arrangement of heat exchangers and cooling stages in the Kapitza liquefier (after Kapitza 1934).

Kapitza (1934) which operated in the Mond Laboratory, Cambridge, for nearly two decades. Fig. 21 indicates the flow circuit in which helium gas, compressed to about 30 atm, is pre-cooled by liquid nitrogen, after which the majority of the gas is expanded in the engine E and returns through the heat exchangers C, B, and A. The minor fraction of compressed gas (about 8 per cent) passes on through exchangers to a Joule–Thomson expansion valve 4.

While this exemplifies the general principles of helium liquefaction discussed above (§ 2 and Fig. 7), it is also a direct development of the Claude process of air liquefaction with the addition of the liquid nitrogen pre-cooling. Kapitza used an unlubricated engine with a clearance of about 0·002 in between the piston and cylinder, the small gas loss past the piston serving as a lubricant. The part of the liquefier below the nitrogen vessel is surrounded by a radiation shield thermally anchored to the liquid-nitrogen container, and the whole assembly is suspended in a highly evacuated metal vessel. This liquefier produced about 1·7 l/h at an overall liquefaction efficiency of 4 per cent.

Two later versions used in Moscow had respective liquefaction rates of 5 l/h (Kapitza and Danilov 1961) and 18 l/h (Kapitza and Danilov 1962). The larger of these machines uses a second expansion engine rather than liquid nitrogen; the flow of gas at 20–25 atm pressure is split three ways with roughly equal amounts going through each of the two engines and the Joule–Thomson stage. The liquefaction yield is about 5 per cent of the gas passing through.

Collins liquefier (*Arthur D. Little, Inc.*)

Closely related in principle of operation to Kapitza's liquefier is that developed by Professor Collins (1947; see also Latham and McMahon 1949) at the Massachusetts Institute of Technology. This liquefier and its commercial development by the Arthur D. Little Corporation has contributed to the present wide development of low-temperature research and to the familiar references to the two eras of low-temperature physics as 'before and after Collins'. As illustrated in the flow diagram of Fig. 22 there is one obvious difference from the Kapitza model, in that there are two expansion engines, the first acting in place of the liquid-nitrogen cooling stage used by Kapitza. Other important features are:

(i) A very hard nitrided nitro-alloy steel for the engines allows the use of a very small clearance (0·0004 in) between piston and cylinder wall and reduces the gas leakage to a negligible amount, without causing appreciable wear on either piston or cylinder.

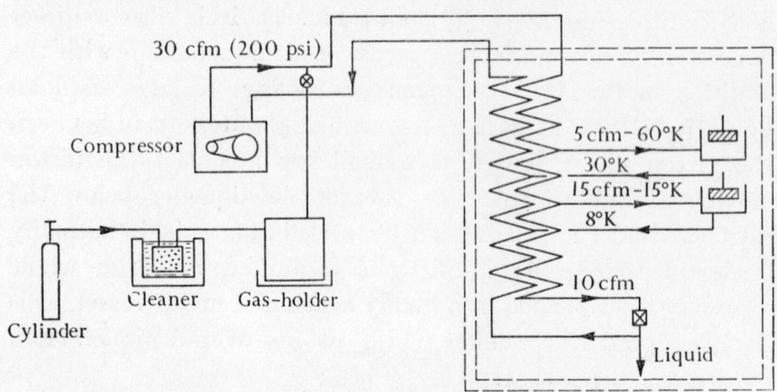

FIG. 22. Simplified diagram of a Collins helium liquefier. Flow rates are typical of the original Collins machine and the earlier commercial models.

(ii) In the heat exchanger (Plates I a and I b,† facing p. 32) the high-pressure helium at about 15 atm flows down through copper-finned cupro-nickel tubes of $\frac{1}{4}$-in diameter, and the low-pressure gas returns past these fins through the annular space between two cylindrical walls: the finned tubing is wound tightly on the inner slightly tapered cylinder and the outer jacket cylinder (similarly tapered) is forced up to cover the windings. A flexible cord is interwound with the exchanger tubing to force the return gas flow into more intimate contact with the copper fins, themselves solder-bonded to the cupro-nickel tubing. This exchanger is of high efficiency and gives a comparatively small pressure drop for the high helium flow rates used in this liquefier. The final section of the heat exchanger carrying gas to the expansion valve is a Linde concentric-tube interchanger and may be clearly seen in Plate II (between pp. 32 and 33).

† I am very grateful to Dr. Howard McMahon of Arthur D. Little, Inc., for photographs from which Plates I–III are taken.

(*a*)

(*b*)

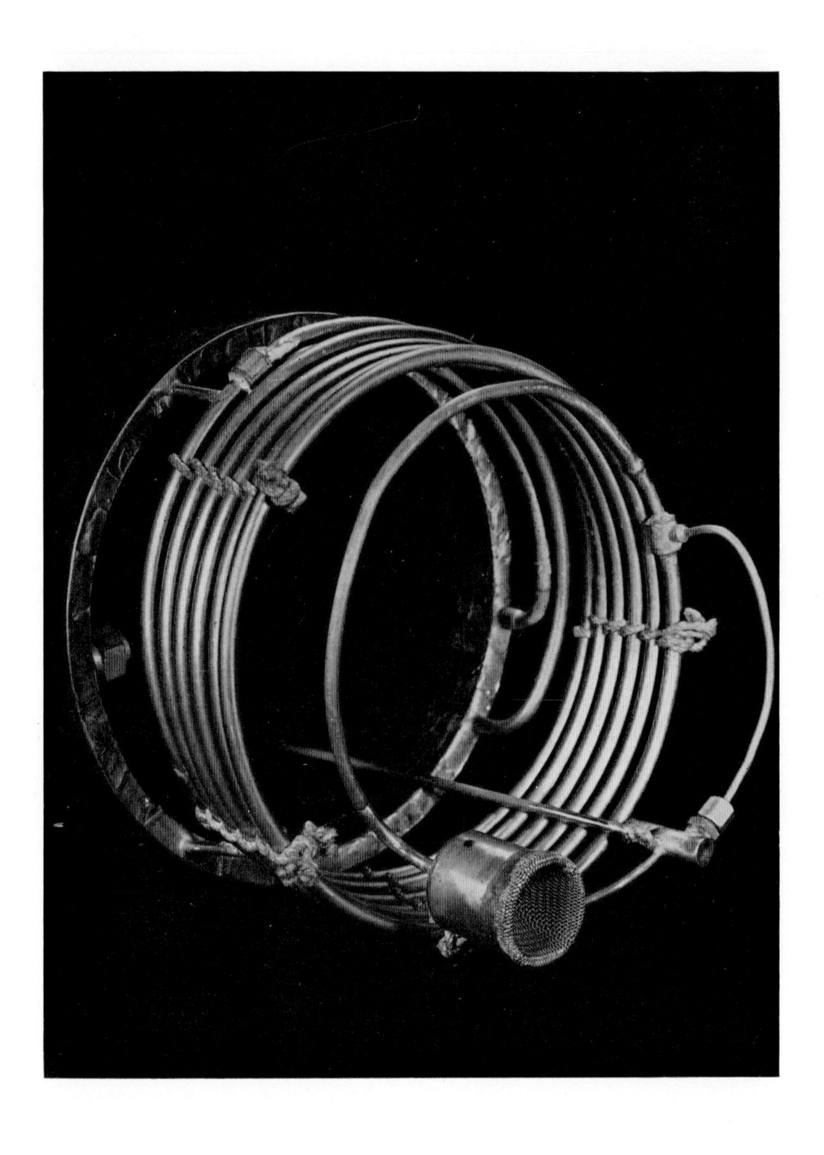

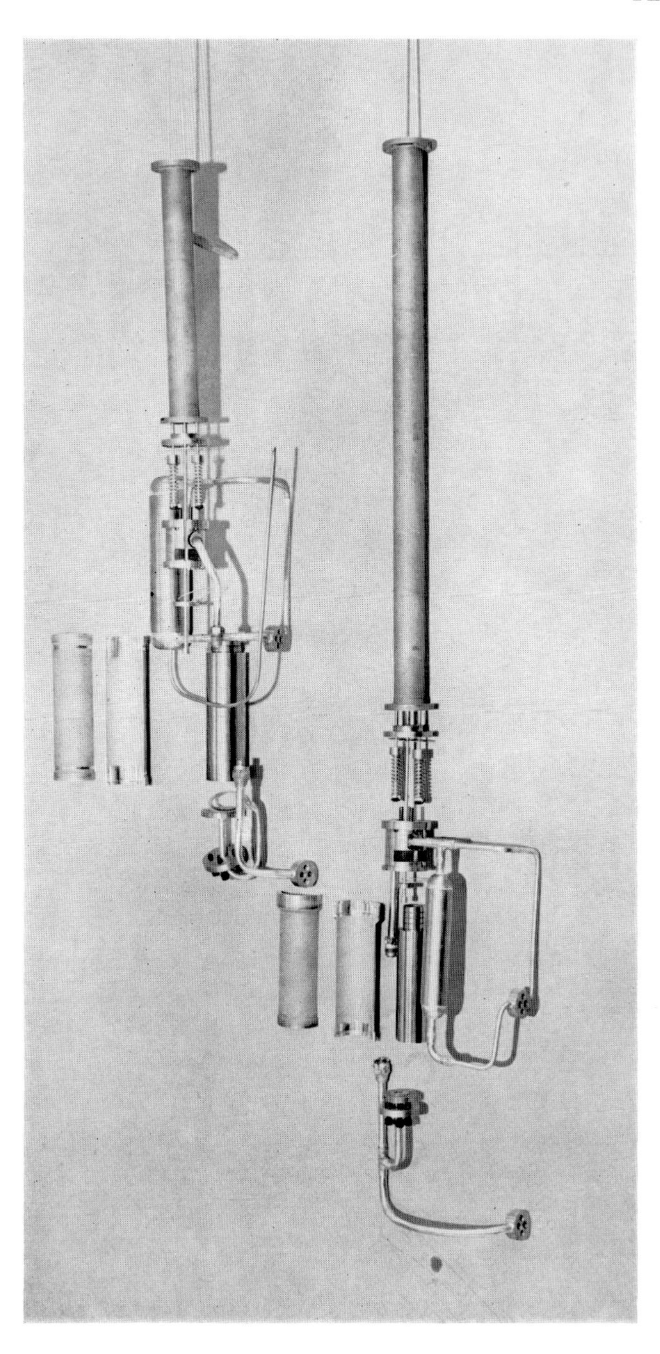

PLATE IV

(iii) In the Collins machine the piston and valve rods are always under tension rather than compression which allows the use of relatively thin stainless steel rods and makes the heat inflow through these small (Plate III) (between pp. 32 and 33).

(iv) Finally and most important from the maintenance viewpoint is the fact that the working parts are surrounded by an atmosphere of helium, rather than a high vacuum. The whole assembly enclosed by the heat exchanger fits inside a stainless steel dewar vessel and the possibility of serious vacuum troubles is considerably lessened by having the engines, helium flow tubes, and couplings, etc., in a static helium atmosphere; there is a change of temperature from about 295° K at the top to 4·2° K at the bottom, and into this region any small leakage of gas is of no great consequence.

With the flow rates shown in Fig. 22 and with no liquid nitrogen cooling, the liquefaction rate is between 2 and 3 l/h. This can be increased to over 4 l/h and the starting time shortened from 4 to 2 hours by using liquid nitrogen to pre-cool the incoming helium and the radiation shield. The production can be further increased to 8 l/h by using two compressors and either increasing the engine speed from 360 to 600 rev/min or using a larger second engine. Another optional feature, which is available on the Arthur D. Little–Collins liquefier, is to allow the Joule–Thomson expansion to occur in an external storage vessel by making the transfer siphon part of the final Linde exchanger.

Other changes from the original liquefier described by Collins (1947) have been made in the method of coupling rods to the crossheads and extracting power. In the present crossheads, the available power is used to drive a small oil pump. Thermal isolation has been improved by using superinsulation (multiple layers of aluminium foil and glass-fibre (see § 12.4) between the walls of the metal dewar). The storage capacity of the liquefier itself has been increased to about 15 litres. Starting from normal working conditions, liquefaction takes about two hours with liquid air cooling. This could be reduced considerably by slight modifications; for example, allowing some gas expanded through the Joule–Thomson valve to bypass the return section of the heat exchanger. Paradoxically the high efficiency of the exchanger prevents cooling from taking place too rapidly. Over 300 of these liquefiers have been made by Arthur D. Little, Inc., for use in research laboratories around the world. Their success has not induced many institutions to make their own due to the difficulties in machining crossheads, nitriding and lapping the pistons and cylinders, making the large vacuum jacket and heat exchanger. Apart from the Massachusetts Institute of Technology, only the National Standards Laboratory in Sydney and the Royal Radar Establishment have constructed them. Collins has shown that the engine construction could be simplified, for example by using steel pistons with plastic or leather rings in place of nitrided nitro-alloy.

Meissner liquefier (Linde-A.G.)

Another liquefier which uses only one expansion engine and has been commercially produced is that of Meissner (1942). Fig. 23 shows the form made by the Linde-A.G. (Meissner, Schmeissner, and Wiedemann, 1957; Meissner, Schmeissner, and Doll 1959). Liquid air is used for pre-cooling and thermal isolation is achieved by the surrounding metal dewar as in the Collins. The two heat exchangers operating between room temperature and 25° K are of the Hampson type and the other two are of the Linde type. The production rate is between 3 and 4 l/h.

Other engine liquefiers

An interesting development in expansion engines was the first use of flexible bellows by Long and Simon (1953). They made a small helium liquefier with brass bellows, which due to metal fatigue had rather a limited life of about half a million cycles. This corresponds to only about 60 h of operation but could be extended considerably with the development of bellows from a more suitable material.

Other forms of expansion cycle are also useful for cooling helium to 10 or 15° K prior to Joule–Thomson expansion, for

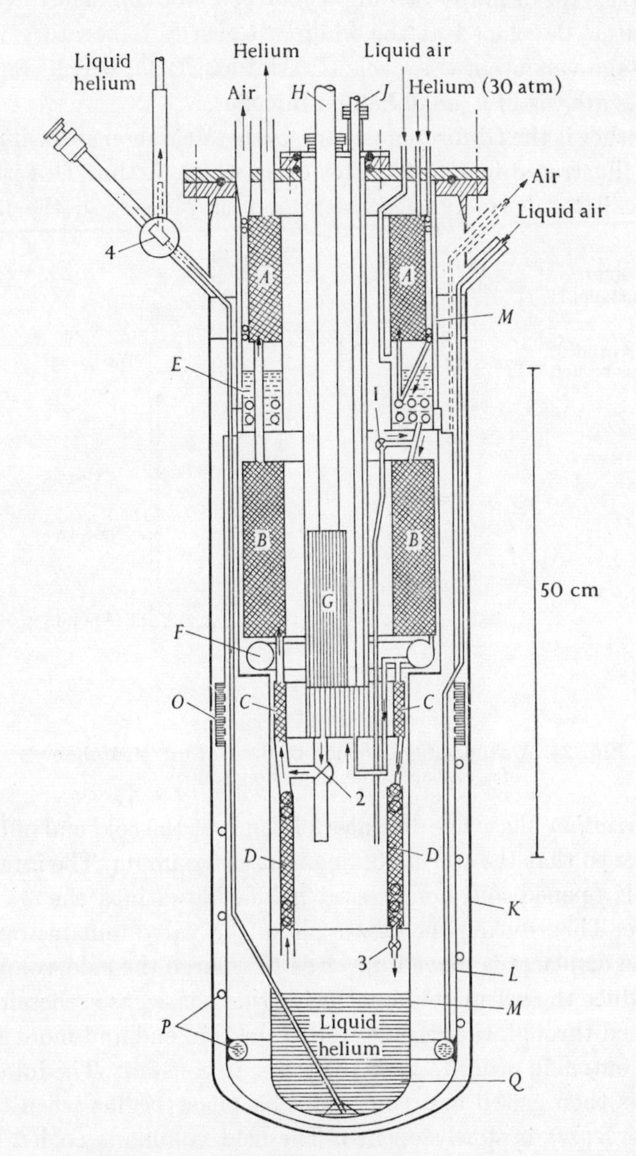

FIG. 23. Diagram of the Meissner helium liquefier, now made by Linde-A.G. (after Meissner, Schmeissner, and Wiedemann, 1957).

example, the Philips–Stirling cycle (§ 3 above). The 'Cryo-generator' developed at the Philips Research Laboratory is a two-stage version of this cycle (Prast 1963, 1965) which can be used as a basis of a small helium liquefier.

Another is the Gifford–McMahon cycle (McMahon and Gifford 1960) illustrated in Fig. 24. This cycle which Arthur D. Little, Inc., call the 'Cryodyne' consists of four phases: in the first

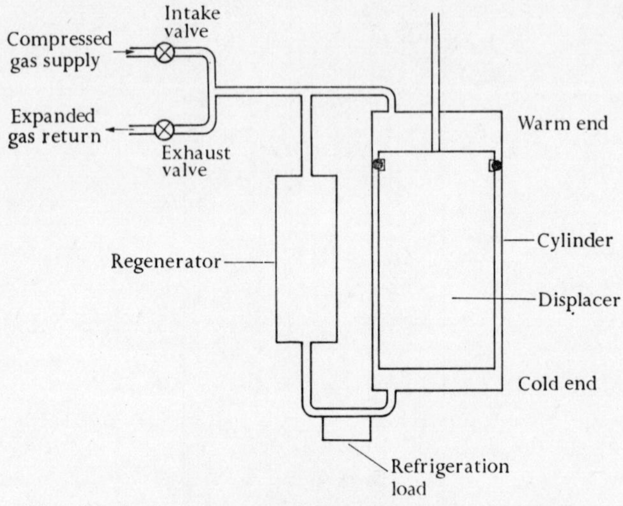

Fig. 24. A single-stage version of the Gifford–McMahon refrigerator (after McMahon 1960).

pressurization phase the displacer piston is at the cold end of the cylinder so that the warm volume is at a maximum. The intake valve is opened and compressed helium flows into the warm volume. Then during the *intake* phase the valve remains open and the displacer is moved upwards to enlarge the cold volume and reduce the warm volume. During this phase gas is therefore displaced through the regenerator to the cold end and more gas slowly enters in order to keep the pressure constant. The intake valve is then closed and the *expansion* phase begins when the exhaust valve is slowly opened. The cold volume is cooled by the expansion. The final exhaust phase occurs when the displacer is moved downwards to displace the remaining cold gas. After this the exhaust valve is closed and the cycle is repeated.

The three-stage or three-cylinder version of this can be used either as a refrigerator or for pre-cooling the helium supply to a Joule–Thomson liquefier as shown schematically in Fig. 25 (McMahon 1960; see also Gifford 1961, and Hoffman 1963).

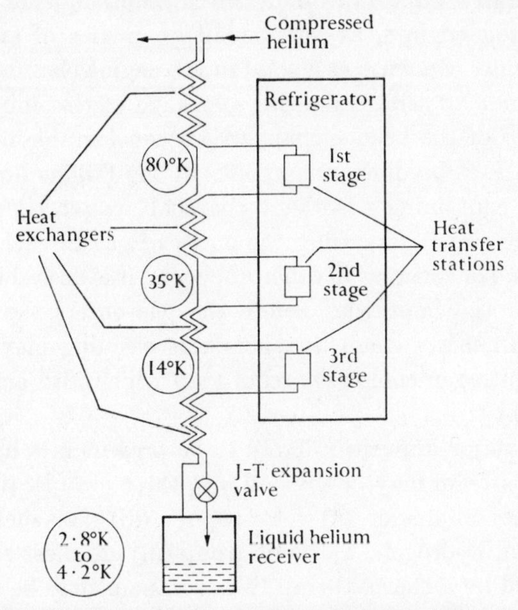

FIG. 25. A helium-liquefaction cycle containing three stages of Gifford–McMahon cooling and a final Joule–Thomson expansion (after McMahon 1960).

Stuart and Hogan (1965) have also described a two-stage model produced by Arthur D. Little, Inc., which uses liquid nitrogen and produces nearly 1 l/h of liquid helium.

Another refrigerator capable of reaching about 15° K and operating continuously is that described by Chellis and Hogan (1964) and developed from a cooling cycle first described by K. W. Taconis (see Yendall 1960). Arthur D. Little, Inc., produce this refrigerator which they call the 'Creacher'; it uses liquid nitrogen as a pre-coolant but requires no external compressor and so is a compact cooling source. It does not appear to have been used yet as a cooling stage in a helium liquefier.

6. Gas Purification

It is generally necessary that the gas stream entering a liquefier should be free of significant amounts of any impurity which may condense in the working parts of the liquefier. This is particularly important in a 'circulation liquefier' employing expansion engines, as very small quantities of any impurity which may condense as a solid in the engine chamber can cause the engine to seize. Likewise, blockage of the Joule–Thomson valve or of the heat exchanger is a considerable nuisance and may in fact be quite dangerous. In the Collins liquefier small vessels containing adsorbent charcoal are placed in the high-pressure circuit preceding the entry of gas to each of the two engines. However, as in other liquefiers, it is desirable to remove most of the impurities before the gas enters the pre-cooling stage. In many small liquefiers this cleaning may occur in a liquid-nitrogen-cooled charcoal trap which also acts as a pre-coolant.

The major impurities likely to be present in a high-pressure helium stream may be divided into three groups: (i) oil vapour from the compressor, (ii) water vapour, (iii) gases such as oxygen, nitrogen, hydrogen. Those of group (iii) are most conveniently removed by a charcoal trap; water vapour may be removed by this if it is not present in such a quantity as to saturate the charcoal. If oil vapour is present in appreciable quantities this will rapidly saturate the charcoal and make it permanently ineffective as a gas adsorbent; compressors should be used which do not 'carry over' much oil vapour into the gas stream or an effective oil filter must be placed in the high-pressure exit line. Such a filter may be a vessel containing copper or steel wool and of sufficient dimensions that the velocity of the gas stream in it is not too great; this allows the oil particles to collect on the copper wool and then oil may be drained or blown out of the vessel at intervals through a drain-valve in the bottom.

In most small liquefiers the total amount of water vapour present in the gas being circulated is relatively small and may be removed together with air and other gases by the charcoal trap. However, in large installations or in a hydrogen liquefier using electrolytic hydrogen, it is usual to place a silica gel drying column in the gas stream or to use refrigeration drying. When circulating hydrogen it is also usual to remove oxygen by a separate purifier—a 'Deoxo' unit which is a palladium catalyst supported on silica gel; these 'Deoxo' purifiers are commercially available from Baker and Company, Inc., of New Jersey.

A common form of charcoal cleaner is shown in Fig. 27 (§ 2.1) but a convenient small cleaning trap may be easily made from brass tubing. One end of the tube is sealed by a brass plate 'sweated' or silver-soldered to it and the entry and exit tubes for the gas stream are soldered through this plate; it is preferable to wrap fine metal gauze around the open ends of these two tubes to prevent charcoal dust being carried over by the gas. The main brass tube is then packed with coconut charcoal of adsorbent grade and the bottom end is sealed with a brass plate. The charcoal may then be activated by connecting the entry or exit tube to a mechanical pump and warming the unit with a gas flame or electrical heater to a temperature of 100°–200° C for a few minutes. It is placed in a convenient glass dewar which is kept filled with liquid nitrogen during the cleaning operation.

The amount of gas required to saturate activated charcoal at 77° K depends somewhat on the gas and varies with the type of charcoal used. Generally, charcoals have a specific surface in the range 100–1000 m^2/g and will adsorb 0·2–0·25 cm^3 of gas (s.t.p.) per m^2 of surface (J. A. Morrison, private communication). Activated coconut charcoals of adsorbent grade have specific surfaces approaching 1000 m^2/g; for example, Brunauer (1943) in his book on physical adsorption quotes figures of 170–230 cm^3/g for the adsorption of N_2, A, O_2, CO_2, on coconut charcoal at about 90° K and gives values of 700–900 m^2/g for the surface areas of this charcoal.

Other adsorbent materials such as activated alumina, Zeolite, Chabazite, Cullite, Sepiolite (see, for example, Beher 1957) appear to be equally effective in absorbing impurities. The Union Carbide Corporation produce adsorbent aluminosilicates (called 'molecular sieves') in powder or pellet form, suitable for cleaning gas streams or acting as a backing pump. They are often more dust-free but perhaps not so readily available as charcoal.

It should also be emphasized that high-pressure cleaning is more effective than low-pressure cleaning, i.e. passing an impure gas at 100 atm pressure through a cooled charcoal cleaner will yield a product with a lower percentage of impurity than by passing the same gas at 1 atm over the charcoal. In either case the adsorption pressure of the impurity at the surface of the charcoal is a function of the temperature and of the volume of impurity already adsorbed, and not of the total gas pressure; hence increasing the total pressure by a factor of 100 decreases the percentage of impurity that is retained by the gas stream by a like factor.

In liquefiers where there is a danger of the cleaning stages becoming ineffective due to passage of large amounts of impurities, a meter may be installed to monitor the impurity; such instruments for determining impurity contents in hydrogen and in helium are made by Gow-Mac Instrument Company, New Jersey.

7. References

Among the more general references from which the writer has drawn the material for this chapter are the following.

COLLINS, S. C. (1956). 'Helium liquefiers and carriers', *Handb. Phys.* **14**, 112.

—— and CANNADAY, R. L. (1958). *Expansion machines for low temperature processes*. Oxford University Press.

CROFT, A. J. (1961). 'Helium liquefiers', *Prog. Cryogen.* **3**, 1.

DAUNT, J. G. (1956). 'The production of low temperatures down to hydrogen temperature', *Handb. Phys.* **14**, 1.

DAVIES, M. (1949). *The physical principles of gas liquefaction and low temperature rectification*. Longmans Green.

HOARE, F. E., JACKSON, L. C., and KURTI, N. (1961). *Experimental cryophysics*. Butterworths, London.

KEESOM, W. H. (1942). *Helium*. Elsevier, Amsterdam.

RUHEMANN, M. and B. (1937). *Low temperature physics*. Cambridge University Press. This includes an historical account of the development of low-temperature physics and discusses the thermodynamic cycles involved in different methods of gas liquefaction.

SCOTT, R. B. (1959). *Cryogenic engineering*. Van Nostrand, Princeton.

—— DENTON, W. H., and NICHOLLS, C. M. (1964). *Technology and uses of liquid hydrogen*. Pergamon, London.

Van Lammeren, J. A. (1941). *Technik der tiefen Temperaturen.* Springer-Verlag, Berlin.

Other references

Ahlberg, J. E., Estermann, I. and Lundberg, W. D. (1937). *Rev. scient. Instrum.* **8**, 422.
Ambler, E. (1951). *Proc. 8th Int. Cong. Refrig.*, London, Commission I, p. 98.
Ashmead, J. (1950). *Proc. phys. Soc.* B**63**, 504.
Baldus, W. and Sellmaier, A. (1965). *Adv. cryogen. Engng* **10**B, 13.
Beher, J. T. (1957). *Ibid.* **2**, 182.
Brewer, D. F. and Edwards, D. O. (1956). *J. scient. Instrum.* **33**, 148.
Brunauer, S. (1943). *The adsorption of gases and vapours*, vol. 1, *Physical adsorption.* Princeton University Press.
Chellis, F. F. and Hogan, W. H. (1964). *Adv. cryogen. Engng* **9**, 545.
Chester, P. F. and Jones, G. O. (1953). *Proc. phys. Soc.* B**66**, 296.
Clusius, K. (1953). *Z. Naturf.* **8a**, 479.
Collins, S. C. (1947). *Rev. scient. instrum.* **18**, 157.
Cooke, A. H., Rollin, B., and Simon, F. E. (1939). *Ibid.* **10**, 251.
Croft, A. J. (1952). *J. scient. Instrum.* **29**, 388.
—— (1964). *Cryogenics* **4**, 143.
—— and Simon, F. E. (1955). *Bull. int. Inst. Refrig.* Annexe 1955–2, p. 81. Paris.
Dash, J. G., Cook, D. B., Zemansky, M. W., and Boorse, H. A. (1950). *Rev. scient. Instrum.* **21**, 936.
Daunt, J. G. and Johnston, H. L. (1949). *Ibid.* **20**, 122.
—— and Mendelssohn, K. (1948). *J. scient. Instrum.* **25**, 318.
Debye, P. (1926). *Annls Phys.* **81**, 1154.
DeSorbo, W., Milton, R. M., and Andrews, D. H. (1946). *Chem. Rev.* **39**, 403.
Fairbank, H. A. (1946). *Rev. scient. Instrum.* **17**, 473.
Giauque, W. F. (1927). *J. Am. chem. Soc.* **49**, 1864.
—— and MacDougall, D. P. (1933). *Phys. Rev.* **43**, 768.
Gifford, W. E. (1961). *Prog. Cryogen.* **3**, 49.
de Haas, W. J., Wiersma, E. C., and Kramers, H. A. (1933). *Physica* **1**, 1.
Hercus, G. R. and White, G. K. (1951). *J. scient. Instrum.* **28**, 4.
Hoffman, T. E. (1963). *Adv. cryogen. Engng* **8**, 213.
Johnson, V. J. and Wilson, W. A. (1955). *Proc. 1954 Cryogenic Engng Conf., N.B.S. Report No. 3517*, p. 246.
Jones, G. O., Larsen, A. H., and Simon, F. E. (1948). *Research* **1**, 420.
Kapitza, P. (1934). *Proc. R. Soc.* A**147**, 189.
—— and Danilov, I. B. (1961). *Soviet Phys. tech. Phys.* **6**, 349.
—— —— (1962). *Ibid.* **7**, 333.
Keesom, W. H. (1933). *Leiden Commun. Suppl.* **76a**.
Keyes, F. G., Gerry, H. T., and Hicks, J. F. G. (1937). *J. Am. chem. Soc.* **59**, 1426.

Kohler, J. W. L. (1960). *Prog. Cryogen.* **2**, 43.
—— and Jonkers, C. O. (1954). *Philips tech. Rev.* **16**, **69**.
Kurti, N. and Simon, F. E. (1935). *Proc. R. Soc.* A**149**, 152.
Lacaze, A. and Weil, L. (1955). *Proc. 9th Int. Cong. Refrig.*, Paris, Commission I, p. 1028.
Latham, A. and McMahon, H. O. (1949). *Refrig. Engng* **57**, **549**.
Long, H. M. and Simon, F. E. (1953). *Appl. scient. Res.* A**4**, 237.
McMahon, H. O. (1960). *Cryogenics* **1**, 65.
—— and Gifford, W. E. (1960). *Adv. cryogen. Engng* **5**, **354**.
Mann, D. B., Sixsmith, H., Wilson, W. A., and Birmingham, B. W. (1963). *Ibid.* **8**, 221.
Meissner, W. (1928). *Phys. Z.* **29**, 610.
—— (1942). *Ibid.* **43**, 261.
—— Schmeissner, F., and Doll, R. (1959). *Kältetechnik* **11**, **317**.
—— —— and Wiedemann, W. (1957). *Ibid.* **9**, 194.
Mendelssohn, K. (1931). *Z. Phys.* **73**, 482.
Parkinson, D. H. (1954). *J. scient. Instrum.* **31**, 178.
—— (1960). *Cryogenics* **1**, 17.
Pickard, G. L. and Simon, F. E. (1948). *Proc. phys. Soc.* **60**, 405.
Prast, G. (1963). *Cryogenics* **3**, 156.
—— (1965). *Philips tech. Rev.* **26**, 1.
Rollin, B. V. (1936). *Proc. phys. Soc.* **48**, 18.
Ruhemann, M. (1930). *Z. Phys.* **65**, 67.
Schallamach, A. (1943). *J. scient. Instrum.* **20**, 195.
Scott, R. B. and Cook, J. W. (1948). *Rev. scient. Instrum.* **19**, 889.
Simon, F. E. (1930). *Ergebn. exakt. Naturw.* **9**, 222.
—— (1932). *Z. ges. Kälteindustr.* **39**, 89.
—— (1956). 40th Guthrie lecture, *Phys. Soc. Lond. Yearb.* p. 1.
Spoendlin, R. (1954). *J. Res. C.N.R.S.* **6**, No. 28, p. 1.
Starr, C. (1941). *Rev. scient. Instrum.* **12**, **193**.
Stuart, R. W. and Hogan, W. H. (1965). *Adv. cryogen. Engng* **10B**, 62.
Woolley, H. W., Scott, R. B., and Brickwedde, F. G. (1948). *J. Res. natn. Bur. Stand.* **41**, 379.
Yendall, E. F. (1960). *Adv. cryogen. Engng* **2**, 188.
Zeldovitch. A. G. and Pilipenko Yu. K. (1960). *Cryogenics* **2**, **101**.

CHAPTER II

STORAGE AND TRANSFER OF LIQUEFIED GASES

1. Dewar vessels

Introduction

ALTHOUGH this section might be more appropriately termed 'containers for liquefied gases' than merely 'dewar vessels', Sir James Dewar's contribution to low-temperature research by the invention of the evacuated double-walled vessel can scarcely be over-emphasized.

The efficiency of storage of any liquefied gas is related to its latent heat of vaporization and normal boiling-point. The latter factor partially determines the extraneous heat inflow from the surroundings and the former governs the evaporation rate as a function of that inflow.

TABLE IV

Latent heat of vaporization and boiling-point at standard atmospheric pressure of some common gases

Gas	CO_2†	O_2	A	N_2	Ne	H_2	^4He
Latent heat (cal/g)	137	51·0	37·9	47·8	20·8	106·8	5·2
Latent heat (cal/cm³)	223	58·1	53·5	38·6	25	7·56	0·65
Boiling-point (° K)	194·6	90·2	87·4	77·3	27·2	20·4	4·2

† Solid.

In designing a container the problem is that of reducing the heat inflow to a minimum, determined by difficulties of construction and the allowable loss of liquid per unit time. The chief sources of heat are radiation to the walls of the containers, radiation down the neck of the storage vessel, heat conduction down the neck or walls, convection in the vapour above the liquid surface, and heat conduction through the residual gas in the surrounding vacuum space. Most of these factors are discussed in detail in Chapter VII and in various papers on the design of liquid-helium containers; for example, Wexler (1951) (see also Wexler and Jacket 1951) and Sydoriak and Sommers (1951) examined the problem of designing suitable containers for storing liquid helium in the laboratory; recently Scott (1957) has considered the thermal design of large storage vessels for liquid helium and liquid hydrogen.

Liquid-air dewars

Glass dewar flasks (Figs. 26 (a) and 26 (b)), both silvered to reduce radiant heat inflow and unsilvered, are available commercially in a wide range of sizes. Usually of Pyrex, they can be made by any competent glass-blower with a suitable glass-blowing lathe. Frequently a special shape, length, or diameter of

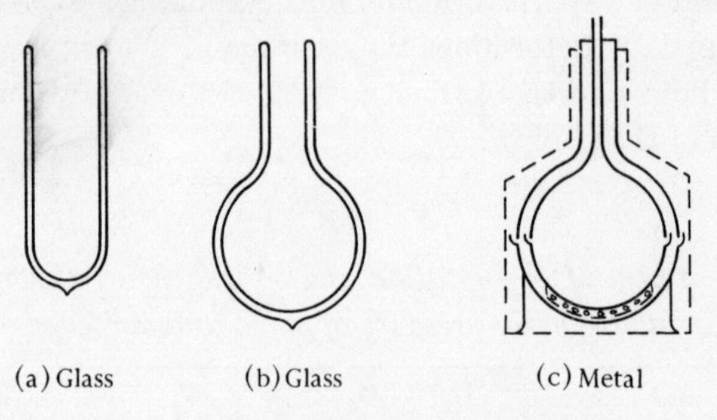

FIG. 26. Common types of dewar flask.

dewar is needed and this may have to be made. A description of the silvering and evacuation procedure has been given by Scott, Cook, and Brickwedde (1931). They found that after silvering by a modified Brashear process, the Pyrex dewar gave best results if slowly raised in temperature to 400° C, then baked for a short period at 550° C and later cooled to about 400° C before sealing. The resulting evaporation rates for liquid air and liquid hydrogen showed that the emissivity of the silver coating is about 0·03. Quite satisfactory dewars also result if the baking temperature is only carried up to about 450° C, and then the danger of distorting the glass walls, which will occur for Pyrex at temperatures slightly in excess of 550° C, is eliminated.

In the spun metal variety (Fig. 26 (c)) the difficulty of complete outgassing makes it necessary to have an internal adsorbent trap. For this purpose a few ounces of coconut charcoal or other adsorbent should be attached in a wire-gauze cage to the inner (cold) wall of the vacuum space, so that it is well exposed to the vacuum system. Such metal flasks, usually of spun copper, with clean polished surfaces facing the vacuum space, and with an inner neck of a low-conductivity alloy (monel,

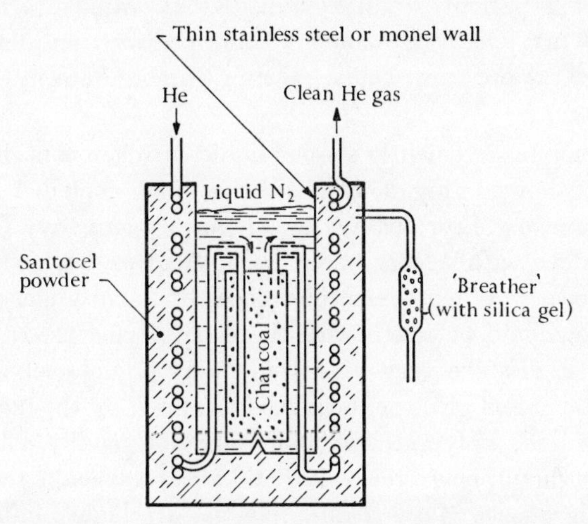

FIG. 27. A charcoal purifying trap cooled by liquid nitrogen.

stainless steel, inconel, etc.), can be 'home-made'. However, the difficulty and cost of having patterns made for the spinning and of the spinning itself usually make it advantageous to buy the commercial article (see Appendix for list of suppliers). The normal evaporation rate of liquid air from such a metal dewar is about 2·5 l/day for the 25-litre size and 10 l/day for the 250-litre size. If they have superinsulation, i.e. multiple layers of aluminium foil and plastic or glass mat, between the evacuated walls the evaporation may be reduced to less than 1 per cent of the capacity per day; such dewars are also more robust.

Powder- and foam-insulation can be very useful in making liquid-air containers. One example is shown in Fig. 27. If the size and construction do not allow evacuation, or if the vacuum that can be maintained is only of the order of a few millimetres of mercury, a powdered insulator such as Santocel (Monsanto Chemical Company) provides good insulation by reducing conduction, convection, and radiation across the space between the walls. Some data for heat inflow through Santocel (White 1948; Fulk 1959) are as follows: at an average temperature of about 180° K, i.e. for walls respectively at room temperature and liquid-nitrogen temperatures, the mean thermal conductivity is about 0·2 mW/cm degK at atmospheric pressure, 0·06 mW/cm degK at about 1 mm pressure, and 0·016 mW/cm degK at pressures below $\frac{1}{100}$ mm (further data in Chapter XII below).

Small conveniently shaped liquid-nitrogen containers such as that shown in Fig. 28 can be cut from an 'exploded' polystyrene plastic, e.g. Styrofoam (Dow Chemical Company). Cutting may be done with a 'cork-borer' type of cutter, hacksaw blade, hot wire, etc. When cementing pieces of foamed plastic together, care should be taken with the type of glue used: polystyrene foams dissolve very readily in many organic solvents so that water-based glues or those recommended by the manufacturers are used. Polyurethane foams are not readily soluble. Great care should be exercised if liquid air or oxygen is to be put into such vessels. Many foamed plastics are highly inflammable in such a situation, although *some* polyurethane foams are relatively safe.

The thermal conductivity of foamed plastics is not very different from that of Santocel or some other powder insulators at atmospheric pressure (see § 12.4). For example, Styrofoam has a conductivity of *ca.* 0·4 mW/cm degK at room temperature and 0·2 mW/cm degK at a mean temperature of 170° K, which is similar to that of Santocel. By contrast, superinsulation in a high vacuum has an effective conductivity of less than 10^{-3} mW/cm degK.

Liquid-hydrogen containers

Apart from its inflammable nature and small latent heat, liquid hydrogen poses an additional problem when storage for any appreciable period of time is desired. Hydrogen contains molecules of two different kinds, *ortho*-hydrogen and *para*-hydrogen, in which the spins of the two protons are respectively parallel and antiparallel. Due to the different energies of the two states, the equilibrium concentration varies from 100 per cent *para*-hydrogen at or below 20° K to about 25 per cent *para*-hydrogen at room temperature. At liquid-nitrogen temperature the states are almost equally probable.

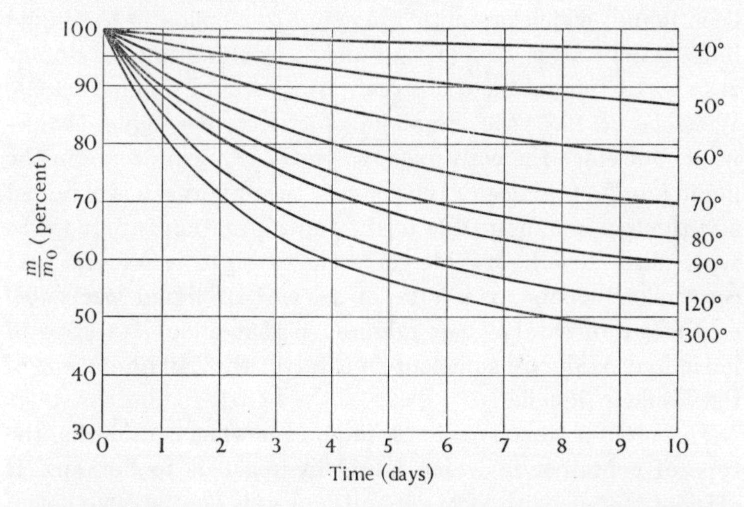

Fig. 28. Effect of *ortho-para* conversion: graphs show the mass of liquid hydrogen m remaining after days of storage, as a percentage of the initial mass m_0 (after Larsen, Simon, and Swenson 1948).

In the absence of a catalyst the rate of conversion is relatively slow with a time constant of the order of 50 h, so that normal hydrogen gas cooled from room temperature, liquefied, and transferred to a storage container may contain nearly 75 per cent *ortho*-hydrogen when first stored. Over a period of days, *ortho-para* conversion occurs with a heat of transformation of 310 cal/mole (see Larsen, Simon, and Swenson, 1948, for determination of rate of conversion and evaporation) to be compared with a heat of vaporization of 216 cal/mole. After 2 days nearly 30 per cent of the liquid has been lost purely due to this internal process. Fig. 28 illustrates the rate of loss for various initial equilibrium concentrations, e.g. where the liquid hydrogen stored had an initial *ortho*-hydrogen concentration corresponding to the equilibrium concentration at 300° K, 120° K, 90° K, etc.

Obviously *ortho-para* conversion is important in large installations where economy is dependent on efficient liquefaction and storage (e.g. see Scott, Denton, and Nicholls 1964). As a result of investigations on the large hydrogen liquefier at the National Bureau of Standards (Boulder), more efficient catalysts have been found, which promote *ortho-para* conversion in the liquid immediately after liquefaction and before transfer to storage vessels. In this method (Barrick, Weitzel, and Connolly 1955, Weitzel *et al.* 1957) the overall liquefaction rate is reduced somewhat, but since the conversion to *para*-hydrogen occurs in the liquid (20° K) practically 100 per cent *para*-hydrogen is obtained and the loss in storage due to the transformation is reduced to zero. With ferric hydroxide gel as the catalyst, it is found that conversion occurs at a rate of 2·6 cm^3 of liquid per cubic centimetre of catalyst per minute, so that about 1·5 litres of ferric hydroxide are sufficient to catalyse the 240 l/h output of the Boulder liquefier.

This *ortho-para* conversion is important when considering the type of container in which liquid hydrogen is to be kept. If efficient storage with low evaporation loss is needed, conversion to the *para* state should be nearly complete and the liquid should be kept in a suitable low-loss dewar vessel (for example, the helium and hydrogen containers with liquid nitrogen shielding) with evaporation loss rates of about 1 per cent per day. If, however, the liquid initially contains a large percentage of *ortho*-hydrogen, considerable evaporation loss is inevitable and storage in an ordinary unprotected dewar vessel—a metal liquid-air container—may be suitable. As the vapour pressure of air is negligible at liquid-hydrogen temperatures, liquid hydrogen creates a high vacuum when placed in the dewar vessel so that the major external heat inflows are radiation across the walls and heat coming down the neck of the dewar by radiation and conduction. Hence in the standard narrow-necked metal dewar, normally used for liquid air, radiation across the vacuum space is the major source of heat leak and this will be, at worst, comparable with that due to *ortho-para* conversion. However, it has been generally observed that metal dewars with liquid-nitrogen-cooled shields do have a rather larger heat leak when used with liquid hydrogen than with liquid helium, probably because of the slow release of dissolved hydrogen gas from the metal walls into the vacuum chamber (cf. Scott 1957).

Liquid-helium dewars for research cryostats

Various research investigations require dewars of various shapes and sizes as will be seen in Chapter VI which discusses cryostat design. The evaporation rate of the refrigerant is not usually a prime consideration provided it is not so high as to require frequent refilling. Metal dewars may be chosen if the geometry is complicated by a need for X-ray or optical windows, narrow tails, etc. Conventional shapes can also be made in metal but may be more easily and cheaply provided in glass.

Glass dewars do pose one problem. Helium gas diffuses through glass, the diffusion rate in the hard boro-silicate glasses being quite high (see Keesom 1942, for references and data; also Rogers, Buritz, and Alpert 1954). Whereas the diffusion rate at room temperature for soft glass and for Jena 16 III measured in cubic centimetres diffusing per second through 1 cm^2 of wall 1 mm thick with 1 atm pressure difference is about $0{\cdot}4 \times 10^{-12}$, it amounts to $0{\cdot}5 \times 10^{-10}$ for Pyrex. Thus in a Pyrex dewar in which 100 cm^2 of glass at room temperature are exposed to helium gas, the thickness of the glass being 2 mm and the volume of the vacuum space being 1 litre, the time required for the helium pressure to reach 10^{-5} mm Hg will be approximately 30 min.

Fortunately the diffusion rate decreases very rapidly with temperature and most of the inner wall of a helium dewar is well below room temperature, much of it at 4° K. It is found in practice that a figure of $\sim$100 h or 6–10 days of research operation are quite feasible with a Pyrex dewar before it becomes sufficiently 'soft' that the evaporation rate is markedly higher and therefore needs re-evacuation. The English Monax glass and its near equivalent in the United States, Kimball N51A glass, are both much less permeable, but are not always readily available in the sizes required for making experimental dewar vessels. The relative merits of these different glasses for dewar vessels have been discussed by Giauque (1947), Lane and Fairbank (1947), and Desirant and Horvath (1948) among others.

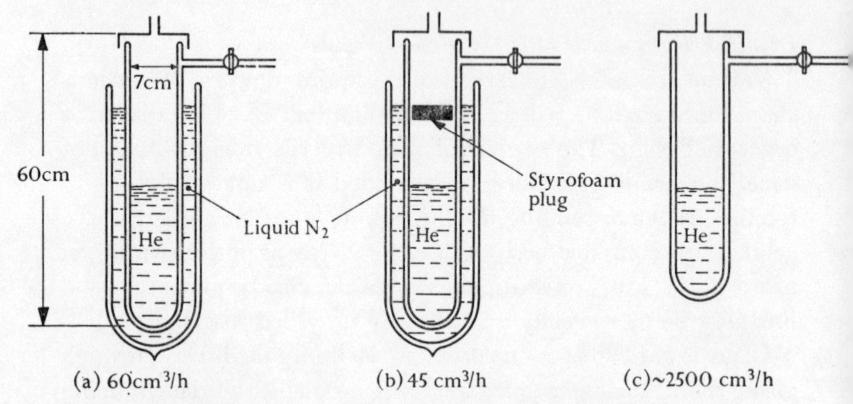

Fig. 29. Rate of evaporation of liquid helium from a glass dewar.

To illustrate the helium evaporation rates which may be encountered in experimental research, Fig. 29 shows a rather typical glass helium dewar pumped to a high vacuum and surrounded by a dewar vessel filled with liquid nitrogen. The helium dewar is 60 cm in length and 7 cm internal diameter; it is approximately half-filled with liquid helium and is (a) silvered except for a 1-cm-wide viewing slit along its length, (b) as before but having a loose-fitting plug of styrofoam, 2 in thick, suspended about 12 cm from the top of the dewar, (c) silvered except for the slit but not surrounded by liquid nitrogen. Some approximate figures for evaporation rate in cubic centimetres of liquid helium per hour are shown on the figure.

To simplify the repeated re-evacuations necessary in Pyrex dewars, the writer has found the following arrangement satisfactory. A short glass side-tube is provided near the top of the dewar, to which a small stop-cock is attached by Tygon (flexible plastic) tubing. The flexible connexion reduces the chance of breakage; the vacuum space is flushed with air and re-evacuated for a few minutes on a mechanical pump or a diffusion pump after every 6 to 7 days of experimental use. In other cases an extended glass 'pip' may be put on the tail of the dewar for repumping but should be sufficiently extended that it can be broken and resealed without introducing strains into the bottom of the dewar. Unless the dewar is also to be used with liquid air, the degree of vacuum is not important as the liquid helium rapidly freezes out any air left in the vacuum space. Indeed, it is often convenient to leave about 1 mm air pressure to assist pre-cooling by the liquid air in the outer dewar.

Metal dewars do not suffer from this 'diffusion' problem. However, they are more difficult to make in a completely vacuum-tight condition. After sealing, slow evolution of gas from the metal walls makes it necessary to have either a charcoal adsorber cooled by liquid air or cryopumping by the contained liquid helium. Materials normally used in construction are 18/8 stainless steels for the inner walls which need to be of low conductivity, high-conductivity copper or aluminium for the one (or more) radiation shields, and brass or stainless steel for the outer vacuum jacket. A liquid-air chamber is usually provided to cool the radiation shield but effective use of the evaporating helium gas can make this unnecessary; for example Fradkov (1962) has described a metal dewar which has no liquid air but has two radiation shields attached at suitable levels in the neck and an evaporation rate of only 60 cm^3/h. The use of superinsulation should make the avoidance of liquid air more commonplace.

Metal dewars of straight cylindrical shape or with narrow tails, windows, rotating bottoms, demountable bottoms, etc., can now be bought in standard sizes or made to specific dimensions by a number of firms (see Appendix). Fig. 30 illustrates schematically two dewars which are demountable in that their tail sections can be removed and replaced by others. In some commercial dewars the surfaces exposed to the vacuum are gold-plated to preserve a low emissivity and low radiative heat inflow. Addition of aluminium foil or superinsulation can assist this.

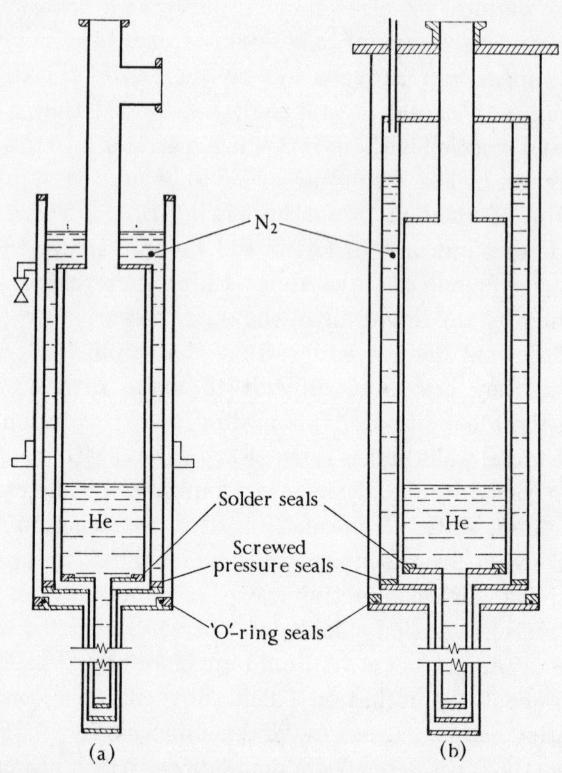

Fig. 30. Schematic diagrams of two common types of metal cryostat. Both have demountable tail sections. (a) is similar to a dewar made by Janis Research Company, Mass. and (b) is similar to dewars produced by Andonian Associates Inc., Mass. Indium O-rings replace the solder seals in some dewars.

Liquid-helium storage dewars

Obviously the relatively high evaporation rates make dewars of the type shown in Fig. 29 unsuitable for long-term storage. Even when fully silvered and surrounded by liquid nitrogen, the large heat inflows down the neck of a cylindrical dewar produce evaporation rates which over a 24-hour period leave little liquid helium.

In 1951 at the Westinghouse Research Laboratories, Wexler and his collaborators analysed the problem of heat inflow; their analysis led to the design and construction of a spherical metal dewar of 12·5-litre capacity which when surrounded by liquid nitrogen gave an evaporation rate of only 100 cm^3 of liquid helium per 24 h (Wexler 1951, Wexler and Jacket 1951). This very low evaporation was obtained using a 20-cm length of $\frac{5}{8}$ in diameter inconel tube (0·010 in wall) as the dewar neck, blackening the inner surface of the tube to prevent any 'funnelled' radiation reaching the helium, and also inserting a metal test-tube down the neck as a further radiation trap. They showed that this residual evaporation was due principally to radiation from the liquid-nitrogen-cooled outer sphere to the inner highly polished copper sphere. Thermal conduction down the narrow inconel tube is quite small as fairly efficient heat exchange occurs between the evaporating cold helium gas and the tube wall.

These evaporation rates led Wexler to a figure of 0·0069 for the emissivity of the polished copper vessel at 4·2° K for 77° K black-body radiation, a figure much higher than that calculated from the electrical resistance by classical theory but in fair agreement with the value based on the theory of the anomalous skin effects (for infra-red absorption data see Chapter VII below).

As a result of this work by Wexler, many manufacturers now produce liquid-helium storage vessels of 10–250-litre capacity like that in Fig. 31, with evaporation rates of *ca.* 1 per cent of their capacity per day. For example, a 25-litre container from Superior Air Products evaporated 200–300 cm^3 per 24 h. Evaporation of liquid nitrogen or air from the shield vessel makes it necessary to refill this about once a week. Storage dewars are also available with superinsulation in place of the liquid nitrogen; these have helium evaporation rates of 2–3 per cent of capacity per day; such vessels are much lighter and more suitable for transport by air.

In Plate IV (facing p. 33) a 25-litre liquid-helium container is shown resting on a convenient fork-lift dolly which enables it to be raised by a hydraulic pump to various heights at which it is required for 'siphoning' liquid helium into experimental cryostats. A steel tube of $\frac{1}{2}$-in diameter for 'reaming' out any solid air in the neck of the dewar and a rubber diaphragm level-indicator (see § 2.3 below) are attached to the dolly. The dewar

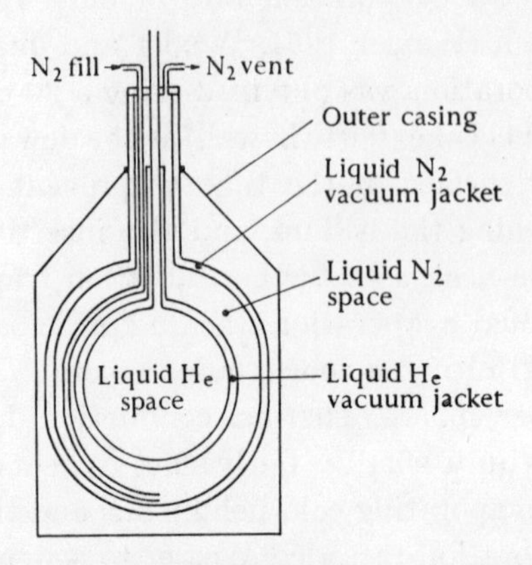

FIG. 31. A commercial metal dewar for the storage of liquid helium or liquid hydrogen.

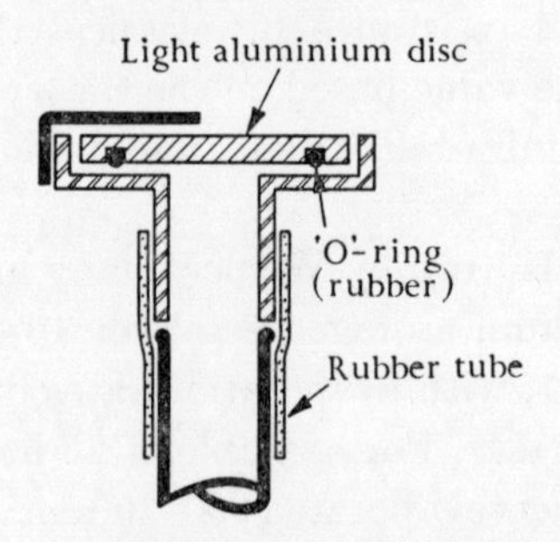

FIG. 32. Simple non-return valve to prevent air or water vapour condensing in the helium-storage dewar.

is normally closed by the non-return valve (Fig. 32) which prevents air or moisture condensing in the neck of the dewar and keeps the liquid under a very slight overpressure. Similarly, condensation in the liquid-nitrogen vents is prevented by the length of plastic tubing, which has a 2-in slit in it to allow nitrogen gas to escape. The liquid-nitrogen space is refilled every

1 to 2 weeks; to avoid the necessity of recooling the inner helium container from 77° to 4° K (requiring several litres of liquid helium) it is desirable to leave a small quantity of liquid helium

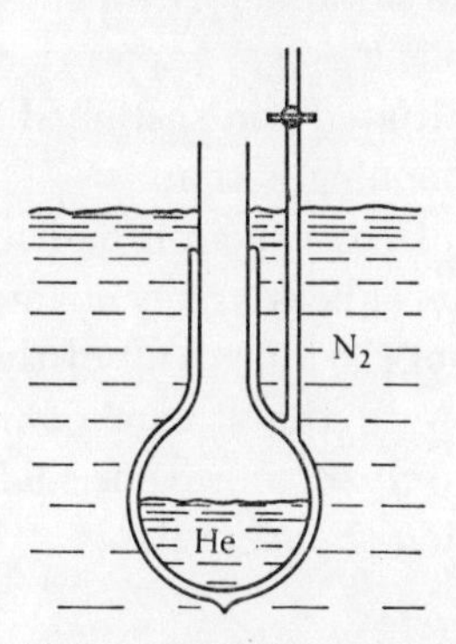

FIG. 33. A Pyrex dewar modified for storing liquid helium (Sydoriak and Sommers 1951).

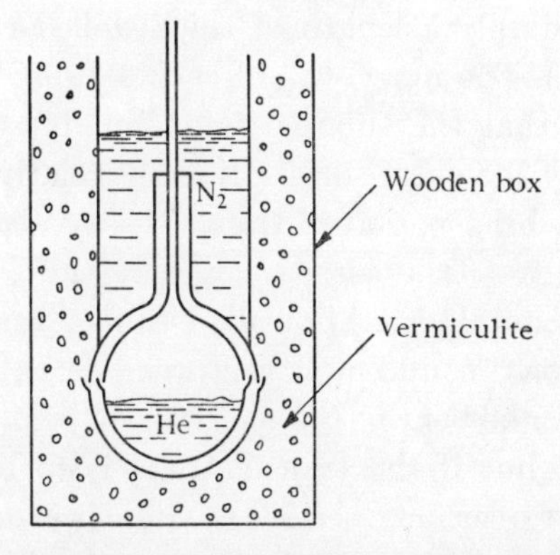

FIG. 34. A commercial copper (liquid-air) dewar modified for storing liquid helium (Gonzalez, White, and Johnston 1951).

in the dewar. Of course this is no longer true when there is an interval of weeks between successive transfers of liquid helium.

Sydoriak and Sommers (1951), and also Gonzalez, White, and Johnston (1951), have described low-evaporation-loss containers which can be constructed by modifying standard liquid-air dewars.

2. Transfer siphons

Introduction

In low-temperature physics, the term 'transfer siphon' has come to mean a tube suitable for transferring low-boiling-point liquids without serious loss by evaporation, although the actual transfer is usually initiated and sustained by a gas overpressure rather than by a siphoning action.

Apart from ease of construction and mechanical reliability, the design of a transfer tube is largely governed by loss considerations; the total heat of the inner wall of the transfer device and the total heat inflow to the stream of liquid during transfer should be small in comparison with the latent heat of vaporization of the liquid being transferred.

Liquid air

If liquid air, nitrogen, or oxygen is being transferred a distance of a few feet in a quantity of a litre in 10 or 15 s, then the tube can be simply a length of thick-walled rubber tube ($\frac{1}{4}$ in i.d., $\frac{3}{4}$ in o.d.). The inner wall of such a rubber tube rapidly freezes hard so that the tube is rigid, but due to its poor heat conductivity the outer surface remains relatively warm and pliable during a brief period of transfer—say about 1 min required to fill a 4-litre flask with an overpressure of $\frac{1}{4}$ to $\frac{1}{2}$ atm on the storage container; thin-walled rubber tubing will freeze hard, become brittle, and may fracture. Alternatively, any metal tube which can be bent to the required shape is suitable for liquid-air transfer, but if the transfer tube is to be left in the storage container over long periods of time, it is desirable that the tube in the neck of the container should be a low-heat-conductivity alloy—monel, stainless steel, cupro-nickel. If the liquid-air transfer is to be rather slower and occupy many minutes, then a wrapping of glass wool, asbestos tape, etc., may be sufficient to make the losses negligible.

When the transfer is of long duration and/or over an appreciable distance, more efficient insulation is desirable. A very convenient covering for the metal tube is a flexible foamed plastic such as Armaflex (Armstrong Cork Company,

Pennsylvania). In the form of tube this slips onto the metal pipe easily and seals against condensation of moisture. Other foam or powder insulators are thermally efficient but are often more difficult to attach to the transfer tubing.

Double-wall metal vacuum siphons are laborious to construct and are therefore more commonly reserved for liquid hydrogen and helium (see below), but can be used for liquid air provided that there is some adsorbent charcoal exposed in one or other

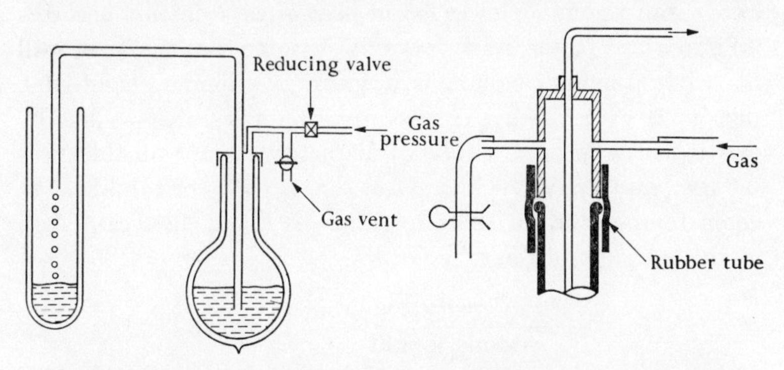

FIG. 35. The transfer of a liquefied gas; inset shows a simple brass coupling for use with a narrow-necked metal dewar.

end of the vacuum enclosure to act as a cryopump (Jones and Larsen 1948). The tubes and charcoal should be well warmed with a flame to outgas them during evacuation and prior to sealing off. With the aid of petrol (or refrigeration) unions a number of such tubes, either straight or bent, can be joined together to carry liquid air from storage flasks to various points where it is required.

In the case of liquid air it is most common to use compressed gas from cylinders or supply lines, to force the liquid through the transfer siphons. If such transfers are fairly frequent, the process of pressurizing and depressurizing the container at each transfer both wastes and frequently contaminates the liquid. In cases where cost and purity are important it is convenient to leave the container sealed by a spring-loaded blow-off valve—designed to blow-off at 3–4 lb/in^2; by means of a valve in the transfer tube, liquid can be drawn off under this pressure head when desired. A useful sealing device of this type has been described by Wexler and Corak (1950). Commercial firms produce liquid-discharge tubes fitted with either manually or electrically controlled valves.

Hydrogen and helium

In designing a transfer tube for liquid helium (or liquid hydrogen) the same considerations apply that have been discussed above, but the much lower latent heat of vaporization becomes an important factor. As an example, assume that the inner wall of a liquid-helium siphon is a metal of atomic weight 100, density 10 g/cm^3, Debye characteristic temperature $\theta_D \sim 300°$ K, and is in the form of a tube of diameter 3 mm, wall thickness 0·5 mm, and length 100 cm. Then the enthalpy or total heat at room temperature of this inner tube is about 1200 cal/g mol, so the enthalpy of the tube is

$$\frac{1200 \times \text{density} \times \text{volume}}{\text{atomic weight}} \simeq 600 \text{ cal.}$$

This indicates that if only the latent heat of liquid helium is used to pre-cool the siphon, about 1 litre of liquid would be required (the latent heat of helium is about 0·65 cal/cm^3 of liquid). Fortunately, however, the heat capacity of the evaporated gas also cools the tube wall; each cubic centimetre of liquid produces helium gas with a heat capacity at constant pressure of 0·15 cal deg^{-1}, representing an enthalpy at room temperature of about 40 cal/cm^3 of liquid equivalent. If we can hope to use about half the cooling content of the evaporated gas, then only about 30 cm^3 of liquid would be required to pre-cool the siphon. In practice the wall thickness of the inner tube is usually rather less than suggested in this example so that the total heat content at room temperature of the inner tube of a typical helium siphon (e.g. see Fig. 37 below) might be 500 calories and require about 50 cm^3 or less of liquid helium for pre-cooling.

Fig. 36 illustrates a simple method of constructing a transfer siphon (Dowley and Knight 1963). Star-shaped spacers of Teflon separate the two tubes of well-annealed stainless steel. They are bent using ice as a filler to stop crimping; alternatively Cerrobend or acetamide can be used as fillers during the bending. Dowley and Knight did not evacuate their siphon but relied on cryopumping. A rather similar siphon, produced by the Oxford

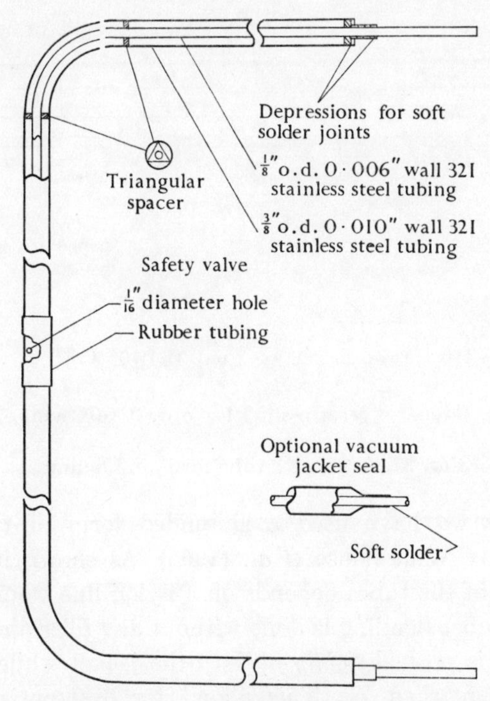

FIG. 36. Details of a transfer tube for liquid helium (after Dowley and Knight 1963).

Instrument Company, has a small brass side-arm for evacuation; in this side-arm is a $\frac{1}{4}$-in diameter steel ball-bearing resting on a rubber O-ring to act as a one-way valve.

Another method of construction which the writer has found satisfactory is that illustrated in Fig. 37. The numbers on the figure indicate the usual order in which the various joints—made with 'Easy-Flo' silver solder except where shown otherwise—are made. The most important joints are those marked 1 and 2, 5, and 6 which must be high-vacuum-tight to avoid a helium leak into the vacuum space. As in the case of a liquid-air siphon, frost on the outer wall indicates any cold spots where there is a serious contact between the two tubes; small Plexiglass stars or similar spacers near the elbows and near the middle of the long horizontal section avoid this trouble.

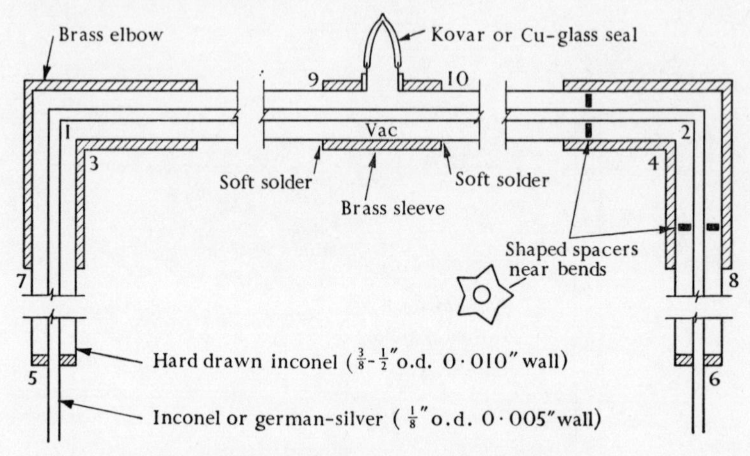

Fig. 37. A transfer tube for liquid helium.

Recently we have used a simplified form of the siphon described by Whitehouse *et al.* (1965). As shown in Fig. 38, separation of the tubes depends on a nylon line wound around the inner tube. Bending is done without any filler and then the inner tube is pushed *tightly* up into the jacket while soldering the points marked 'S'. This allows for differential thermal contraction without using any bellows in the side-arm.

Other siphons have been described by Stout (1954), Croft and Jones (1950), Jacobs and Richards (1954), Serin (see Rose-Innes 1964, p. 12). Sometimes it is useful to have one which is more flexible than those just mentioned. They can be made with flexible bellows or rubber or plastic in the outer jacket-tube (e.g. Mathewson 1955, Swim and Clement 1960, Kimber 1963) and are also commercially available. Generally, however, conventional siphons like those in Figs. 36–38 can be made with quite long side-arms and, if used with different extension pieces on the sides and sliding O-ring seals on the storage dewar and/or cryostat dewar, can be accommodated to a variety of differences in height between the two dewars.

If the transfer of liquid helium takes only a few minutes then the amount of radiant heat reaching the liquid stream is not serious. If as in the example above, we assume an inner tube of

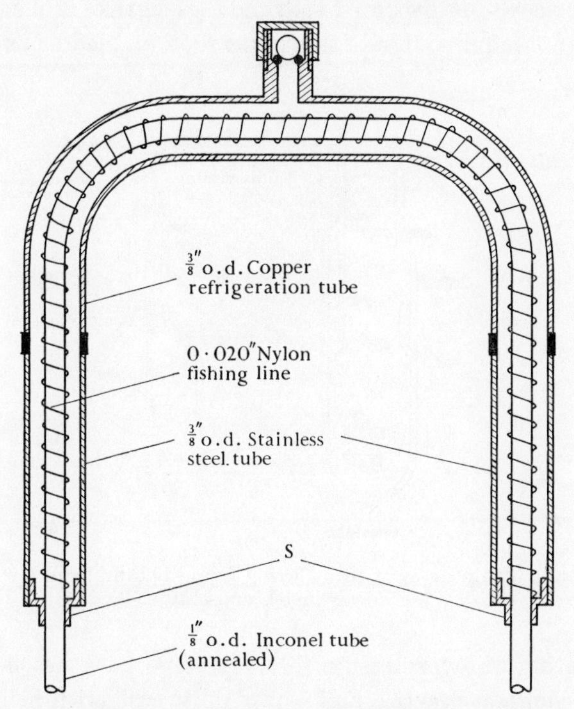

FIG. 38. A transfer tube for liquid helium, based on that described by Whitehouse *et al.* (1965).

3 mm diameter, 100 cm long, and further assume that its emissivity $\varepsilon \sim 0{\cdot}05$ and that it is exposed to room temperature radiation from a surrounding black body, then the heat inflow

$$\begin{aligned}\dot{Q} &= \sigma A\varepsilon(T_1^4 - T_2^4)\\ &\simeq 5{\cdot}7\times 10^{-12}\times 100\times 0{\cdot}05(300)^4\ \mathrm{J/s}\\ &= 0{\cdot}23\ \mathrm{J/s}\\ &\simeq 3\ \mathrm{cal/min},\end{aligned}$$

which is sufficient to evaporate about 5 cm³ of liquid helium per minute.

A method of transfer applicable to both liquid helium and liquid hydrogen is shown schematically in Fig. 39 and in the photograph of Plate IV. Provision for collecting the evaporated gas is shown, but if this is unnecessary, the 'sealing' of the siphon into the cryostat—either through a rubber bung or a rubber sleeve—is dispensed with and the length of rubber by-pass tube is not required. Use of the rubber football bladder as a means

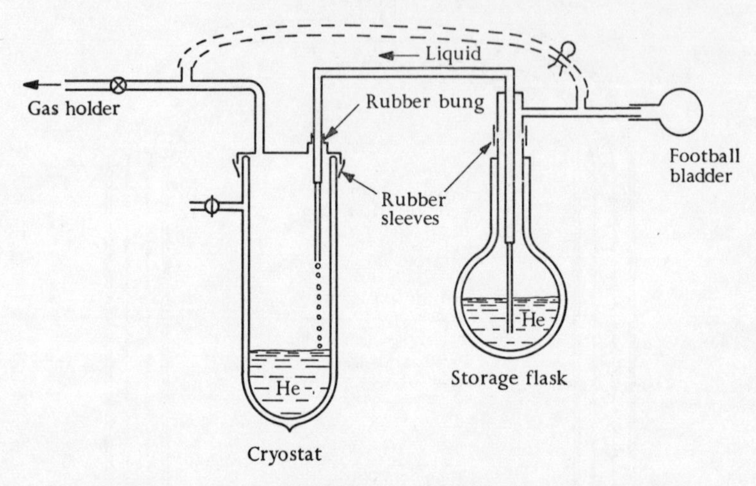

Fig. 39. The process of transferring liquid helium from a storage dewar into a cryostat.

of creating an overpressure in the storage flask has become a fairly common practice and works quite well with helium and hydrogen; a fluctuating pressure applied with the fingers to the 'bladder' causes an oscillation of warm gas into the flask and maintains quite easily the pressure of about $\frac{1}{4}$ to $\frac{1}{2}$ lb/in^2 required for transfer. In cases where a substantially higher overpressure is required, e.g. where the pressure in the gas holder (and therefore in the gas-return pipe and cryostat dewar) is relatively high, a gas cylinder with reducing valve can supply the overpressure required to force the liquid through the transfer tube.

Considerable care should be taken when first transferring liquid helium into either a warm dewar or into a partially filled dewar, otherwise litres of liquid can be evaporated needlessly.

If the dewar is warm (say 80° K) and particularly if it contains equipment of large heat capacity, a slow initial transfer allows the cooling capacities of the liquid and vapour to be used more fully. If liquid is already in the dewar, a slow build-up of pressure in the storage flask (helped by closing the relief clip in Fig. 39 a few minutes beforehand) is essential to avoid having a warm stream of helium gas hitting the liquid surface. The novice is well advised to use a flow-rate meter or monitor on the gas recovery system to check the progress of the transfer. A little care can result in liquid collecting after, say, 10–15 ft^3 of gas are evaporated (equivalent to less than $\frac{1}{2}$ litre of liquid helium) rather than 30 or 40 ft^3.

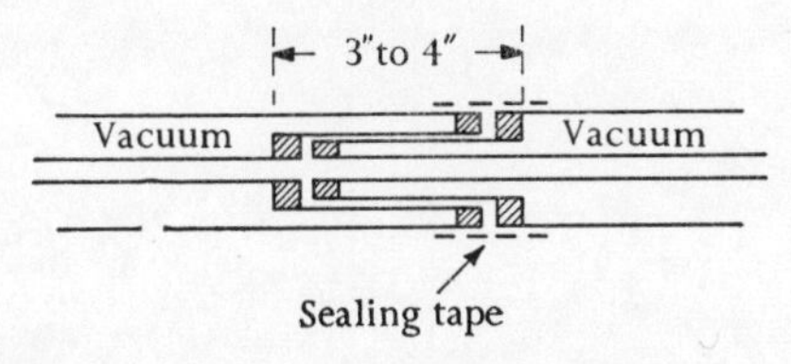

FIG. 40. A coupling for liquid helium (or hydrogen) transfer tubes.

A junction in a transfer tube is useful sometimes; for example when two parts of the siphon are permanently attached to the respective dewars from which and to which the liquid is being transferred. Fig. 40 shows a sleeve coupling which can be sealed against leakage by a strip of adhesive tape.

A valve is sometimes required in a transfer siphon, for example when drawing off liquid helium from a liquefier. The example in Fig. 41 shows a stainless-steel needle (total taper 10°–20°) seating in cylindrical bush of brass or some other metal softer than the stainless-steel needle; this bush also seals the end of the double-walled vacuum siphon. The valve is actuated by a knurled brass piece which screws on to two brass sleeves, one soldered to the upper outside transfer tube and the other to the low-conductivity metal tube which moves the valve needle; one of the threads is right hand and the other is left hand, and a metal pin prevents the lower sleeve from turning as the knurled head is turned. Descriptions of such needle valves and also ball valves used for the same purpose have been given by Croft and Jones (1950), Scott and Cook (1948), and in the manuals of the Arthur D. Little Corporation which describe their Collins liquefier.

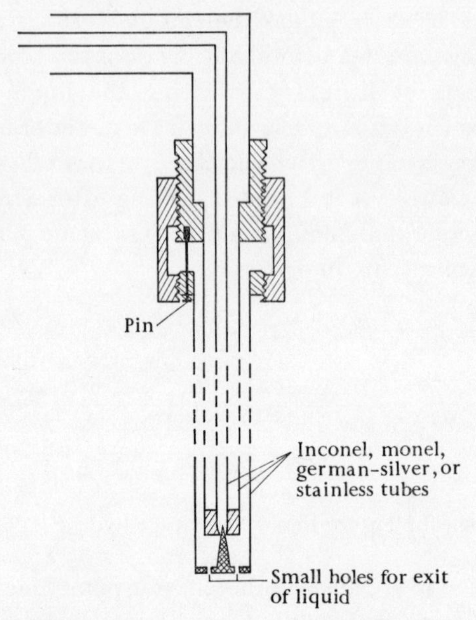

FIG. 41. A valve on a transfer tube.

3. Liquid-level indicators

Introduction

A number of devices for determining the level of a liquefied gas in a dewar have been described in the past few years; some of these have been suitable for only one or two liquids and others for most liquefied gases; some have been essentially discontinuous in nature, i.e. only registering when the liquid reaches a certain level, and other have been continuous recorders of level; some also have been used to actuate a transfer of liquid into the dewar. Many indicators are thermometric in principle, e.g. the change in vapour or gas pressure with temperature, or the change in electrical resistance or expansion of a bimetallic strip are used in level detectors which depend on the change in temperature at or near the liquid surface.

Rather than discuss the details of all these level indicators, let us consider some examples of the types which seem to satisfy most requirements.

Examples

The devices most commonly used to give a continuous indication of liquid level or liquid depth depend on measurement of (a) hydrostatic pressure difference between bottom of liquid and vapour above it, (b) electrical resistance of a wire partially immersed in liquid, (c) electrical capacitance between coaxial cylinders immersed in the liquid and vapour, or (d) flotation.

In the hydrostatic pressure-head type of depth gauge, an oil-filled U-tube or some other form of differential manometer records the difference in pressure between the base of the column of liquid in the container and the pressure above the surface. These are universally applicable to liquids, but they are rather insensitive in the case of liquefied gases of very low density such as hydrogen. Magnehelic pressure gauges (e.g. from F. Dwyer Manufacturing Co., Indiana) have been used successfully for this purpose by Trujillo and Marino (1964), Lyon and Gillich (1965), Pope and McLaughlin (1966), etc.

Electrical resistance wires which give a continuous recording most conveniently are of materials which are superconducting in the liquid (helium) but have a relatively high resistance when they are in the normal state, that is above the liquid level. The fine wire can be wound onto a threaded rod (e.g. Rasor 1954) which extends to the bottom of the dewar. Tantalum wire (critical temperature $T_C \sim 4{\cdot}3°$ K) was first used for this purpose (Feldmeier and Serin 1948, Rovinskii 1961) but other wires with slightly higher values of T_C have often been found to be more suitable. For example, Ries and Satterthwaite (1964) have used manganin wire of 0·008 cm diam. plated with a thickness of *ca.* 3×10^{-5} cm of a tin $+40$ per cent lead alloy; Figgins, Shepherd, and Snowman (1964) used a similar coating ($T_C \simeq 7°$ K), tinned onto constantan wire and they also used vanadium wire ($T_C \simeq 5{\cdot}3°$ K) successfully.

Capacitance gauges depend on the difference in dielectric constant of liquid and vapour. Dash and Boorse (1951), Williams and Maxwell (1954), and Meiboom and O'Brien (1963) have described their construction. Fig. 42 illustrates one which

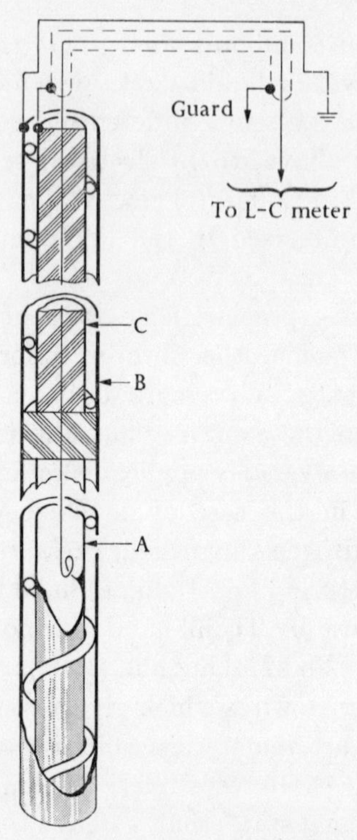

FIG. 42. Cylindrical capacitor depth gauge made from stainless-steel tubes and a nylon fishing line as a spacer (after Meiboom and O'Brien 1963).

can be simply made from a $\frac{1}{8}$-in o.d. stainless-steel tube in a $\frac{3}{16}$ in o.d. tube, separated by nylon fishing line. The capacity between ground B and the sensitive electrode, tube A, is measured by connecting it to a Tektronix type 130 LC-meter with a coaxial cable, whose shield is attached to C and the 'guard' circuit on the meter.

Float gauges are another means of showing changes in liquid level. In the 'old' Oxford hydrogen liquefier (Jones, Larsen, and

Simon 1948) sealed bulbs of thin-walled German silver float in each of the three liquid-air dewars; light German silver tubes are attached to the 'bulbs' and these indicate the level by their height in glass viewing tubes at the top of the liquefier.

Styrofoam also makes a very suitable float as it has a density even less than that of liquid hydrogen; Babiskin (1950) mentions its use in liquid-helium dewars, the Styrofoam having a light balsa wood stick attached above it as an indicator.

Allen (1960) describes the use of Styrofoam beads floating on the surface of liquid helium as a convenient means of indicating the surface more clearly and incidentally of reducing the heat inflow. Optical dipsticks have also been used (Kitts and Harler 1954, Geake 1954).

When there is no restriction on the vertical movement of the depth gauge so that the level can be found by searching with a 'probe' of some kind, there are very simple devices which can be made. By placing the ear above the end of an open tube (for example, $\frac{1}{4}$-in diameter brass or German silver tube) and lowering it into a dewar, an unmistakable 'bubbling' noise can be heard when the liquid (helium, hydrogen, nitrogen, oxygen) level is first reached; however, this is rather unsatisfactory for detecting liquid helium or hydrogen levels in narrow-neck metal storage flasks. Another remarkably simple device, which uses the human finger as a sensing element and is very suitable for helium and can be used for hydrogen, is the rubber diaphragm level finder (Gaffney and Clement 1955). It is very easily made with a small piece of thin rubber sheet cut from a rubber glove which seals the top of a small brass cup at the end of an open tube (0·1 in diam.). Spontaneous oscillations in the gas column in the narrow tube cause the rubber diaphragm to vibrate with a frequency which falls quite markedly when the lower end of the tube reaches the liquid level. The sensitivity may be increased by suitably enlarging and shaping the funnel near to the diaphragm or by using a stethoscope to detect the oscillations (cf. Trammell 1962).

This diaphragm device (Fig. 43) has proved the simplest and most popular of level-finders for use with liquid helium, but it should not be left sitting in the liquid because of the thermal oscillations which increase the heat inflow to the liquid.

Other useful level-finders have been described, many of them based on the electrical resistance of a carbon resistor or a

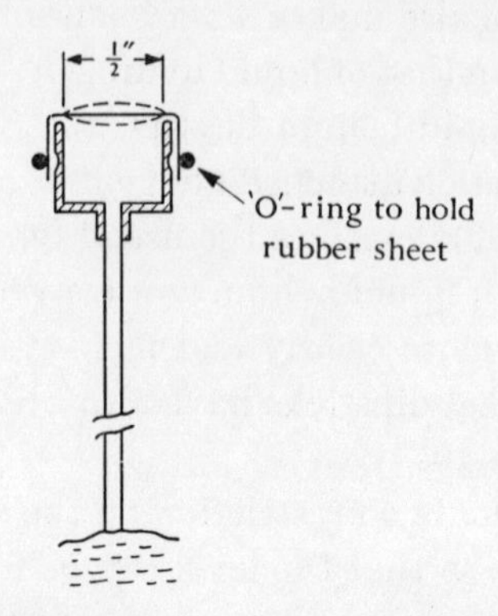

Fig. 43. A rubber diaphragm level finder for liquid helium (Gaffney and Clement 1955).

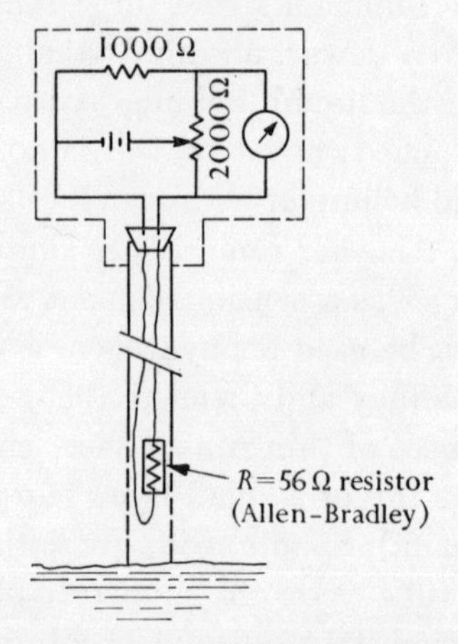

Fig. 44. A carbon resistor used as a liquid-helium level-finder.

thermistor; their resistance varies sharply with a small temperature change and therefore is very different when immersed in liquid where heat dissipation is good and in the vapour just above the liquid. Fig. 44 illustrates a small 'dipstick' type of level-finder which has a carbon resistor at the bottom of a stainless-steel tube and small bridge circuit at the top. This is operated by a 30-V hearing-aid battery and uses a 100-μA meter to indicate when the bridge is balanced (in liquid helium) or unbalanced (in the vapour). Further information on these resistive probes can be found in Maimoni (1956), Sauzade *et al.* (1965), Lewin (1963), etc.

4. Liquid-level controllers

A type of level indicator which can also activate a transfer of liquid is very useful in many low-temperature laboratories. The problem of keeping liquid-air traps filled by some automatic process is often important, e.g. in charcoal purifiers, pre-cooling

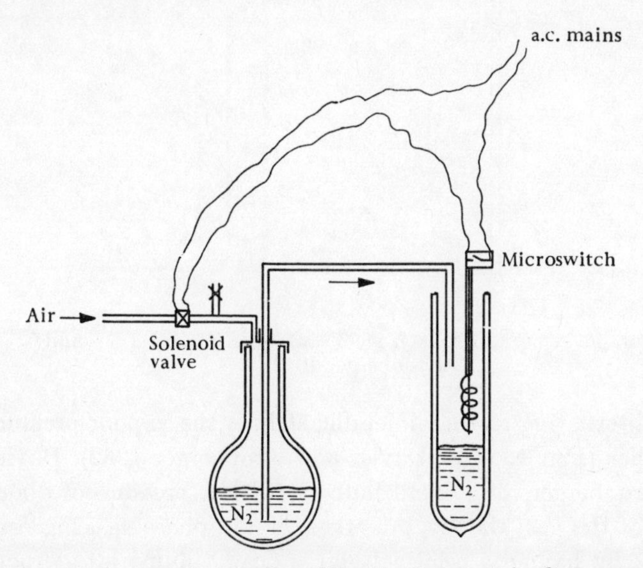

FIG. 45. Automatic transfer of liquid nitrogen (air, O_2) controlled by bimetallic strip.

dewars, etc. Two simple devices for doing this are shown in Figs. 45, 46. In the first (based on that of Lounsbury 1951) a commercial bimetallic strip thermostat activates a micro-switch relay; this in turn operates a solenoid valve allowing compressed air into a storage container, forcing the liquid across into the dewar. When the liquid reaches the bimetallic strip, the relay opens, the solenoid valve closes and the overpressure in the container is slowly released through a 'bleed' valve (e.g. needle valve which is slightly open) or by means of a second solenoid valve, and then the transfer stops.

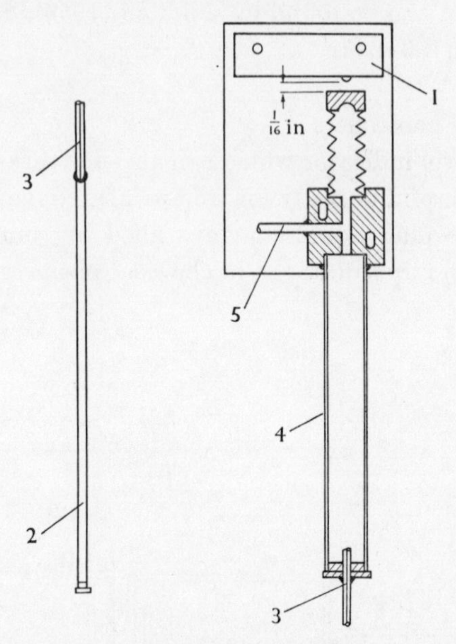

FIG. 46. A liquid-nitrogen-level controller (after Davies and Kronberger 1952).

An alternative to the bimetallic strip is the vapour-pressure controller (Fig. 46 after Davies and Kronberger 1952). Davies and Kronberger fill a small bulb (2) with a pressure of about 2 atm of the gas, the level of whose liquid phase is being controlled; by having a relatively large dead volume (4) at room temperature they ensure that the operation of the microswitch does not depend too critically on the temperature of the vapour above the liquid surface. Jones (1948) developed a rather similar device and used the expansion or contraction of a bellows to operate the liquid-air inlet valve by a direct mechanical linkage.

Similar in principle to these latter devices and particularly simple is a small glass level controller (after Sherwood 1952) in which a platinum–mercury switch is used to control electric power to a heater in a liquid-air container. This heating produces an overpressure in the container sufficient to transfer liquid air to another dewar; when the dewar is nearly filled, the relay opens, heating ceases, and the overpressure is released slowly through a 'bleed' valve.

Many descriptions of other level controllers have appeared in the pages of the *Journal of Scientific Instruments*, *Review of Scientific Instruments*, and *Cryogenics* but most do not differ greatly in principle from those discussed above.

Most of the common sensing elements which were discussed in the previous section can be made to operate a refilling system; that is, carbon resistors (e.g. Davies and Gosling 1963), thermistors (Flinn and Moore 1959), vapour-pressure bulbs (Purser and Richards 1959, Roizen and Gannus 1962), thermocouples (Greaves 1963), floats (Henshaw 1957, Leefe and Liebson 1960), capacitance (Nechaev, 1962) etc. From this sensing element a signal is carried via a mechanical linkage or electrically to valve(s) on the liquid storage flask. The valve(s) may be mechanically or electrically operated and may be situated either in the liquid transfer tube or the gas-pressure supply line (Fig. 45) or in the bleed (relief) valve.

In some instances a requirement for very precise control of level is met and in others an effort is made to avoid the transfer of liquid occurring too frequently.

5. References

Allen, R. J. (1960). *Rev. scient. Instrum.* **31**, 203.

Babiskin, J. (1950). *Ibid.* **21**, 941.

Barrick, P. L., Weitzel, D. H., and Connolly, T. W. (1955). *Proc.* 1954 *Cryogenic Engng. Conf. N.B.S. Report No.* 3517, p. 210.

Croft, A. J. and Jones, G. O. (1950). *Br. J. appl. Phys.* **1**, 137.

Dash, J. G. and Boorse, H. A. (1951). *Phys. Rev.* **82**, 851.

Davies, E. A. and Gosling, D. S. (1963). *J. scient. Instrum.* **40**, 429.

Davies, M. G. and Kronberger, H. (1952). Ibid. **29**, 335.

Desirant, M. C. and Horvath, W. J. (1948). *Rev. scient. Instrum.* **19**, 718.

Dowley, M. W. and Knight, R. D. (1963), Ibid. **34**, 1449.

Feldmeier, J. R. and Serin, B. (1948). Ibid. **19**, 916.

Figgins, B. F., Shepherd, T. A., and Snowman, J. W. (1964). *J. scient. Instrum.* **41**, 520.

Flinn, I. and Moore, P. (1959). *J. scient. Instrum.* **36**, 374.
Fradkov, A. B. (1962). *Cryogenics* **2**, 177.
Fulk, M. M. (1959). *Prog. Cryogen.* **1**, 64.
Gaffney, J. and Clement, J. R. (1955). *Rev. scient. Instrum.* **26**, 620.
Geake, J. E. (1954). *J. scient. Instrum.* **31**, 260.
Giauque, W. F. (1947). *Rev. scient. Instrum.* **18**, 852.
Gonzalez, O. D., White, D., and Johnston, H. L. (1951). Ibid. **22**, 702.
Greaves, C. (1963). *J. scient. Instrum.* **40**, 425.
Henshaw, D. E. (1957). Ibid. **34**, 207.
Jacobs, R. B. and Richards, R. J. (1957). *Rev. scient. Instrum.* **28**, 291.
Jones, G. O. (1948). *J. scient. Instrum.* **25**, 239.
—— and Larsen, A. H. (1948). Ibid. **25**, 375.
—— —— and Simon, F. E. (1948). *Research* **1**, 420.
Keesom, W. H. (1942). *Helium*, p. 131. Elsevier, Amsterdam.
Kimber, R. A. (1963). *Cryogenics* **3**, 104.
Kitts, W. T. and Harler, F. L. (1954). *Rev. scient. Instrum.* **25**, 926.
Lane, C. T. and Fairbank, H. A. (1947). Ibid. **18**, 522.
Larsen, A. H., Simon, F. E., and Swenson, C. A. (1948). Ibid. **19**, 266.
Leefe, S. and Liebson, M. (1960). Ibid. **31**, 1353.
Lewin, J. D. (1963). *J. scient. Instrum.* **40**, 539.
Lounsbury, M. (1951). *Rev. scient. Instrum.* **22**, 533.
Lyon, D. N. and Gillich, J. J. (1965). Ibid. **36**, 1164.
Maimoni, A. (1956). Ibid. **27**, 1024.
Mathewson, R. C. (1955). Ibid. **26**, 616.
Meiboom, S. and O'Brien, J. P. (1963). Ibid. **34**, 811.
Nechaev, Yu. I. (1962). *Cryogenics* **2**, 175.
Pope, W. L. and McLaughlin, E. F. (1966). *J. scient. Instrum.* **43**, 260.
Purser, K. H. and Richards, J. R. (1959). Ibid. **36**, 142.
Rasor, N. S. (1954). *Rev. scient. Instrum.* **25**, 311.
Ries, R. and Satterthwaite, C. B. (1964). Ibid. **35**, 762.
Rogers, W. A., Buritz, R. S., and Alpert, D. (1954). *J. appl. Phys.* **25**, 868.
Roizen, L. I. and Gannus, V. K. (1962). *Cryogenics* **2**, 145.
Rose-Innes, A. C. (1964). *Low temperature techniques.* English University Press, London.
Rovinskii, A. E. (1961). *Cryogenics* **2**, 115.
Sauzade, M., Georges, C., Pontnau, J., and Lesas, P. (1965). *Cryogenics* **5**, 42.
Scott, R. B. (1957). *J. Res. natn. Bur. Stand.* **58**, 317.
—— and Cook, J. W. (1948). *Rev. scient. Instrum.* **19**, 889.
—— —— and Brickwedde, F. G. (1931). *J. Res. natn. Bur. Stand.* **7**, 935.
—— Denton, W. H., and Nicholls, C. M. (1964). *Technology and uses of liquid hydrogen.* Pergamon, London.
Sherwood, J. E. (1952). *Rev. scient. Instrum.* **23**, 446.
Stout, J. W. (1954). Ibid. **25**, 929.
Swim, R. T. and Clement, J. R. (1960). *Adv. cryogen. Engng* **6**, 396.
Sydoriak, S. G. and Sommers, H. S. (1951). *Rev. scient. Instrum.* **22**, 915.

TRAMMELL, A. (1962). *Rev. scient. Instrum.* **33**, 490.

TRUJILLO, S. M. and MARINO, L. L. (1964). *J. scient. Instrum.* **41**, 184.

WEITZEL, D. H., DRAPER, J. W., PARK, O. E., TIMMERHAUS, K. D., and VAN VALIN, C. C. (1957). *Proc. 1956 cryogen. engng conf. N.B.S., Boulder, Colorado*, p. 12. Univ. of Colorado, Boulder.

WEXLER, A. (1951). *J. appl. Phys.* **22**, 1463.

—— and CORAK, W. S. (1950). *Rev. scient. Instrum.* **21**, 583.

—— and JACKET, H. S. (1951). *Ibid.* **22**, 282.

WHITE, J. F. (1948). *Chem. Engng Prog.* **44**, 647.

WHITEHOUSE, J. E., CALLCOTT, T. A., NABER, J. A., and RABY, J. S. (1965). *Rev. scient. Instrum.* **36**, 768.

WILLIAMS, W. E. and MAXWELL, E. (1954). Ibid. **25**, 111.

CHAPTER III

HEAT EXCHANGERS

1. Introduction

WHILE the topic of 'heat exchangers' is a very general one, the use of the word in low-temperature physics is usually confined to systems in which heat is transferred from an entering warm gas stream to a returning cold gas stream; the mathematical formulae developed for dealing with these systems can also treat cases where heat is transferred from a gas stream to a cold liquid bath. The problem of calculating desirable dimensions for countercurrent heat exchangers and for cooling coils is common to both the design of gas liquefiers and such associated equipment as purifying traps.

As was discussed (§ 1.2) above, the heat exchanger is essential in the Joule–Thomson process for liquefying helium; in the many cases where it is not essential, economy of operation makes the role of the heat exchanger very important. Generally we are faced with a task of reducing the wastage of 'cold' by utilizing the cooling capacity of returning gas streams to cool the incoming gas. In a liquefier these returning gas streams may be the cooled (but unliquefied) portion of the gas being circulated, or may be cold gas evaporated from a liquid refrigerant used in the cooling or purifying stages. For maximum efficiency in the interchange process a maximum amount of heat should be transferred from the incoming gas stream to the outgoing streams, so that the temperature of the outgoing gas at its exit approaches closely the temperature of the incoming gas at its entry point into the system. In Fig. 47, a warm gas enters at temperature T_1 and leaves at T_x, exchanging heat with a cold gas entering at T_0 and leaving at T_2.

In the limit of maximum efficiency $T_2 \rightarrow T_1$ and temperature $T_0 < T_x < T_1$, but the value of T_x can only be found if the relative heat capacities of the two gas streams are known, i.e. the mass flow per second in each stream and the specific heat of the gas in each stream. Of course, in the case where the returning stream has a greater heat capacity than the downflowing stream, then $T_x \to T_0$, and T_2 may be appreciably less than T_1, but this is an artificial condition which is scarcely likely to arise in the cases with which we are concerned.

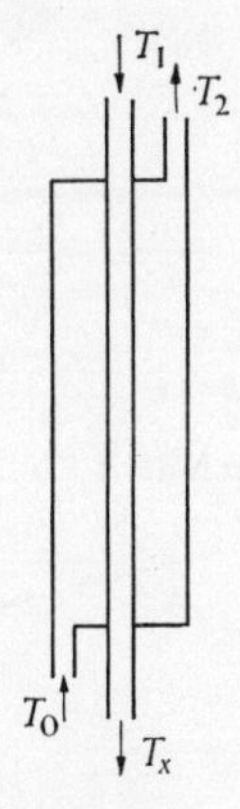

FIG. 47.

Associated with the calculation of efficiency of heat transfer is the calculation of the pressure drop for the gas flowing through the exchanger. It is generally necessary that the pressure drop be small in comparison with the total entry or exit pressure of the gas.

2. Calculation of pressure gradient and heat transfer

Pressure drop

Consider a gas flowing in a circular pipe of diameter D cm and length L cm, a total mass m g passing any section per second. Then if the density of the gas be ρ g cm^{-3} and viscosity η CGS units (poises), classical hydrodynamics and dimensional analysis (see, for example, Jakob 1949, McAdams 1954) indicate that the pressure drop Δp in dyn cm^{-2} is given by

$$\Delta p = \frac{1}{2}\psi\frac{LG^2}{D\rho} \text{ where } G = \frac{4m}{\pi D^2}\,\text{g cm}^{-2}\,\text{s}^{-1}, \tag{1}$$

and ψ is a dimensionless factor which is given by

$$\psi = 64\left(\frac{\eta}{GD}\right) \tag{2}$$

for laminar flow (Poiseuille flow) or

$$\psi = 0{\cdot}316\left(\frac{GD}{\eta}\right)^{-0\cdot25} \tag{3}$$

for turbulent flow (see Fig. 48).

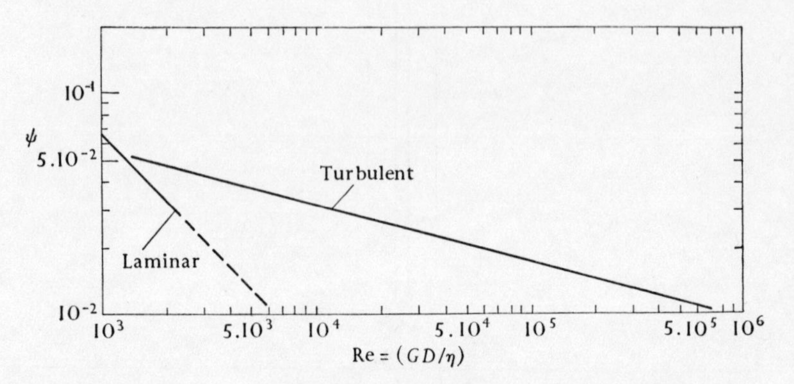

FIG. 48. The dimensionless factor ψ as a function of the Reynolds number.

The dimensionless factor (GD/η) is usually called the Reynolds number, denoted by Re. In the case of a straight pipe of circular section, the condition for turbulence is that Re > 2300, but this critical Reynolds number varies considerably for different shapes of the carrying tube. In most applications to low-temperature equipment, Re $\gg$ 2300 and the flow will be turbulent, a fact that is assumed in the calculation of heat transfer coefficients given below.

For generality the equation for Δp may be extended to the case of non-circular section tubes, by defining a hydrodynamic diameter D_h given by

$$D_h = 4 \times \frac{\text{cross-sectional area of tube considered } (A)}{\text{total perimeter of surfaces in contact with gas stream } (P)}; \tag{4}$$

then G is the mass flow per cm²/s, i.e. $G = m/A$, and

$$\Delta p = \psi L G^2 / 2\rho D_h. \tag{5}$$

Heat transfer coefficient

If the gas flowing through the tube is at a temperature different by an amount ΔT from the temperature of the tube wall, heat will be transferred at a rate

$$\dot{Q} = h\Delta T,$$

where the coefficient h is the 'heat transfer coefficient'. For turbulent flow, the temperature is sensibly constant across the section of the tube and the major temperature difference occurs across the thin layer of gas at the tube surface. By dimensional arguments Nusselt (1909) related the heat transfer coefficient h to the thermal conductivity λ of the gas, the effective diameter

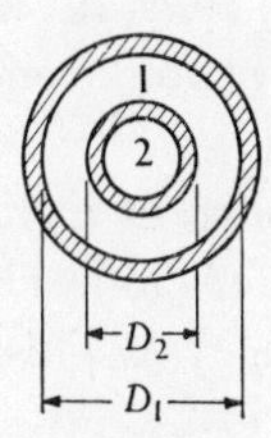

FIG. 49

D_e, the dimensions of the tube, viscosity of the gas η, and its heat capacity C_p. The effective diameter D_e is now defined by

$$D_e = 4 \times \frac{\text{cross-sectional area of tube considered}}{\text{perimeter of surface to which the gas stream transfers heat}} \tag{6}$$

so that in some cases D_e may be quite different from D_h. For example, in the simple counter-current exchanger of two concentric tubes (in Fig. 49), the gas stream flowing in the annular space 1 will suffer a viscous drag due to surfaces of

$$\text{perimeter} = \pi D_1 + \pi D_2;$$

for our purposes the gas transfers heat only to the inner tube of perimeter πD_2.

Therefore to calculate Δp, we use

$$D_h = 4\frac{(\pi D_1^2/4 - \pi D_2^2/4)}{\pi D_1 + \pi D_2} = (D_1 - D_2),$$

and to calculate h_1, we use

$$D_e = 4\frac{(\pi D_1^2/4 - \pi D_2^2/4)}{\pi D_2} = \frac{(D_1^2 - D_2^2)}{D_2}.$$

Nusselt's equation for h in the dimensionless form is

$$\text{Nu} = \text{const. } (\text{Re})^x (\text{Pr})^y (D_e/L)^z,$$

where Nusselt's number $\text{Nu} = hD_e/\lambda$,

Reynolds number $\text{Re} = GD_h/\eta$,

and Prandtl's number $\text{Pr} = \eta C_p/\lambda$. (7)

Experimental values for the constant and for the indices x, y, z have been obtained under various conditions. z is found to be extremely small so that $(D_e/L)^z \simeq 1$; it is also found that $x = 0{\cdot}8$, $y \simeq 0{\cdot}4$, and the constant has a value of about 0·02.

In the form given by McAdams (1954) for turbulent flow $hD_e/\lambda = 0{\cdot}023(D_e G/\eta)^{0{\cdot}8}(C_p \eta/\lambda)^{0{\cdot}4}$. This can be reduced to the more useful dimensional form

$$h = \frac{0{\cdot}023}{(\text{Pr})^{0{\cdot}6}} C_p \frac{G^{0{\cdot}8}\eta^{0{\cdot}2}}{D_e^{\;0{\cdot}2}}, \tag{8}$$

where h is in cal/cm² s degK if λ is in cal/s cm degK, and C_p in cal/g degK.

General equations for heat interchange

By applying considerations of conservation of energy to a section of the exchanger shown in Fig. 50, an equation relating the heat transfer coefficients h_1 and h_2 of the two gas streams and their heat capacities $m_1 c_1$, $m_2 c_2$ to the dimensions and temperature distribution may be obtained.

Here c_1 and c_2 are the specific heats of the gases in cal/g degK, T_{L_1} is the entry temperature of gas stream no. 1 and T_{0_1} its exit temperature; similarly T_{0_2} and T_{L_2} are the entry and exit temperatures respectively of the second gas stream (upflowing stream). If this countercurrent exchanger consists of two concentric tubes, and the inner tube has an internal diameter D_1

and external diameter D_2 then (Mandl 1948)

$$L = \frac{\alpha}{\gamma}\ln\frac{T_{L_1}+\beta/\gamma}{T_{0_1}+\beta/\gamma}, \tag{9}$$

where

$$\alpha = m_1 c_1\left(\frac{1}{h_1 S_1}+\frac{\Lambda}{\lambda' S'}+\frac{1}{h_2 S_2}\right), \tag{10}$$

$$\beta = \left(\frac{m_1 c_1}{m_2 c_2}T_{0_1}-T_{0_2}\right), \tag{11}$$

$$\gamma = 1-m_1 c_1/m_2 c_2, \tag{12}$$

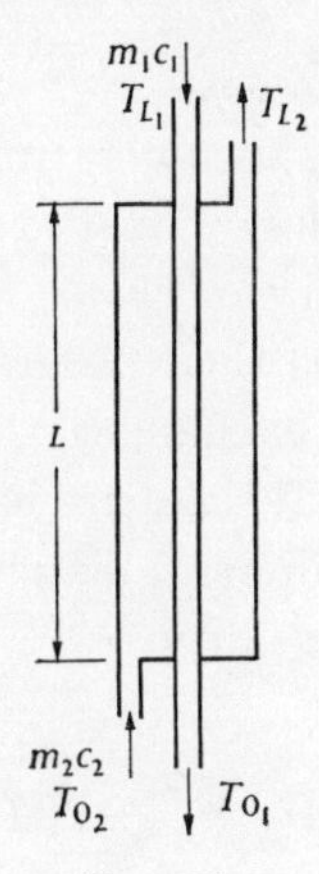

Fig. 50

the heat-transfer perimeters

$$S_1 = \pi D_1,$$

$$S_2 = \pi D_2,$$

$$S' = \pi(D_2-D_1)/\ln(D_2/D_1),$$

and the thickness of the wall of the inner tube is

$$\Lambda = (D_1-D_2)/2;$$

λ' = thermal conductivity of material of the inner tube.

An important and common case is when $m_1 c_1 = m_2 c_2$, i.e. when the same quantity of the same gas flows in each stream, then (9) reduces in the limit as $m_1 c_1 \to m_2 c_2$, to

$$L = \frac{\alpha}{\beta}(T_{L_1}-T_{0_1}). \tag{13}$$

In a more general case where the downflowing stream flows in n identical parallel tubes (each of inner and outer diameter D_1 and D_2 respectively) and each carries m_1/n g/s of gas, (9) and (13) are modified ,to become

$$L = \frac{\alpha}{\gamma n} \ln \frac{T_{L_1} + \beta/\gamma}{T_{0_1} + \beta/\gamma} \tag{14}$$

and

$$L = \frac{\alpha}{\beta n}(T_L - T_0). \tag{15}$$

Physical data for gases and some approximate formulae

Before summarizing the formulae and discussing examples of construction and calculation of heat exchangers, it is interesting to see what simplification can be made to the above formulae without seriously affecting their accuracy.

When designing a heat exchanger, the approximate temperature ranges are fixed so that the density ρ and viscosity η are not affected by altering the dimensions or the flow rate. We may note then that the pressure drop

$$\Delta p \propto \frac{G^{1 \cdot 75}}{D_h^{\,1 \cdot 25}} \propto \frac{m^{1 \cdot 75}}{A^{1 \cdot 75} D_h^{\,1 \cdot 25}}, \tag{16}$$

where A is the cross-sectional area of the tube; for a circular cross-section $\Delta p \propto m^{1 \cdot 75}/D^{4 \cdot 25}$, so that the pressure drop decreases very rapidly with increase in tube diameter, and increases somewhat less rapidly with increase in the mass flow of gas.

The accompanying Figs. 51 and 52 give representative data on the viscosity and thermal conductivity of air, N_2, O_2, H_2, and He. Sources of data are *Helium* by W. H. Keesom (1942, Elsevier, Amsterdam), the Landolt–Bornstein tables, and *Tables of Thermal Properties of Gases* (*N.B.S. Circular* 564, 1955. U.S. Government Printing Office, Washington).

At pressures below about 200 atm, these properties are not very sensitive to change in pressure so that in most heat-exchanger problems any change with pressure may be neglected, particularly as the pressure drop Δp and the heat transfer coefficients h are rather insensitive to changes in η or λ.

Similarly, the specific heat data (shown in Table V together with normal densities) may be used fairly generally without considering the effect of pressure or temperature.

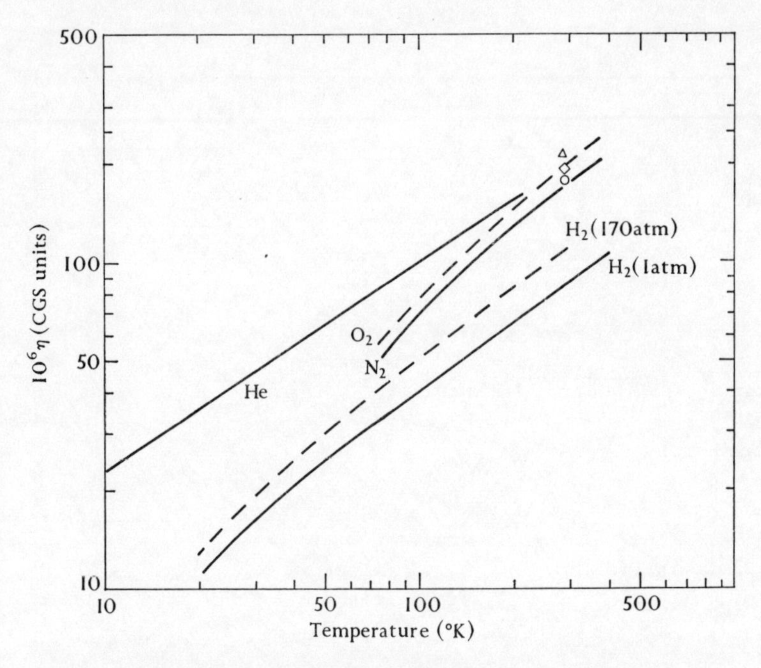

Fig. 51. Viscosity of some gases as a function of temperature; the experimental points represent values for air at 20° C and 160 atm (△), 80 atm (◇), and 1 atm (○).

Table V

	Air	N_2	O_2	H_2	He
C_p (cal/g degK) at 0° C	0·240	0·248	0·219	3·40	1·25
ρ (g/l at s.t.p.)	1·292	1·250	1·428	0·090	0·179

In Table VI are given some values for Prandtl's number ($\text{Pr} = C_p\,\eta/\lambda$) calculated for He, H_2, and air at different temperatures from the data in Figs. 51, 52, and Table V. The relative constancy of Pr and more particularly of $(\mathrm{Pr})^{0\cdot6}$ suggests that a simpler form for calculating h can be used in practice.

Thus, assuming $(\mathrm{Pr})^{0\cdot6} \simeq 0\cdot85$, eqn (8) reduces to

$$h \simeq 0\cdot027 C_\mathrm{p}\, G^{0\cdot8}\eta^{0\cdot2}/D_\mathrm{e}^{0\cdot2}.$$

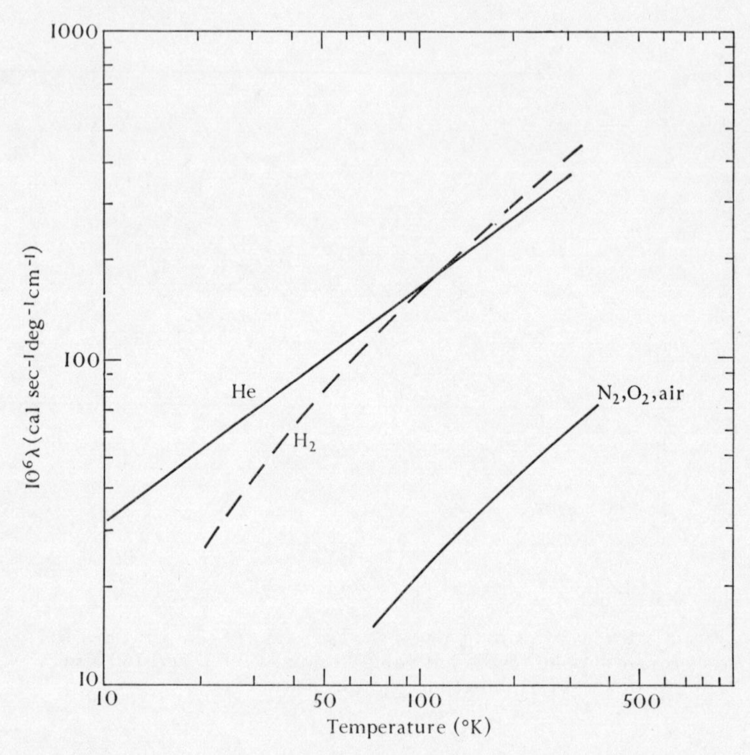

FIG. 52. Variation of thermal conductivity of gases with temperature.

TABLE VI

	T, °K	Pr	$(\mathrm{Pr})^{0\cdot6}$	$(\eta)^{0\cdot2}$
He	10	0·93	0·96	0·12
	50	0·81	0·88	0·146
	100	0·76	0·85	$0\cdot15_9$
	180	0·72	0·82	$0\cdot17_2$
H_2	50	0·72	0·82	0·12
	180	0·72	0·82	0·143
Air	180	0·75	0·82	$0\cdot16_7$

Also, since η and D_e appear in the form $\eta^{0\cdot 2}$ and $D_e^{0\cdot 2}$ a further rough approximation that may be used is to assume that $\eta^{0\cdot 2} \simeq 0\cdot 15$ (see Table VI) and that $D_e^{0\cdot 2} \simeq 1$ since $D_e \sim 1$ cm in many instances.

Hence $$h \simeq 0\cdot 004 C_p G^{0\cdot 8}, \tag{17}$$

and we have $$h \propto G^{0\cdot 8} = (m/A)^{0\cdot 8}, \tag{18}$$

which for a circular tube becomes $h \propto m^{0\cdot 8}/D^{1\cdot 6}$.

Equations (16) and (18) are useful when considering the suitability of varying the tube diameter or mass flow of gas through a heat exchanger. Obviously, for a maximum value of $h/\Delta p$, the hydrodynamic diameter D_h and cross-sectional area A should be large, but must eventually be limited by the requirement of turbulence, and by difficulties of construction.

As shown by eqns (13) and (14) the length L required to give any particular temperature distribution is proportional to α, which in turn depends (from (10)) on the sum of the reciprocals of the effective heat transfers. The second term in (10) involving the heat conducted across the wall of the exchanger tube is given by $\Lambda/\lambda' S'$. For a circular pipe, this term reduces to $2t/\lambda'\pi(D_1+D_2)$ provided that D_2/D_1 approaches 1, i.e. provided that the wall thickness t is small in comparison with the diameter;

then $$\alpha = m_1 c_1\left(\frac{1}{h_1 S_1}+\frac{t}{\lambda'\pi\bar{D}}+\frac{1}{h_2 S_2}\right). \tag{19}$$

Also, the thermal conductivity of the metal wall is usually much greater than the coefficients h for heat transfer from the gas stream to the wall so that this term $t/\lambda'\pi\bar{D} = t/\lambda' S'$ is negligible in comparison with

$$\frac{1}{h_1 S_1}+\frac{1}{h_2 S_2}.$$

Summary of formulae

$$\Delta p = \psi L G^2/2\rho D_h, \qquad \{(5)\}$$

where $$\psi = 64/\mathrm{Re}\ \text{(laminar flow)} \qquad \{(2)\}$$

or $$\psi = 0{\cdot}316(\text{Re})^{-0{\cdot}25} \text{ (turbulent flow).} \qquad \{(3)\}$$

$$h = \frac{0{\cdot}023}{(\text{Pr})^{0{\cdot}6}} C_{\text{p}} \frac{G^{0{\cdot}8}\eta^{0{\cdot}2}}{D_{\text{e}}^{0{\cdot}2}} \quad (\text{Pr} = C_{\text{p}}\,\eta/\lambda) \qquad \{(8)\}$$

$$\simeq 0{\cdot}004 C_{\text{p}}\, G^{0{\cdot}8}. \qquad \{(17)\}$$

$$L = \frac{\alpha}{\gamma n} \ln \frac{T_{L_1} + \beta/\gamma}{T_{0_1} + \beta/\gamma} \quad (m_1 c_1 \neq m_2 c_2) \qquad \{(14)\}$$

$$L = \frac{\alpha}{n\beta}(T_{L_1} - T_{0_1}) \quad (m_1 c_1 = m_2 c_2), \qquad \{(15)\}$$

where $$\beta = \frac{m_1 c_1}{m_2 c_2} T_{0_1} - T_{0_2}, \qquad \{(11)\}$$

$$\gamma = 1 - m_1 c_1 / m_2 c_2, \qquad \{(12)\}$$

and $$\alpha = m_1 c_1 \left(\frac{1}{h_1 S_1} + \frac{\Lambda}{\lambda' S'} + \frac{1}{h_2 S_2} \right), \qquad \{(10)\}$$

$$\simeq m_1 c_1 \left(\frac{1}{h_1 S_1} + \frac{1}{h_2 S_2} \right), \qquad (20)$$

where $S_1 = \pi D_1$, $S_2 = \pi D_2$.

Cooling of a gas stream by a liquid bath

The cooling of a gas by passing it through a coil immersed in a cold liquid is common; eqns {(14)} and {(11)} are easily simplified to give the cooling efficiency as a function of length. It is assumed that heat transfer both from the wall to the liquid and across the tube wall are very good in comparison with that from the gas stream to the wall of the carrying tube; then the second and third terms in the bracket on the right-hand side of {(10)} can be neglected. If the bath temperature be T' and a stream of m g/s of gas of specific heat c cal/g degK enters a tube at temperature T_L and leaves the tube (of length L cm) at temperature T_0, then

$$L = \frac{mc}{hS} \ln \left(\frac{T_L - T'}{T_0 - T'} \right), \qquad (21)$$

where h is defined by eqn {(8)} and S is the inner perimeter of the tube. For a given efficiency, that is for given values of

$(T_L - T')/(T_0 - T')$, the required length L is proportional to m/hS; so that, for a circular tube of diameter D,

$$L \propto m/hD \propto m/G^{0\cdot8}D \propto D^{0\cdot6}m^{0\cdot2}.$$

As the diameter is increased, the pressure drop in the gas stream decreases very rapidly (as $D^{-4\cdot25}$) and the required length increases slowly with diameter (as $D^{0\cdot6}$).

3. Methods of construction

Introduction

Many types of heat exchanger have been made but most appear to be derivatives of the two basic patterns mentioned in Chapter I—those of Linde and Hampson. In considering their relative merits, a number of factors must be borne in mind: (i) the relative difficulties of construction, (ii) the difficulty of mathematical analysis, (iii) the efficiency of heat transfer that can be obtained while keeping the pressure drop within reasonable bounds, (iv) the total mass and hence total heat capacity of the assembly. As far as factor (ii) is concerned, the Linde pattern lends itself to mathematical treatment, while in the Hampson pattern it is usually extremely difficult to make any theoretical estimate of the pressure drop or heat transfer in the low-pressure gas stream; thus in the latter case a combination of experience and experimental trial are needed to produce an efficient exchanger. The factor (iv) concerning heat capacity is particularly important in most laboratory liquefiers where a fairly short cooling-down time is desired; the time required to reach the equilibrium temperature distribution increases with the mass of the heat exchanger.

The Linde type of exchanger

Some examples of Linde-pattern exchangers are shown in Fig. 53 (see also Plate II between pp. 32 and 33). In (a), (b), and (c) the tubes are concentric and the outer wall contributes seriously to the pressure drop in the gas stream occupying the outer annular space, but assists very little in raising the heat transfer coefficient, since there is merely a touch contact between the inner tube and the outer tube. Frequently the central tube(s) is used for the high-pressure stream, and the low-pressure stream occupies the annular space between the tubes; however, there are often advantages in a construction such as (c) where a large inner tube may be used for the low-pressure stream to minimize $\Delta p/h$ and the high-pressure flow is through a narrow gap between the tubes.

Those shown in (d) and (e) are solder-bonded parallel-tube exchangers which are also amenable to mathematical analysis and are efficient, provided that the solder-bonding between the tubes is adequate. These are most simply made by tying the tubes together at intervals of a few inches with a twist of wire and then drawing the long bundle of tubes slowly through a molten bath of soft solder. The solder bath may be contained in a length of channel iron supported above bunsen burners; care should be taken to clean the tube surfaces of dirt or grease and to wet them thoroughly with soldering flux before they enter the solder bath.

The Fig. 53 (f) shows a section of a twisted-tube exchanger in which the inner tube is flattened and twisted into a 'corkscrew' shape before insertion into the outer jacket tube. This twisted tube exchanger was devised by Nelson about 45 years ago and first described by Bichowsky (1922).

In concentric tube exchangers, heat transfer in the outer space can be improved by winding a copper wire helically around the inner tube, and bonding it with soft solder. This increases the heat transfer surface and the turbulence, but also increases the pressure gradient slightly.

Hampson exchangers

The Linde exchanger is at its greatest disadvantage when one gas stream is at low pressure—allowing only a small pressure gradient—and the flow rate is large. The Hampson-type exchanger largely avoids this by shortening the low-pressure return path of the gas stream. As mentioned before it is much more difficult to treat mathematically with any degree of accuracy and also must be constructed accurately with uniform spacing between the tubes that build up the spiral pancake pattern. The main heat exchanger used by Collins (1947) in his helium liquefier (see Chapter I and Plates I (a) and I (b)) is a particularly efficient type of Hampson exchanger in which the high-pressure tubing ($\frac{1}{4}$-in diameter cupro-nickel) is 'finned' with copper ribbon and wound in a helix in the annular space between two co-axial cylinders of low thermal conductivity

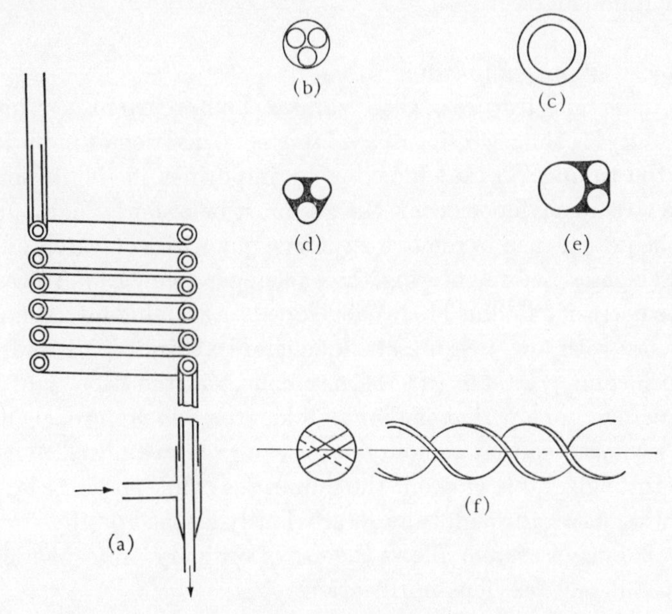

Fig. 53. Some heat exchangers of the Linde pattern.

metal. The copper ribbon 0·10 × 0·040 in is edgewound and solder-bonded to the high-pressure tubing, and with the aid of the cotton cord interwinding produces a good heat transfer between the low-pressure returning helium stream and the high-pressure tube. Finned copper tubing for heat exchangers is now produced commercially by Imperial Chemical Industries Ltd. and called Integron; the helical fin is rolled up out of the thickness of the tube.

In another variation of the Hampson spiral described by Nicol *et al.* (1953), the high-pressure tube of $\frac{1}{4}$-in o.d., 0·030-in wall copper is effectively finned by threading the copper tube in a die, mounted in a lathe. The threads, 28 per inch are cut 0·015 in deep and the tube is then annealed and wound in layers on a central tube; cotton cords are interwound with the tubing to guide the low-pressure gas stream into close contact with the finning. Another interesting form of Hampson exchanger is the coiled-coil construction used by Parkinson (1955) in making small liquefiers.

Condensation of impurities

An obvious feature of these various countercurrent heat interchangers is the necessity of avoiding condensation of impurities in the tubing. So that blocking by impurities in the incoming gas stream shall not occur, the gas must be of fairly high purity or be pre-cleaned to remove any large quantities of water, oil, or other gases before entering the exchanger (see § 1.6). To avoid this necessity Collins (1946) developed a reversing interchanger for use with low-pressure air liquefiers. In this, impurities from the incoming stream (no. 1) may collect in the tube A of the exchanger for a period of about 3 min, after which the exchanger is 'reversed' so that the outgoing clean gas stream (no. 2) flows up through A, blowing out the impurities, while stream 1 is now flowing down through tube B, previously cleaned out by stream no. 2. Such a system allows the use of ordinary atmospheric air without any previous purification.

Some general remarks

The material used for heat-exchanger tubing is largely dictated by the availability of tube of the required dimensions and by the fact that it must be bent, for most low-temperature equipment, into a helix or a spiral without splitting or crimping. Both copper and copper-nickel (annealed) bend easily and if a trial bend shows there is a danger of their crimping they may be filled with Cerrobend, Wood's metal, acetamide, or ice.

Copper may be an undesirable material in some instances because of its high heat conductivity. The heat flow along the exchanger due to thermal conduction should be negligible in comparison with the heat content of the gas stream. The formulae and the analyses which we indicated in § 2 are only valid if the exchange of thermal energy between the heat exchanger and the surroundings is very much smaller than the energy exchanged between the gas streams; this also implies that the exchanger should be in an evacuated chamber or in a stagnant gas atmosphere within a dewar vessel.

In the Linde type of heat exchanger, a rough calculation will often show that copper is quite acceptable as a tube material due to the long length of path which the heat must travel; in a Hampson exchanger the major heat conduction will occur along the walls of the two cylinders which enclose the annular space, and this usually necessitates the use of stainless steel, monel, or inconel for these walls.

4. Examples of heat-exchanger analysis

(*a*) *Cooling of hydrogen gas by liquid nitrogen*

Consider cooling a stream of compressed hydrogen gas in a bath of liquid nitrogen; the gas enters at a temperature of 120° K, a pressure of 100 atm, and at a rate equivalent to 300 l (s.t.p.)/min. The temperature of the pumped liquid nitrogen is 64° K. The problem is to find suitable dimensions for the immersed cooling coil such that the hydrogen gas will be cooled to within 1 deg K of the bath temperature.

$$T_L = 120^\circ\text{ K},\quad T' = 64^\circ\text{ K},\quad\text{and}\quad T_0 \simeq 65^\circ\text{ K}.$$

Therefore average temperature of gas $\bar{T} = 92{\cdot}5^\circ$ K.

Hence we obtain the viscosity $\eta \simeq 43\times 10^{-6}$ CGS units,

$$\lambda \simeq 150\times 10^{-6}\ \text{cal/cm s degK},$$

and the density
$$\rho = 0{\cdot}090\times 10^{-3}\times 100\times\frac{273}{92{\cdot}5}$$
$$= 0{\cdot}0266\ \text{g/cm}^3.$$

The mass flow
$$m = \frac{300\times 0{\cdot}090}{60} = 0{\cdot}45\ \text{g/s}.$$

Assume, as a trial, a coil of copper tubing of internal diameter

0·4 cm, and length L cm, then the Reynolds number

$$\begin{aligned} \mathrm{Re} &= GD/\eta \\ &= \frac{0\cdot45\times0\cdot4}{\pi(0\cdot2)^2\times43\times10^{-6}} \quad (G = 3\cdot57\ \mathrm{g/s\,cm^2}) \\ &= 33\ 200. \end{aligned}$$

Therefore the flow will be turbulent as required.

Pressure drop

$$\begin{aligned} \Delta p &= \psi LG^2/2\rho D \\ &= \psi\times L\times12\cdot7/(0\cdot4\times0\cdot0266\times2), \end{aligned}$$

and obtaining

$$\psi = 0\cdot026 \quad \text{from Fig. 47,}$$

$$\Delta p = 15\times L\ \mathrm{dyn/cm^2/cm} \simeq 1\cdot5\times10^{-3}\ \mathrm{atm/m}.$$

From equation {(8)}, heat transfer coefficient

$$h = \frac{0\cdot023}{(\mathrm{Pr})^{0\cdot6}}C_\mathrm{p}\frac{G^{0\cdot8}\eta^{0\cdot2}}{D_\mathrm{e}^{0\cdot2}}.$$

Now $\quad C_\mathrm{p}\,\eta/\lambda = 3\cdot4\times43\times10^{-6}/150\times10^{-6} = 0\cdot97$

$$(C_\mathrm{p} = 3\cdot4\ \mathrm{cal/g\,degK}),$$

therefore $\quad (\mathrm{Pr})^{0\cdot6} = 0\cdot98.$

Hence

$$\begin{aligned} h &= \frac{0\cdot023}{0\cdot98}\times3\cdot4\times\frac{2\cdot8\times0\cdot14}{0\cdot833} \\ &\doteq 0\cdot0376\ \mathrm{cal/cm^2/degK}. \end{aligned}$$

From equation (21), desirable length

$$\begin{aligned} L &= \frac{mc}{hS}\ln\left(\frac{T_L-T'}{T_0-T'}\right) \\ &= \frac{0\cdot45\times3\cdot4}{0\cdot038\times1\cdot26}\ln\left(\frac{120-T'}{T_0-64}\right) \\ &= 73\log_{10}\left(\frac{120-T'}{T_0-64}\right) \\ &\simeq 73\log_{10}\frac{56}{\Delta T}\ \mathrm{cm}, \end{aligned} \tag{22}$$

where $$T_0 = 64+\Delta T.$$

Thus as $\Delta T \to 0$, the efficiency $\to$ 100 per cent, and for $\Delta T = 1$ deg K, i.e. about 98 per cent efficiency, we require $L = 128$ cm, to which corresponds a pressure gradient

$$\Delta p = 0{\cdot}002 \text{ atm.}$$

Fig. 54 shows a graphical solution of (22) above. It is apparent from the low value of Δp that a much smaller diameter tube

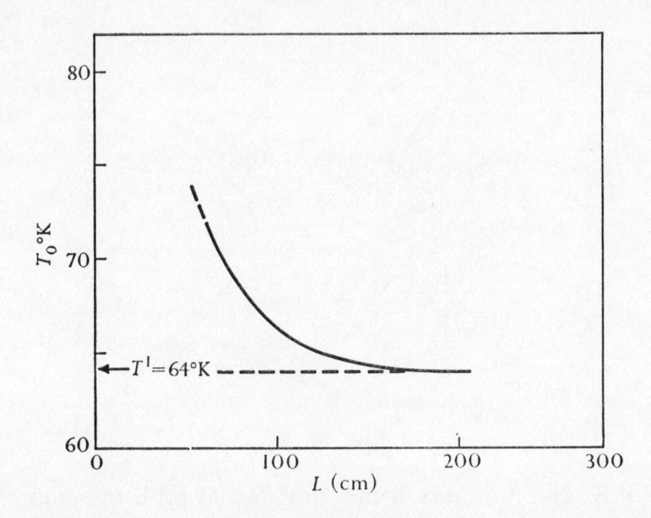

Fig. 54. Exit temperature of hydrogen gas as a function of the length of the cooling coil.

could be used without making Δp excessive. For example, reducing the tube diameter to one-half of its assumed value would increase Δp about thirty times, but would increase h nearly four times and decrease the required length (for the same efficiency) by a factor of nearly 2.

In Chapter I the efficiency of a heat exchanger was defined as the heat transferred divided by the heat available for transfer. In the example just given this is seen to be equivalent to a first approximation to

$$\text{efficiency} = \frac{T_L - T_0}{T_L - T'} = 1 - \frac{\Delta T}{(T_L - T')}.$$

(*b*) *Helium heat exchanger operating between liquid oxygen and liquid hydrogen temperatures*

In this example from a Linde liquefier illustrated by Fig. 55, helium at 20 atm pressure enters the heat exchanger at temperature $T_L = 90°$ K, flowing at 8 m³/h. The returning gas stream at a pressure of 1 atm enters the exchanger at temperature

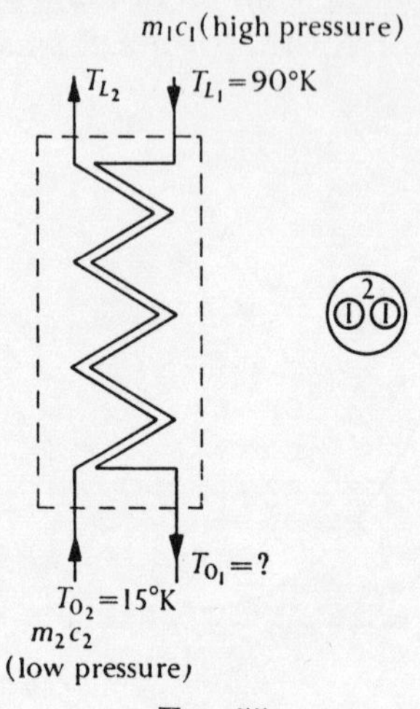

Fig. 55

$T_{0_2} = 15°$ K, the flow rate being 7 m³ (at s.t.p.)/h (assuming a liquefaction efficiency $\mathscr{E}$ of 0·14). At the average temperature

$$\bar{T} = 52{\cdot}5°\,\text{K},$$

$$\text{viscosity } \eta = 70 \times 10^{-6} \text{ CGS units},$$

$$\lambda = 110 \times 10^{-6} \text{ cal/cm/degK},$$

$$C_p = 1{\cdot}25 \text{ cal/g degK},$$

$$\rho \text{ (s.t.p.)} = 0{\cdot}179 \times 10^{-3} \text{ g/cm}^3.$$

Hence $\quad m_1 = 0{\cdot}4$ g/s, $\quad \rho_1 = 1{\cdot}8 \times 10^{-2}$ g/cm³;

$\qquad\quad m_2 = 0{\cdot}35$ g/s $\quad \rho_2 = 0{\cdot}9 \times 10^{-3}$ g/cm³.

High-pressure circuit

Assume the compressed gas flows through two inner tubes ($n = 2$) of copper, 0·15 cm inner diameter in parallel.

Since $D_h = 0{\cdot}15$ cm, $G_1 = 11$ g/cm^{2}s,

therefore $\mathrm{Re} = 2{\cdot}4 \times 10^4$ (turbulent flow),

$$\psi = 2{\cdot}5 \times 10^{-2},$$

and so, from (5), $\Delta p = 0{\cdot}056$ atm/m length.

Also Prandtl's number

$$\mathrm{Pr} = 0{\cdot}80$$

and $D_e = 0{\cdot}15$ cm.

Therefore, from (8), the heat transfer coefficient

$$h_1 = 4{\cdot}8 \times 10^{-2} \text{ cal/cm}^2\text{/degK}.$$

Low-pressure circuit

The jacket tube of 0·5 cm i.d. cupro-nickel encloses the two copper tubes (0·24 cm o.d., 0·15 cm i.d.), so that the low-pressure stream flows through an annular gap of

$$\text{area} = \tfrac{1}{4}\pi\{(0{\cdot}5^2) - 2 \times (0{\cdot}24^2)\},$$

and the total perimeter is given by $P = \pi(2 \times 0{\cdot}24 + 0{\cdot}5)$.

Then $D_h = 4A/P = 0{\cdot}138$ cm,

and proceeding as before,

$$\Delta p = 0{\cdot}15 \text{ atm/m}.$$

In the case of heat transfer, the effective perimeter is simply the circumference of the two inner tubes, i.e. the outer jacket is neglected.

Then $D_e = 0{\cdot}28$ cm,

and we obtain $h_2 = 1{\cdot}64 \times 10^{-2}$ cal/cm^2/degK.

Length L as a function of T_{0_1}

Assuming $\mathscr{E} = 0{\cdot}14$,

so that $m_1 c_1/m_2 c_2 \simeq 1{\cdot}14$,

then from {(11)}, {(12)}, and {(10)}

$$\beta = 1{\cdot}14 T_{0_1} - 15,$$

$$\gamma = -0{\cdot}14,$$

and $$\alpha = 0{\cdot}4\times 1{\cdot}25\left(\frac{100}{4{\cdot}8\times 0{\cdot}47}+\frac{100}{1{\cdot}64\times 0{\cdot}75}\right)$$

(using $S_1 = 0{\cdot}47$ cm, $S_2 = 0{\cdot}75$ cm).

Note that the second term in equation (10), $t/\lambda'\pi\bar{D}$, has a value of about 0·08 (assuming $\lambda' \simeq 1$ cal/cm s degK for copper), which is negligible in comparison with those terms involving h_1 and h_2.

Thence $$L = \frac{\alpha}{\gamma n}\ln\left(\frac{T_{L_1}+\beta/\gamma}{T_{0_1}+\beta/\gamma}\right)$$

$$= -514\log_{10}\frac{1{\cdot}14T_{0_1}-27{\cdot}6}{T_{0_1}-15},$$

which for $T_{0_1} = 25°$ K gives $L = 537$ cm,

26° K ,, ,, 380 cm,

28° K ,, ,, 247 cm,

30° K ,, ,, 183 cm,

35° K ,, ,, 109 cm.

If we require exchanger efficiency η of 0·96,

$$\eta = 0{\cdot}96 = 1-\frac{T_{0_1}-24{\cdot}2}{90-24{\cdot}2},$$

therefore $T_{0_1} = 26{\cdot}8°$ K, requiring $L \simeq 300$ cm.

For $L = 300$ cm,

$$\Delta p \text{ (high pressure)} = 0{\cdot}17 \text{ atm},$$

$$\Delta p \text{ (low pressure)} = 0{\cdot}45 \text{ atm}.$$

5. References

In this chapter the author has made extensive use of the mathematical treatment of heat exchangers by Mandl (1948), but detailed discussion of the problem has been given in many more accessible texts such as Jakob (1949) and McAdams (1954).

Helpful accounts of heat exchangers have been written by Daunt (1956) and Scott (1959, p. 18). Also Starr (1941) treated the problem with particular reference to small hydrogen liquefiers. An article by Jacobs and Collins (1940) described some results of experimental investigations of the efficiencies of various types of exchanger. Other useful general references are Bosworth (1952) and Hausen (1950).

Bichowsky, F. R. (1922). *J. ind. Engng Chem.* **14**, 62.

Bosworth, R. C. L. (1952). *Heat transfer phenomena.* Wiley, New York.

Collins, S. C. (1946). *Chem. engng* **53**, 106.

—— (1947). *Rev. scient. Instrum.* **18**, 157.

Daunt, J. G. (1956). *Handb. Phys.* **14**, 1.

Hausen, H. (1950). *Wärmeübertragung im Gegenstrom, Gleichstrom und Kreuzstrom.* Springer, Berlin.

Jacobs, R. B. and Collins, S. C. (1940). *J. appl. Phys.* **11**, 491.

Jakob, M. (1949). *Heat transfer,* Vol. i. Wiley, New York.

McAdams, W. H. (1954). *Heat transmission,* 3rd edn. McGraw-Hill, New York.

Mandl, F. (1948). Thesis. Oxford.

Nicol, J., Smith, J. S., Heer, C. V., and Daunt, J. G. (1953). *Rev. scient. Instrum.* **24**, 16.

Nusselt, W. (1909). *Z. Ver. dt. Ing.* **53**, 1750.

Parkinson, D. H. (1955). *Vacuum* **4**, 159.

Scott, R. B. (1959). *Cryogenic engineering.* Van Nostrand, Princeton.

Starr, C. (1941). *Rev. scient. Instrum.* **12**, 193.

CHAPTER IV

TEMPERATURE MEASUREMENT

1. Introduction

According to the zeroth law of thermodynamics any two bodies which are in thermal equilibrium with a third body are in thermal equilibrium between themselves; that is, when placed in thermal contact there is no transfer of heat from one to the other. Such bodies are said to be at the same temperature. A temperature scale may then be defined in terms of any of a number of thermometric properties of the system, and the scale and the size of particular temperature intervals (or degrees) may be fixed by reference to such common physical phenomena as boiling-points, melting-points, phase changes, etc. Thus the thermal expansion of a solid, liquid, or gas, the electrical resistance or thermoelectric power of a metal, and the magnetic susceptibility are all physical parameters which in the case of a particular chosen substance could be used to define a temperature scale, with the scale in degrees being a linear (or other chosen function) of the particular parameter for that substance. But, even if the same numbers on each of these scales were used to represent the temperatures of two or more chosen fixed points, there is no obvious reason why the different scales should agree at other temperatures or why any one of them should have a particular fundamental significance. Such a multitude of scales would be obviously rather unsatisfactory and very unsatisfying.

However, useful temperature scales can be based on the physical properties of particular systems and one particularly important scale depends on the properties of gases as determined experimentally: it has been found that at sufficiently low pressures the isotherms of gases are described by Boyle's law, i.e. the product of pressure and volume, pV, is constant. The 'perfect' or 'absolute' gas scale of temperature assigns numerical values to the temperature θ on the basis of the relation $\theta = a(pV)$ in the limit as $p \to 0$.

If it is then agreed that there shall be 100° between the ice point and steam point, the constant a is fixed so that

$$a = \frac{100}{\lim(pV)_s - \lim(pV)_i}$$

and the ice-point temperature,

$$\theta_i = 100\frac{\lim(pV)_i}{\lim(pV)_s - \lim(pV)_i}.$$

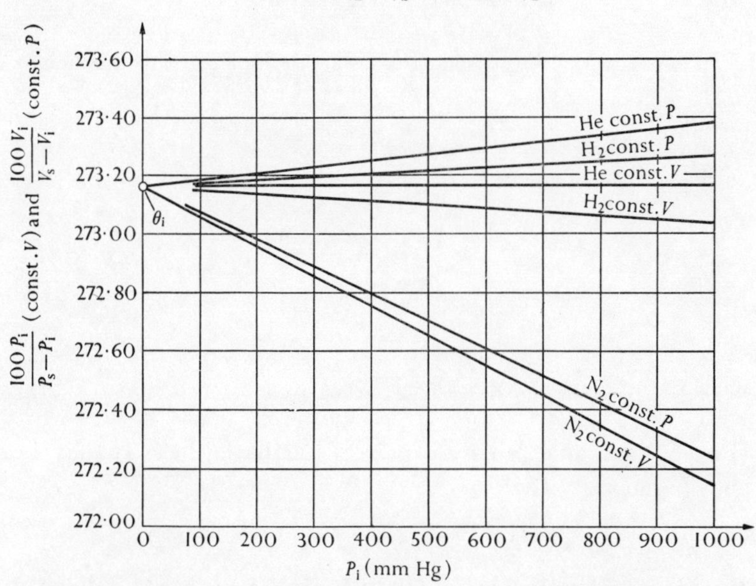

Fig. 56. The absolute ice point determined by graphical extrapolation (from *Heat and thermodynamics*, 2nd edn., by M. W. Zemansky. Copyright, 1943, McGraw-Hill Book Co.).

For such a centigrade absolute gas scale, it has been found experimentally for a number of gases (see Fig. 56) that in the low-pressure limit, θ_i has a value of 273·15°.

The real significance of this scale goes beyond this apparent wide applicability to various gases. The second law of thermodynamics may be stated in the form 'that no heat engine operating in a closed cycle, can transfer heat from a reservoir at a lower temperature to a reservoir at a higher temperature'.

From this follows an alternative statement that no engine can be more efficient than an ideal Carnot engine, which operating between two reservoirs has an efficiency depending only on the temperatures of these reservoirs.

Such a Carnot engine, taking in heat Q_1 at a temperature θ_1 in a reversible isothermal process and rejecting heat Q_2 at temperature θ_2, has an efficiency

$$\eta = 1 - \frac{Q_2}{Q_1} = f(\theta_1, \theta_2).$$

In such a Carnot cycle the two portions of the isotherms

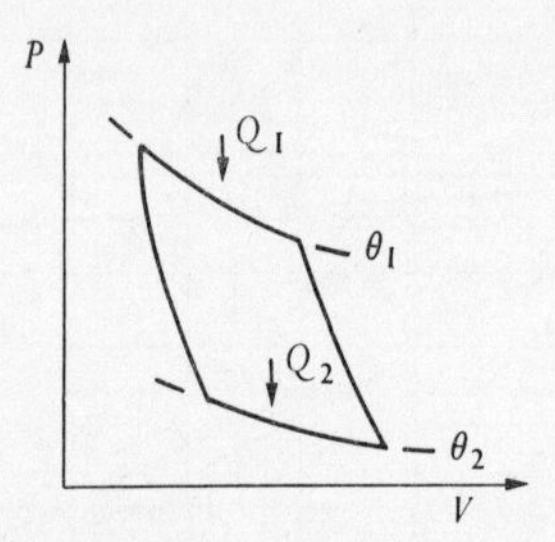

Fig. 57. A Carnot cycle.

(Fig. 57) are bounded by a pair of adiabatic lines. It may be shown generally that

$$Q_1 : Q_2 : Q_3 : Q_4 : \ldots = f(\theta_1) : f(\theta_2) : f(\theta_3) : f(\theta_4) : \ldots.$$

By identifying this ratio $f(\theta_1) : f(\theta_2) : f(\theta_3) : \ldots$ etc., directly with the ratio $T_1 : T_2 : T_3 : \ldots$ etc., a temperature scale independent of the system or of the working substance can be defined. This scale is called the Kelvin scale or absolute thermodynamic scale.

For a perfect gas, or for a real gas inthe limit of low pressures, pV is a function of temperature only and hence the internal energy is only a function of temperature so that such a gas could be used as the working substance in a Carnot cycle. It can be shown that

$$\frac{T_1}{T_2} = \frac{\theta_1}{\theta_2}.$$

If, as in the case of the gas scale, we fix numerical values for the temperature T on the thermodynamic scale by assuming $T_{\text{steam}} - T_{\text{ice}} = 100°$, the two scales will coincide.

On the thermodynamic or Kelvin scale there is a temperature at which the heat Q_1 taken in is zero and this corresponds to a temperature $T_1 = 0$, called the absolute zero of temperature. That such a temperature exists follows from the second law; that such a temperature is unattainable in a finite number of steps follows from the third law, as we discussed at the beginning of Chapter I.

2. Relation of thermodynamic and gas scales to the international scale

So we have a thermodynamic scale of temperature, which in principle at least can be realized by the absolute gas scale; in practice with real gases it can be realized with a degree of accuracy which depends on the accuracy of the gas thermometry and the accuracy of the corrections which are applied to the real gas to account for its non-ideality. However, accurate gas thermometry over a wide temperature range is exceedingly difficult; the necessity of a scale which could be more easily realized and reproduced in practice was met in 1927 when the Seventh General Conference on Weights and Measures adopted the International Temperature Scale. This scale was based on the same fundamental fixed points: ice point and steam point, taking the ice point as °0 Celsius (0° C) and the steam point as 100° Celsius (100° C). The instruments for its realization are the electrical resistance of platinum as a thermometer and at high temperatures ($> 600°$ C) the thermocouple and radiation pyrometers.

To quote from Hall (1956):

the original aim of the international scale was to provide a practical scale of temperature reproducible to very high accuracy and as nearly as possible identical with the thermodynamic Celsius scale. That is to say, it should be identical with the thermodynamic Celsius scale to within the limits of measurement with the gas thermometer, but it should offer the higher reproducibility obtainable with such instruments as the platinum resistance thermometer, not in themselves capable of defining an absolute scale.

A further development in the international Celsius scale has been the replacement in 1948 of the ice point (melting point of ice under one atmosphere pressure) by the triple point of ice, so that although the ice point is still 0° C (Int. 1948), it is defined as being 0·0100 deg C below the triple point of ice. Thus the two fundamental points used in fixing the international scale experimentally were the triple point at 0·0100° C and the steam point at 100° C.

As far as fixing the Kelvin or thermodynamic scale is concerned it would seem an unnecessary constraint to use two fixed points (ice and steam) as there is already a fixed point at the absolute zero. Kelvin pointed this out in 1854 and Giauque returned to the attack in 1938, proposing that the then best known value of the ice point on the absolute gas scale should be adopted permanently to define the scale. Finally in 1954 after examination of the values for θ_i or T_i obtained by precision gas thermometry in Germany (involving recalculation of the earlier results from the Physikalisch-Technische Reichsanstalt), Japan, Holland, and the United States, a value of 273·15° K was adopted by the Tenth General Conference of Weights and Measures. They recommended that the thermodynamic scale should henceforth be defined by one fundamental fixed point, the triple point of water, and that the best value for the temperature of this point is 273·16° K on the thermodynamic scale in use up to the present.

Now we have two scales, one defined at the ice (or triple) point and steam point, and the other at the absolute zero and the ice (or triple) point. Since each of these may be expressed in degrees Celsius or degrees Kelvin we have the situation depicted below as amended by the 11th General Conference in 1960:

International Practical Scales

International practical Temperature	*International practical Kelvin Temperature*
t_{int}	$T_{\text{Int}} = t_{\text{int}} + 273{\cdot}15$
° C (Int. 1948)	° K (Int. 1948)

Thermodynamic Scales

Thermodynamic Celsius Temperature	*Thermodynamic Kelvin Temperature*
$t = T - 273{\cdot}15$	T
°C (therm)	°K

3. Fixed points

In Table VII are listed the fixed points of the international scale or International Practical Temperature Scale (I.P.T.S.) as it has been called since 1960. At the higher temperatures it is now established that this Scale (1948, 1960) differs seriously

TABLE VII
Fixed Points (1960)

Fixed point (under pressure of 1 013 250 *dyn cm*$^{-2}$)	*Temperature,* ° C (Int. 1948)
Boiling point of oxygen	$-182{\cdot}97$
Triple point of water	$+0{\cdot}01$
Boiling point of water	100
Freezing point of zinc	419·505
(or boiling point of sulphur	444·6)
Freezing point of silver	960·8
Freezing point of gold	1063

from the Thermodynamic Scale and it is likely that the 1967 General Conference will authorize its Consultative Committee on Thermometry to amend some of these values. Probable new values would be 419·58° C for the zinc point, 961·93° C for the silver point, and 1064·43° C for the gold point.

For purposes of interpolation and for extrapolation above the gold point the international scale is divided into three regions.

(*a*) $-182{\cdot}97°$–$630{\cdot}5°$ C. The scale is defined here by the platinum resistance thermometer of which the physical and chemical purity are such that $R_{100}/R_0 \geqslant 1{\cdot}392$. In the region above 0° C, the resistance is then represented by the Callendar equation

$$t = 100\frac{R_t - R_0}{R_{100} - R_0} + \delta\frac{t}{100}\left(\frac{t}{100} - 1\right)$$

and the three constants are to be determined by measurement at the triple point of water, steam, and zinc (or sulphur) points.

For temperatures below 0° C, a four-constant equation is employed of which the Callendar–Van Dusen form is

$$t = 100\frac{R_t - R_0}{R_{100} - R_0} + \delta\frac{t}{100}\left(\frac{t}{100} - 1\right) + \beta\left(\frac{t}{100}\right)^3\left(\frac{t}{100} - 1\right),$$

and the additional constant β is ascertained by a measurement at the oxygen point. The departures of the international scale defined thus or I.P.T.S. from the thermodynamic scale are the continuing subject of study by improved gas thermometry.

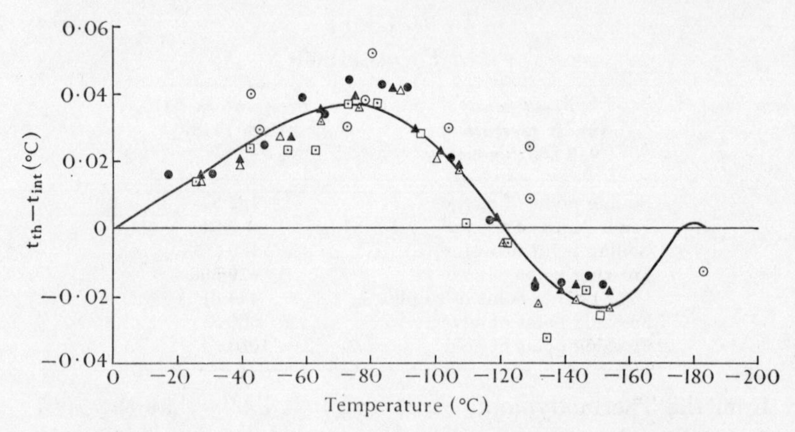

FIG. 58. Comparison of measurements of t (thermodynamic)–t (international). Curve ——— represents work of Barber of Horsford while points cover earlier work of Keesom and Dammers, Heuse and Otto (after Barber and Horsford 1965).

Below room temperature Barber and Horsford (1965) have confirmed that they amount to more than 30 millidegrees (Fig. 58).

(*b*) 630·5°–1063° C. Here the platinum + 10 per cent rhodium against pure platinum thermocouple defines the scale, using the equation

$$E = a + bt + ct^2.$$

a, b, c are determined by measuring the thermoelectric force E at the freezing points of silver and gold at 630·5° C; this latter temperature is to be measured with a platinum resistance thermometer in the case of a particular antimony sample or the thermocouple may be checked directly against a resistance thermometer in the vicinity of 630·5° C.

(*c*) Above 1063° C. A standard optical pyrometer is used to compare the intensity of radiation from a luminous black body of unknown temperature with that from a black body at the freezing point of gold. This range of temperature is of much less concern to the low-temperature physicist.

Below 90° K

At lower temperatures there are no internationally recognized fixed points or scales with the exception of the region from 1–5° K, which is covered by the 1958 ^{4}He and 1962 ^{3}He vapour pressure scales which are discussed in § 5.† Thermodynamic temperatures can be realized above 1 or 2° K by gas thermometry (or acoustic thermometry) and below this by magnetic thermometry but not usually with the sensitivity or reproducibility or ease that is required of practical thermometers. For convenience some national laboratories have calibrated a number of secondary thermometers—platinum resistance—against gas thermometers from 10–90° K and used these as practical sources of reference. The best known is the N.B.S. 1939 scale, established by Hoge and Brickwedde (1939). This was later redefined to be 0·01 deg lower than the 1939 scale and designated the N.B.S. 1955 scale. Scales have been established at the Pennsylvania State University (P.S.U. scale—see Moessen, Aston, and Ascah in *Temperature. Its measurement and control in science and industry*, Reinhold, New York, Vol. 3 (1962), Pt. 1, p. 91), the PhysicoTechnical and Radiotechnical Measurements Institute in Moscow (P.R.M.I. scale—see Borovik-Romanov *et al.* in *Temperature*, Vol. 3 (1962), Pt. 1, p. 113), and the National Physical Laboratory, Teddington (N.P.L. scale—see Barber 1962). Because there is no apparent analytic relation between resistance and temperature below 90° K, these scales exist in tabular form expressing the resistance R_T or reduced resistance

† Efforts are being made to extend the I.P.T.S. down to 13·8° K by the use of platinum resistance thermometers and the CCT64 table discussed below.

$W_T = R_T/R_{273}$ as a function of T. These scales have been intercompared by measuring at N.P.L. and P.R.M.I. a group of eight platinum thermometers for which there were in existence 'calibrations' in terms of national scales. The result (Fig. 59 (a))

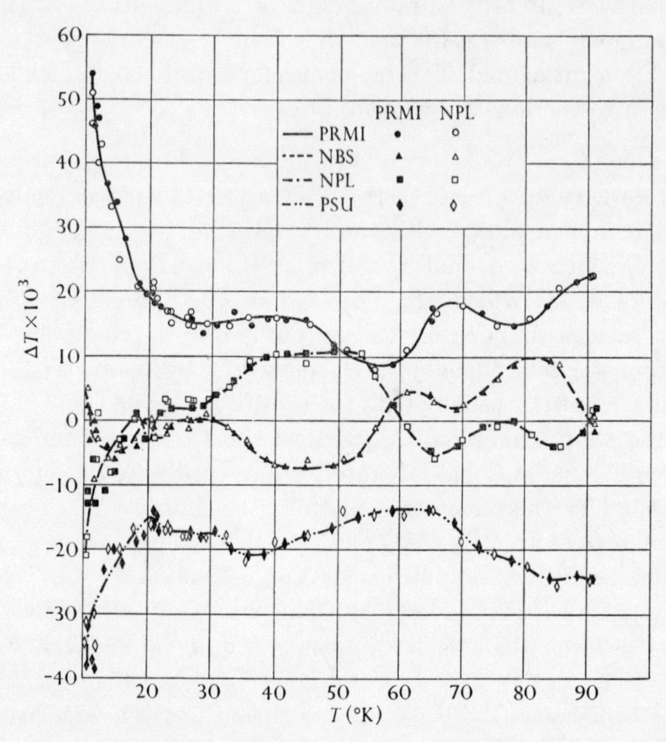

Fig. 59(a). Comparison of 'national' scales which formed the basis of the CCT64 scale, direct intercomparison (after Orlova *et al.* 1966).

confirmed that differences of more than 30 millidegrees exist between the scales. Even when readjusted (Fig. 59 (b)) with a common value of 90·170° K for the oxygen point and 20·384° K for the equilibrium-hydrogen point, discrepancies and irregularities still exist (see also the discussions of irregularities in existing scales between *ca.* 30 and 100° K by Furukawa and Reilly 1965, and Roder 1965). A weighted mean of the $W(T)$ tables has been recommended by the Advisory Committee on Thermometry as a provisional scale—CCT64.

Table VIII has values of 'fixed 'points which are useful when calibrating or comparing thermometers. Note that some of these points may still be in error in terms of thermodynamic temperature by 10 millidegrees or so. This situation may be improved in the near future. More accurately based and more accurately

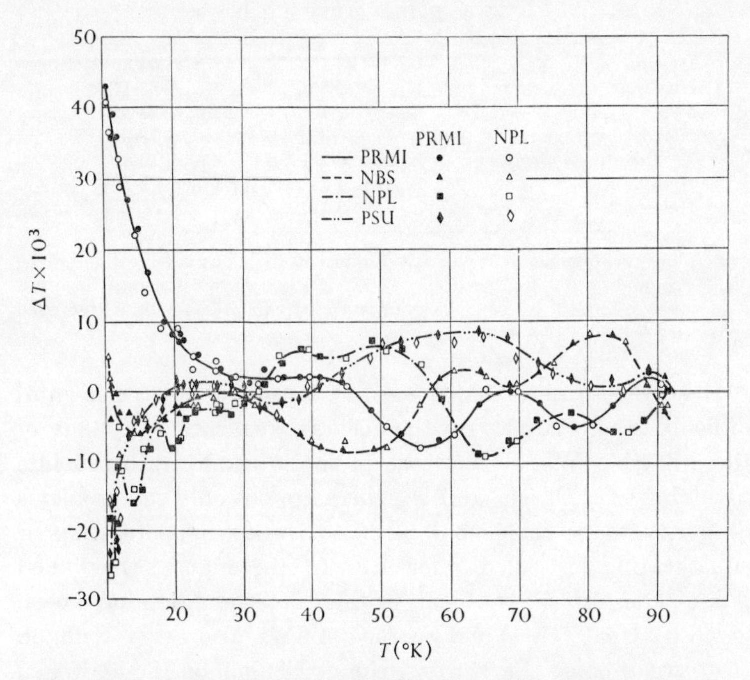

FIG. 59(b). Comparison of 'national' scales which formed the basis of the CCT64 scale, after adjustment to common oxygen point and hydrogen point (after Orlova *et al.* 1966).

realized low-temperature scales should result from the acoustic thermometer and the semiconducting thermometer. The former determines absolute temperatures from the velocity of sound in helium gas with an ultrasonic interferometer (Plumb and Cataland 1965); the latter are highly reproducible resistance elements of germanium which are particularly sensitive at the lowest temperatures and useful for the awkward region from 4° to 10° K. Calibration of these can now be performed at the National Bureau of Standards on an N.B.S. provisional scale

(2°–20° K) which is based on the acoustic thermometer. However, because there is yet no good analytic expression for $R(T)$, these need to be calibrated or compared at many points.

TABLE VIII

'Fixed' points below 90° K

Oxygen	b.p. = 90·188° K†	t.p. = 54·361†
Nitrogen	b.p. = 77·355	t.p. = 63·15
Neon	b.p. = 27·10	t.p. = 24·55
Normal hydrogen	b.p. = 20·39†	t.p. = 13·96
Equilibrium hydrogen	b.p. = 20·280†	t.p. = 13·810
Helium-4	b.p. = 4·215	λ point = 2·172
Helium-3	b.p. = 3·191	

† Values recommended in the 8th Report of the Consultative Committee on Thermometry (September 1967). With the exception of neon, the other values are intended to be consistent with vapour-pressure data discussed below in § 5 (and Table A in Appendix).

It is well to note that there are considerable experimental difficulties associated with the accurate realization of many of these fixed points. In the case of the low-temperature points involving CO_2, O_2, N_2, and H_2, there are not only the problems of purity to be considered but also those of supercooling or superheating. Detailed references to these and other fixed points, and also the problems of their determination have been given by Hoge (1941) and Scott (1941) (cf. also Berry 1962, on the oxygen point; Barber and Horsford 1963, on the hydrogen point; and Barber 1966, on the sublimation point of CO_2 at 194·67° K).

4. Gas thermometry

Introduction

Of all forms of precision thermometry, high-precision gas thermometry would seem the most demanding and has therefore remained the province of a mere handful of laboratories. Apart from the difficulties of correcting for the non-ideality of the gas used and at low pressures for the thermo-molecular pressure difference—both of which are discussed briefly at the end of this section—there are many other corrections and difficulties to be overcome. Details of the Leiden and Berlin gas thermometers are given by Keesom (1942). In the proceedings of the symposia on temperature measurement, entitled *Temperature. Its measurement and control in science and industry* (Reinhold, New York), there are detailed articles by Keyes (Vol. 1, 1941), by Beattie describing the M.I.T. precision gas thermometer (Vol. 2, 1955), and by Moessen, Aston, and Ascah (P.S.U.), Borovik-Romanov *et al.* (P.R.M.I.), and Barber (N.P.L.) in Vol. 3, Part 1 (1962).

Details of precision gas thermometry are not appropriate to the present monograph but a brief account of less accurate and more common types of gas thermometers which are used in low-temperature research may be useful. Gas thermometry in its simpler form not only gives results directly in terms of absolute temperature and has a sensitivity which can be increased by using higher filling pressures at lower temperatures but is unaffected by magnetic fields and is easily adapted to differential thermometry, i.e. measurement of small temperature differences.

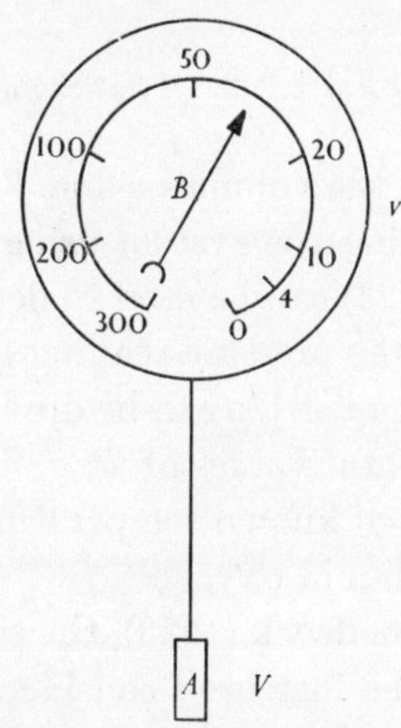

FIG. 60. A simple gas thermometer.

Simon's gas thermometer

Fig. 60 shows the particularly simple form due to Simon (see Ruhemann 1937) in which the thermometer, usually helium gas-filled, is essentially constant volume due to the very small change in volume of the Bourdon spiral in the pressure gauge B. The bulb A which is at the low temperature to be measured, has volume V and is connected by a fine capillary to the gauge of volume v. This is simply a vacuum dial gauge which for more accurate results can be a calibrated standard gauge, capable of reproducing its readings to $\frac{1}{10}$ per cent of full-scale deflexion.

If we assume the thermometric gas is 'perfect' within the required limits of accuracy, then for a constant mass of gas,

$$\frac{pV}{T} = \text{constant.}$$

The system may be filled to a pressure p_0 at room temperature T_0; then, neglecting the capillary volume, the pressure p at a temperature T is given by

$$\frac{pv}{T_0}+\frac{pV}{T} = \frac{p_0 v}{T_0}+\frac{p_0 V}{T_0},$$

whence
$$p = p_0\frac{(v+V)T}{VT_0+vT} \tag{23}$$

or
$$\frac{1}{T}(VT_0)+v = \frac{1}{p}(p_0 v+p_0 V). \tag{24}$$

If one or if neither of the volumes v and V are known, they can be found by calibrating the system using two or three known temperatures. Then (23) can be used to determine the unknown temperature T from the pressure reading p. Alternatively, from (24) a graph of $1/T$ against $1/p$ can be drawn using two (or three as a check) experimental values of T, p obtained by immersing the bulb A in liquids of known temperature, e.g. liquid helium, liquid nitrogen, and liquid oxygen.

As discussed by Woodcock (1938), the sensitivity of this type of thermometer can be increased considerably at low temperatures by making v large in comparison with V. Then only at low temperatures will the majority of the gas be in the bulb A, so that at high temperatures it is very insensitive to change in T (but unfortunately sensitive to changes in room temperature T_0). Therefore the scale on the gauge in Fig. 60 can be expanded at the low-temperature end as much as desired by making v/V sufficiently large. In such cases the thermometer is of little use at higher temperatures, although it may be capable of an accuracy of $\pm\frac{1}{20}$ deg K at temperatures below 30° K.

Note that if a linear dependence of p on T is required, so that the sensitivity is more or less independent of the temperature, then V must be large in comparison with v; this means physically that the major part of the helium gas is in the variable temperature bulb A at all temperatures and therefore to a first approximation p is proportional to T.

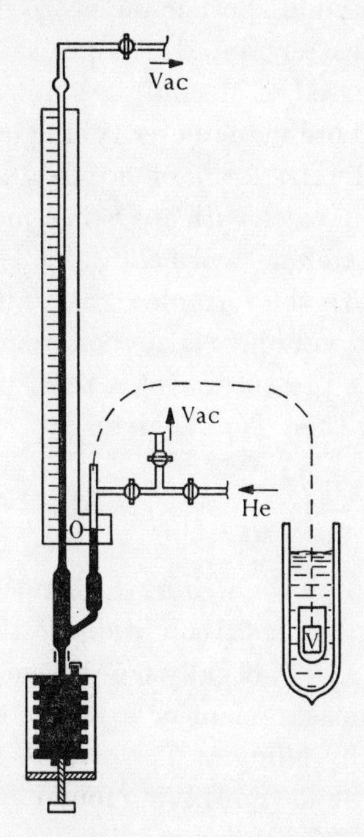

FIG. 61. A constant-volume gas thermometer.

Oil or mercury manometer

A more accurate variant of this latter situation (where $V \gg v$) is shown in Fig. 61; a thermometer bulb (volume V, temperature T) is connected by 0·5-mm o.d. (0·3-mm i.d.) German-silver capillary tubing to a glass manometer. This constant volume manometer is filled with mercury or butyl phthalate and the liquid is adjusted to the fiducial mark 'O' in the right-hand arm by means of the liquid reservoir below. Butyl phthalate, which has a density of about 1·04 g/cm^3, is very suitable and allows the use of a brass sylphon bellows for the reservoir and capillary tubing (say, 1·5 mm bore) in the manometer. Due to the effects of surface tension in these capillary tubes, it is advisable to use precision bore tubing (e.g. 'Veridia' from Chance Bros. (England) or precision bore Pyrex from H. S. Martin, Illinois, or Wilmad Glass Company, New Jersey) particularly for differential thermometry when a second short manometer arm is added to the system. Mercury has certain advantages as a manometer fluid as it is easily outgassed of air and does not absorb helium, but it requires a wider bore manometer tubing (at least 10 mm i.d.) and a slightly different form of adjustable reservoir, e.g. a stainless-steel piston sealed with an O-ring, moving in a stainless-steel cylinder or stainless-steel bellows.

We may generalize the formulae given above to include the case where the dead volume v is at room temperature t, and the system is filled to a pressure p_0 at a fixed point T_0 other than room temperature; then $T(p)$ is given by

$$\frac{T}{p} = \frac{T_0}{p_0}\left(\frac{1+vT/Vt}{1+vT_0/Vt}\right) = \frac{T_0}{P_0}\frac{(1+\Delta)}{(1+\Delta_0)}. \tag{25}$$

In the cryostat used by the writer, the copper bulb of volume V is limited by the experimental conditions to about 5 cm^3 capacity and then v is made small, usually 0·5–1·0 cm^3. The value of V is found by volume measurement or from the construction, and v is found from (25) by filling at $T_0 = 273°$ K or $T_0 = 295°$ K and then cooling to 90° or 77° K. Having found v, and assuming that t always remains 295° K (an air-conditioned room), a convenient table of $(1+\Delta)/(1+\Delta_0)$ is drawn up for various temperatures, e.g. for $T_0 = 295°$ K, $T_0 = 77{\cdot}3°$ K, $T_0 = 4{\cdot}2°$ K. Finally the pressure p is read by means of a reflecting glass scale marked from 0 to 100 cm with lines 1 mm apart; such scales are commercially available.

Used in this way, the thermometer gives readings which appear to be correct to $\pm\frac{1}{20}$ degK in the range 90°–55° K over which it can be checked against the vapour pressure of liquid oxygen. Again, from 4·2° to 1·8° K it is correct to $\frac{1}{100}$ deg K. By checking with a calibrated platinum resistance thermometer, the scale is correct within the limit of reading from 90° up to about 140° K and from 4° up to about 20° K.

However, if the thermometer bulb is near, say, 40° K and (Fig. 61) part of the connecting capillary is in liquid helium, then with small bulbs a substantial fraction of the gas is in the capillary (even if it is only 0·3 mm i.d.); corrections should be applied if the temperature is to be known to an accuracy of 2 per cent or better. This, of course, depends on the relative dimensions of the bulb and capillary; the use of larger thermometer bulbs reduces such corrections. Between room temperature and about 100° K it is advisable to correct for the contraction of the thermometer bulb as the total change in bulb volume amounts to about 1 per cent in cooling from room temperature to 100° K.

A differential manometer

For measurement of thermoelectric power or thermal conductivity, small temperature differences are measured and a gas thermometer of the type just discussed is easily adapted (see Fig. 62). For this, the absolute temperature T_A and T_B of two thermometer bulbs A and B are read as before by measuring the height h in centimetres of oil with respect to the fiducial mark 'O'. Then

$$\frac{T}{p_{(\text{cm oil})}} = \frac{T_0}{p_{0(\text{cm oil})}}\frac{1+\Delta}{1+\Delta_0}.$$

If the cross-sectional area of the bore of the capillary tube is A cm², it is easily shown that

$$\frac{\delta T}{\delta p} = \frac{T}{p}\left(1+\frac{vT}{Vt}\right)+\frac{AT^2}{Vt} \text{ where } \delta T = T_A - T_B \text{ and } \delta p = \delta h_{\text{cm oil}},$$

whence
$$\frac{\delta T}{\delta p} = \frac{T_0(1+\Delta)^2}{P_0(1+\Delta_0)}+\frac{AT^2}{Vt}. \tag{26}$$

By tabulating $(1+\Delta)^2/(1+\Delta_0)$ for the different values of T_0

normally used and also tabulating AT^2/Vt, $\delta T/\delta p$ can be quickly read off the table and the conversion from δh to δT made. The small difference δp is usually read with a cathetometer although we have found that where $\delta h \leqslant \frac{1}{2}$ in a geodetic survey level, e.g. Wild N3 level, can be used more speedily and gives δh to $\pm\frac{1}{100}$ mm with less trouble in levelling.

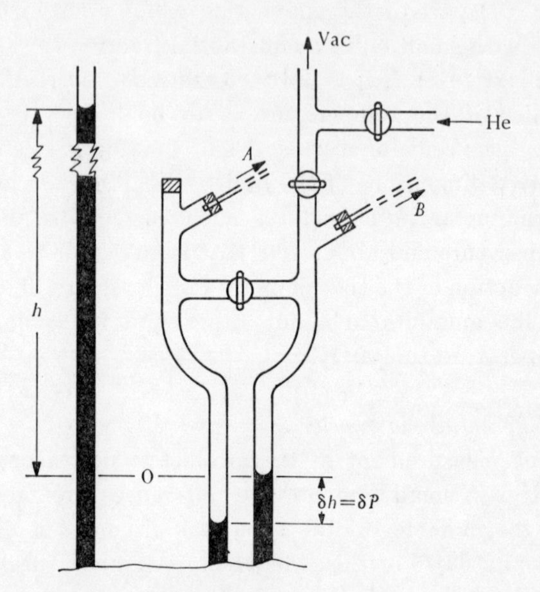

Fig. 62. A differential manometer.

Butyl phthalate manometers similar to the type shown in Fig. 62 were described by Hulm (1950). Note that such gas thermometers have the merit that by refilling to a suitable pressure at 4·2° K they can be used with a high sensitivity in the liquid-helium temperature range as well as at higher temperatures. Their sensitivity as a differential thermometer may be increased by using mercury for the total pressure measurement (h) and oil for the difference measurement (δh). A variant of this was described by Mendelssohn and Pontius (1937) in which the mercury manometer and differential oil manometer are separate. In either case, the use of mercury allows a total pressure to be used which is about ten times larger than is possible with an oil column of the same height, but the lower density liquid, oil, is still used for measuring small differences in temperature.

TABLE IX

Example of table for use with gas thermometer

(Dead volume $v = 1{\cdot}29$ cm³, $V = 4{\cdot}46$ cm³, $t = 295°$ K, and 1·5 mm i.d. glass capillary is used in manometer)

For $T_0 = 77{\cdot}3°$ K				*For* $T_0 = 4{\cdot}2°$ K			
T	$\dfrac{1+\Delta}{1+\Delta_0}$	$\dfrac{(1+\Delta)^2}{(1+\Delta_0)}$	$\dfrac{AT^2}{Vt}$	T	$\dfrac{1+\Delta}{1+\Delta_0}$	$\dfrac{(1+\Delta)^2}{(1+\Delta_0)}$	$\dfrac{AT^2}{Vt}$
50	0·979	1·022	0·034	2	0·998	1·000	—
60	0·984	1·042	0·048	5	1·001	1·006	—
70	0·993	1·061	0·066	10	1·006	1·016	0·001
80	1·001	1·079	0·086	15	1·011	1·026	0·003
90	1·011	1·099	0·109	20	1·016	1·036	0·005
100	1·020	1·119	0·135				

Summarizing the precautions to be observed in this form of practical gas thermometry, we note:

(i) Correction for dead volume v, included in the formulae above.

(ii) Requirement that the temperature of the manometer (usually room temperature) be kept fairly constant, particularly when vT is comparable with Vt; at very low temperatures $Vt \gg vT$ and the problem is not significant.

(iii) Necessity for uniform bore tubing to avoid capillarity corrections.

(iv) Contraction of thermometer bulb which may be important at temperatures above that of liquid nitrogen.

(v) Correction for volume of gas in the connecting capillaries which may be in a temperature gradient. This is a more difficult correction to make and is largely avoided by making V very large in comparison with the volume of the capillary. In cases where the capillary is partially at 4·2°K and the thermometer bulbs are at a much higher temperature, this can cause a substantial error.

However, temperatures can be measured with an inaccuracy of a few parts per thousand over a quite wide temperature range by these methods without any arduous precautions and purely by reference to two or three easily available fixed points.

Corrections for a non-perfect gas and for thermo-molecular pressure difference

Before leaving the subject of gas thermometry it is pertinent to illustrate the magnitude of the corrections that may arise from non-ideality of the gas and from thermo-molecular pressure differences.

If the equation of state of a gas is expressed as

$$pv = A(1+Bp+\text{terms in } p^2, \text{ etc.}),$$

the virial coefficients B, etc., can be found experimentally and can be used to correct the real gas scale to the true thermodynamic scale. Whereas the third and fourth virial coefficients are very small, the second virial coefficient B has values (see Table X from Keesom 1942; see also Kilpatrick, Keller, and Hammel 1955) which even for helium can produce a significant correction to the gas scale, particularly at high pressures and low temperatures.

TABLE X

T° K	2·6	4·0	14·0	22·0	50	100	300
$10^3 B$	−5·30	−3·61	−0·549	−0·101	+0·338	+0·492	+0·508

The deviations of the standard constant-volume helium gas scale (using $p_0 = 1$ m Hg at 0° C) from the thermodynamic scale are shown in Fig. 63 below (after Hoge 1941).

The other correction due to thermomolecular pressure difference arises when the mean free path of the gas molecules is sufficiently large to be comparable with the width of the capillary connecting the thermometer bulb to the manometer; that is, at very low pressures there is a difference in pressure between the ends of the capillary tube when a temperature difference exists between the bulb and the manometer. Our knowledge of the magnitude of this effect is largely a result of experimental work done at Leiden (see Keesom 1942, for summary and references). This work indicates that for helium in the hydro-dynamic region, i.e. when the diameter of the capillary $2R$ is greater than the mean free path, the thermomolecular pressure difference Δp is given by

$$\frac{\Delta p}{p} = 724{\cdot}2\frac{1}{(Rp)^2}\left\{1-\left(\frac{T_2}{273{\cdot}1}\right)^{2{\cdot}294}\right\}. \quad (27)$$

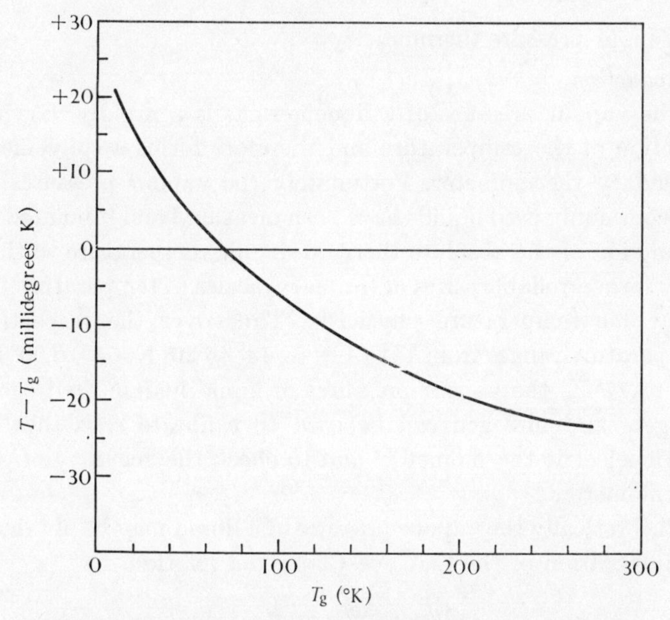

FIG. 63. Deviations of the helium-gas scale (T_g) from the thermodynamic scale (T) for $p_0 = 1$ m Hg.

This assumes that the temperature at the hot (manometer) end is 273·1° K, T_2 is the temperature at the cold end, R is in centimetres and p in microbars. Thus, if $T_2 = 4{\cdot}2°$ K, $R = 1$ mm, and $p \simeq 1$ cm Hg ($\simeq 13\,000\ \mu$b), then $\Delta p/p \sim 4\times 10^{-4}$.

In most thermometry applications this correction is negligible, but in some vapour-pressure measurements where the total pressure p is of the order of $\frac{1}{10}$ mm Hg or less, the mean free path for helium atoms may be comparable with the tube width. At

$\frac{1}{10}$ mm pressure, the mean free path in helium is about $\frac{1}{100}$ mm at 4° K, $\frac{3}{10}$ mm at 80° K, and $\sim$ 1 mm at room temperature; in some of these cases the correction Δp becomes considerable and the 'hydrodynamic region' formula above is no longer applicable (see Keesom 1942) For further data on the 'intermediate' region—between the hydrodynamic and the Knudsen regions—the reader is referred to Keesom (1942) or Roberts and Sydoriak (1956). The latter include data for ^{3}He.

5. Vapour-pressure thermometry

Introduction

The vapour pressure of a liquefied gas is a rapidly varying function of the temperature and therefore forms a convenient secondary thermometer. Fortunately the vapour pressures of the commonly used liquids have been measured and tabulated as a function of the absolute thermodynamic temperature so that they form a reliable series of 'primary' scales of temperature for many low-temperature physicists. Thus over the respective temperature ranges from 1° to 4·2° K, 14° to 20° K, 55° to 90° K, 63° to 77° K, the vapour pressures of liquid helium, hydrogen, oxygen, and nitrogen can be used to calibrate resistance or thermoelectric thermometers and to check the accuracy of gas thermometers.

Theoretically the vapour pressure of a liquid may be obtained by integration of the Clausius–Clapeyron relation

$$\frac{dP}{dT} = \frac{\Delta S}{\Delta V} = \frac{L}{T\Delta V},$$

where ΔV is the change in volume at vaporization and L is the latent heat.

If L is constant, $$\log P = \frac{A}{T} + B,$$

and more generally, if L is a linear function of temperature,

$$L = L_0 + aT;$$

therefore $$\log P = \frac{A}{T} + B \log T + c. \tag{28}$$

Although equations of this type can be fitted quite closely to the experimental values of vapour pressure, it is usually convenient to tabulate P as a function of T and use such a table to find $T(P)$ from experimental values of the vapour pressure.

Some typical experimental situations in which vapour pressure is measured are sketched schematically below (Fig. 64). In (a) a liquid is boiling under atmospheric or some controlled pressure and the vapour pressure above the surface is measured by a tube which leads to a pressure gauge. In such an instance

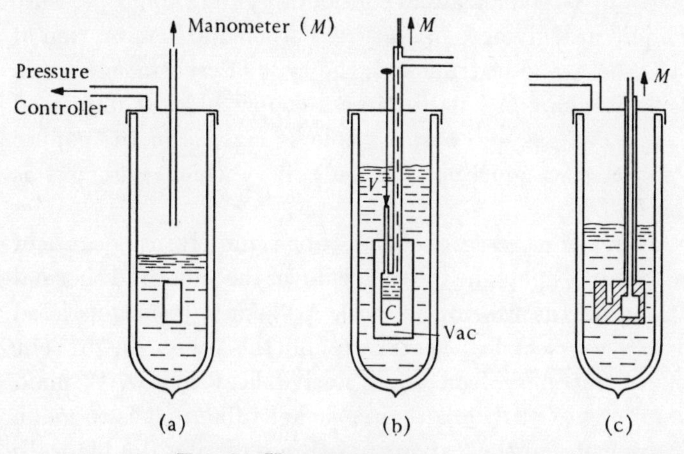

Fig. 64. Vapour-pressure measurement.

temperature inhomogeneities in the liquid may make the recorded vapour pressure a rather poor indication of the true temperature at points within the liquid. Efforts have been made to avoid or allow for these inhomogeneities in oxygen and helium baths. Berry (1962) has used absorbent cotton in oxygen as a means of preventing superheating and achieving a stability of better than a millidegree. In liquid helium Swim (1960) has shown that gradient of a few millidegrees exist near the liquid surface. These cannot be correlated with hydrostatic pressure effects (see also Hoare and Zimmerman 1959).

The second case (b) shows a fairly common type of cryostat in which the inner space C has a small chamber attached which can be filled with liquid from the dewar through the valve V. A tube connected to the pressure gauge at one end is taken down to a point close to the surface of the liquid in the small chamber. This liquid may be boiling under atmospheric or under reduced pressure. If this chamber is of a good heat-conducting metal (e.g. copper), and there are no points along the sampling tube which are colder than the liquid in the chamber, then the vapour pressure will form a good indication of the liquid temperature. However, in either of these cases (a) or (b) a trace of impurity in the liquid will affect the result markedly, e.g. a trace of oxygen in nitrogen or vice versa alters considerably the vapour pressure of the liquid at a given temperature. For accurate calibration of other thermometric instruments, the type of cryostat shown in Fig. 64 (c) is commonly used. Here a copper block is immersed in the liquefied gas and a small hole or capsule in the copper block forms a vapour-pressure bulb into which pure gas is condensed.

The form of the pressure sensing tube is important because of dangers of 'cold' spots, heat leakage down the tube, and thermal oscillations (in the case of helium). A vacuum jacket, assisted perhaps by an electrical heater around the tube, can prevent cold spots but may lead to excessive heat inflow. A good compromise is to provide an outer jacket tube, that is to make the sensing tube to the vapour-pressure bulb of a double wall, but to allow vapour into this space. An interesting account of measuring helium pressure has been given by Cataland, Edlow, and Plumb (*Temperature*, Vol. 3, Pt. 1, p. 413, 1962).

There are special problems in measuring the vapour pressure of liquid helium-4 below the λ-point because of He II film flow up the tube wall. This film evaporates higher up at a warmer level and refluxes back to the bulb causing a steady heat inflow and a temperature difference between the liquid and the wall of the bulb; this difference is accentuated by the Kapitza thermal resistance at the boundary. Montgomery and Pells (1963) have shown that these discrepancies can reach 10 or 15 millidegrees at $1{\cdot}5^\circ$ K: to avoid or reduce such errors it seems best either (i) to have a ^{3}He vapour pressure bulb, (ii) to have a pumped ^{4}He chamber, as shown in Fig. 64 (b) for measurements below the λ-point and a bulb above the λ-point, (iii) to solder copper fins or gauze inside the bulb to increase the effective heat-transfer surface, or (iv) to have a constriction ($\frac{1}{2}$ or 1 mm diam.) in the tube above the He II bath to reduce film flow (see also Cataland *et al.* in *Temperature*, Vol. 3, Pt. 1, p. 413).

Liquid hydrogen also presents a special problem because of the existence of the *ortho-* and *para-* forms discussed in Chapter II. Normal hydrogen (75 per cent *ortho*, 25 per cent *para*) boils under standard pressure at 20·39° K, while equilibrium-hydrogen (practically pure *para*-hydrogen in thermodynamic equilibrium at its boiling point) boils at 20·28° K. However, for most purposes, if pure hydrogen gas taken from a room-temperature reservoir is condensed into the bulb, its boiling point will remain quite close to 20·39° K for some hours. If a very precise determination is required, then a small quantity of a suitable catalyst such as of neodymium oxide, ferric hydroxide gel, or chromic anhydride (see Chapter II and Los and Morrison 1951) may be placed in the bulb; then by allowing an hour or so to elapse before measurement, it may be assumed that the liquid present is equilibrium-hydrogen. Ferric hydroxide which is the most commonly used catalyst becomes ineffective after prolonged use or exposure to the atmosphere and water vapour but can be reactivated by outgassing it for a few hours at a temperature of 100–120° C.

For purposes of pressure measurement, a mercury column or accurate dial gauge (e.g. Budenberg standard vacuum gauge or a Wallace and Teirnan gauge) may be used at normal pressures, and an oil manometer, McLeod gauge, or hot-wire gauge for lower pressures. Another useful but more elaborate manometer is the quartz-spiral gauge (Texas Instrument Co.) which is direct reading from 1000 mm Hg down to a few microns.

As was mentioned in the previous section, the effect of the thermo-molecular pressure difference must be taken into account when the pressure to be measured is sufficiently low that the mean free path of the gas molecules becomes comparable with the diameter of the connecting tubes. As a guide to when this condition is liable to arise, it may be noted that at room temperature, the mean free path l for most gases is of the order of 100 cm at 10^{-4} mm Hg; hence, being inversely proportional to pressure, $l \sim 1$ cm at 10^{-2} mm Hg, 1 mm at 10^{-1} mm Hg, and $\frac{1}{10}$ mm at 1 mm Hg pressure; l decreases with temperature as T^m where the exponent m has values of 1·1–1·3 for most gases.

By using tubes $\frac{1}{2}$ cm or more in diameter, the corrections are made negligible unless measurements of vapour pressure are extended to the range well below 1 mm Hg. A detailed discussion of these corrections in the hydrodynamic, intermediate, and molecular (or Knudsen) regions have been given by Keesom (1942) in the case of helium.

Data

Table A in the appendix has data for the vapour pressure of liquid helium-3, helium-4, normal- and equilibrium-hydrogen, nitrogen, and oxygen. The units of pressure, p, are millimetres of mercury at 0° C and standard gravity (980·665 cm/s^2) so that readings taken at room temperature or in places where the gravity is appreciably different must be corrected as discussed in the next section.

The values of ^{3}He are abbreviated from the compilation of Sherman, Sydoriak, and Roberts (1964) entitled 'The 1962 ^{3}He scale of temperatures'. This 1962 ^{3}He scale has been recognized by the International Committee on Weights and Measures as an international standard. In algebraic form, the relation between P (mm Hg, 0° C) and T° K is:

$$\ln P = -2{\cdot}49174/T + 4{\cdot}80386 - 0{\cdot}286001T + 0{\cdot}198608T^2$$
$$-0{\cdot}0502237T^3 + 0{\cdot}00505486T^4 + 2{\cdot}24846 \ln T$$
$$(0{\cdot}2 < T < 3{\cdot}324° \text{K}).$$

A detailed discussion of the scale appears in the review by Roberts *et al.* (1964).

The ^{4}He values are also taken from an internationally approved scale, namely the 1958 ^{4}He scale (Brickwedde *et al.* 1960). This scale resulted from the marriage of the two scales T_{55E} and T_{L55} which were produced respectively from the efforts of Clement and others in Washington and Van Dijk and his colleagues in Leiden. The history of this scale and its deviations from its predecessors are given in an interesting review by Van Dijk (1960).

Although these helium scales are internationally accepted and internally smooth, they may yet be shown to depart from true thermodynamic temperatures and therefore subject to revision. Preliminary measurements by Plumb and Cataland (1965) with acoustic thermometers suggests that the 1958-scale temperature may be a few millidegrees too low at 4·2° K and near 2° K.

The vapour pressures of equilibrium- and normal-hydrogen in the table are from Woolley, Scott, and Brickwedde (1948) because of their convenient tabulation. More recently data for equilibrium-hydrogen have been obtained by Hoge and Arnold (1951) and Barber and Horsford (1963). These only differ from our tabulated values by about 3 millidegrees over the range from 14° to 20° K. Barber and Horsford have expressed their data by the relation

$$\log_{10} P = 13{\cdot}26223 - 80{\cdot}30193/T - 0{\cdot}748084T + 0{\cdot}0302862T^2 - 0{\cdot}000444259T^3$$

$$(13{\cdot}8 < T < 20{\cdot}3^\circ\,\mathrm{K}).$$

The vapour pressure of neon is not included in the table. It has been measured by Grilly (1962) and fitted to the equations:

$$\log_{10} P = 6{\cdot}89224 - 110{\cdot}809/T + 5{\cdot}4348\times 10^{-3}\,T \qquad (20{\cdot}301 < T < 24{\cdot}544^\circ\,\mathrm{K})$$

and

$$\log_{10} P = 7{\cdot}46116 - 106{\cdot}090/T - 0{\cdot}0356616T + 0{\cdot}000411092T^2 \qquad (24{\cdot}544 < T < 44{\cdot}384^\circ\,\mathrm{K})$$

with maximum deviations in P of 0·4 and 0·2 per cent respectively. He found the normal boiling point to be 27·092° K and triple point at 24·544° K (324·8 mm Hg).

The data for nitrogen in Table A are from Armstrong (1954) above the triple point and from the work of Keesom and Bijl (1937) below it. More recent thermodynamic calculations of the vapour pressure by Moussa, Muijlwijk, and Van Dijk (1966) give values of 77·343° K and 63·141° K for the boiling and triple points respectively. These are based on 90·170° K for the boiling point of oxygen whereas Armstrong's values (see also Table VIII above) are based on the N.B.S. 1939 scale.

For oxygen, values are taken from the recent tables of Muijlwijk, Moussa, and Van Dijk (1966). These do not appear to differ significantly from the data of Hoge (1950*a*) at the boiling and triple points but do differ between 70° and 80° K.

Gravity and temperature corrections

Readings of vapour pressure recorded from a 'liquid column' pressure gauge are subject to gravity and temperature corrections. Therefore for precise measurements in cases where the value of the acceleration due to gravity g is substantially different from that at sea-level at latitude 45°, or where the temperature of the mercury column is substantially different from that used in the table, correction is necessary.

Gravity corrections, which are usually extremely small, are given in various physical tables (e.g. *Smithsonian physical tables*, 9th revised edition (1954), p. 608) and are simply proportional to the difference between the acceleration due to gravity g, at the place considered and standard gravity, 980·665 cm s^{-2}.† Hence the correction to barometric height h is

$$\Delta h = +\frac{(g-980{\cdot}665)}{980{\cdot}665}\times h. \tag{29}$$

Similarly, the correction Δh due to a difference in temperature ΔT between the barometric column in question and the standard column referred to in the tabulation, is proportional to ΔT, i.e.

$$\Delta h = -\alpha\Delta T\,.\,h. \tag{30}$$

For a mercury column and glass scale, this correction constant $\alpha = 0{\cdot}000172$ per deg C and for a mercury column and brass scale

† The figure of 980·665 is the standard value adopted in 1901 for use in barometer reductions and is quite close to the value for normal gravity at sea-level at latitude 45°.

$\alpha = 0{\cdot}000163$ per deg C (see, for example, *Smithsonian physical tables*, 9th revised edition (1954), p. 607).

The correction to be applied to a reading of height of a mercury column at sea-level and an ambient temperature of *ca.* 20° C is about minus 1 part in 300 before comparing it with the vapour-pressure table.

6. Electrical resistance of metallic elements

Introduction

In a metal, the free electrons responsible for electrical conduction are scattered by imperfections in the crystal lattice and by the thermal vibrations of the lattice. These processes limit the conductivity and so determine the electrical resistivity ρ. This resistivity is a function of the number of free electrons per atom n, the velocity of the electrons v, the electronic charge e, and the effective mean free path. The mean free path l, being partly limited by thermal vibrations whose amplitude is temperature-dependent, is itself therefore a temperature-dependent quantity. The charge e is a constant and n and v are practically independent of temperature so that this effective mean free path is the principal factor in determining the temperature variation of electrical resistance.

We may define a resistivity ρ_{r} due to the static imperfections —either chemical impurities or physical impurities—and a resistivity ρ_{i} caused by thermal vibrations. Due to the static character of the impurities, ρ_{i} is the quantity which we expect to change with temperature and the total resistivity may be written, assuming the validity of Matthiessen's rule,

$$\rho = \rho_{\mathrm{r}} + \rho_{\mathrm{i}}. \tag{31}$$

The thermometric property with which we are concerned is therefore ρ_{i} and $d\rho_{\mathrm{i}}/dT$ determines the sensitivity of the electrical resistance thermometer. ρ_{i} for most metallic elements is approximately proportional to temperature down to temperatures in the vicinity of $\theta_{\mathrm{D}}/3$; θ_{D} is the Debye characteristic temperature. Below this ρ_{i} decreases more rapidly with temperature and between about $\theta_{\mathrm{D}}/10$ and $\theta_{\mathrm{D}}/50$ (the lower limit of reliable investigation) $\rho_i \propto T^n$, where $3 < n < 5$. Therefore at very low temperature, the sensitivity of electrical resistance as a thermometric element decreases rapidly (Fig. 65 after Dauphinee and Preston-Thomas 1954), and in the case of metallic elements of the highest available purity $\rho_i \ll \rho_r$ at temperatures in the vicinity of $\theta_D/100$. This enables us to determine ρ_r, which experimentally appears to be constant at the very low-temperature end of the scale.†

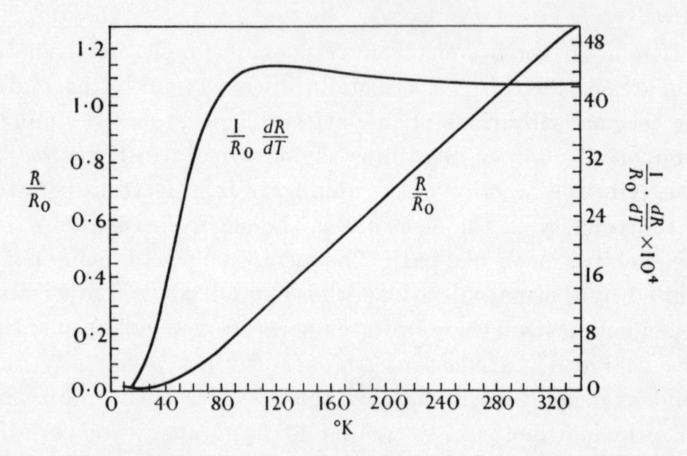

FIG. 65. Variation of resistance ratio and the temperature coefficient of resistance with temperature for a copper resistance thermometer (after Dauphinee and Preston-Thomas 1954).

It is fairly obvious that the ideal metallic element for use as a resistance thermometer should have the following properties:

(i) A resistivity ρ_i which has a variation with T at higher temperatures which is as close to linear as possible; this simplifies the task of interpolation considerably.

(ii) For low-temperature use, the θ_D should be as low as possible so as to preserve a high sensitivity to a low temperature.

(iii) The element should be obtainable in a state of high

† Some metallic elements become superconducting, i.e. their total electrical resistance falls sharply to zero at a critical temperature, and this may prevent an accurate estimate of ρ_r.

purity so that ρ_r will be insignificant over a wide temperature range.

(iv) It should be a metal which is chemically inert and should have a high stability of resistance, so that its calibration is retained over long periods of time and not affected by thermal cycling.

(v) It must be capable of being mechanically worked, i.e. drawn into wire and wound into required forms.

The noble metal platinum fulfils most of these requirements quite adequately although from the low-temperature viewpoint its range of usefulness would be much greater if its Debye temperature were lower. However, metals like lead, bismuth, or gallium which have a low characteristic temperature ($\theta_D \simeq 100°$ K, cf. for Pt, $\theta_D \simeq 225°$ K) are rather unsuitable for other reasons.

Platinum thermometers

In § 4.3 we mentioned the use of the platinum resistance thermometer for realizing the international temperature scale between $-182{\cdot}97°$ and 630° C, using the formulae

$$t = \frac{1}{\alpha}\frac{R_t - R_0}{R_0} + \delta\frac{t}{100}\left(\frac{t}{100} - 1\right) + \beta\left(\frac{t}{100}\right)^3\left(\frac{t}{100} - 1\right) \tag{32}$$

for

$$-182{\cdot}97 < t < 0° \text{ C}$$

and

$$t = \frac{1}{\alpha}\frac{R_t - R_0}{R_0} + \delta\frac{t}{100}\left(\frac{t}{100} - 1\right) \tag{33}$$

$$\text{for } t > 0° \text{ C}.$$

For a precision thermometer used to realize the international temperature scale, $\alpha = \{(R_{100}/R_0) - 1\}/100$ must have a value greater than 0·00392, a figure which corresponds approximately to a ratio of $\rho_r/\rho_{0°\text{C}}$ less than about 15×10^{-4}. The constants δ and β representing the departure of R_t from a linear dependence on temperature are determined by measuring R at the zinc and oxygen points respectively. These and other requirements have been discussed in detail by Mueller (1951), Hall (1955), and Stimson (1955).

Alternatives to the Callendar forms of the resistance equations are the following:

$$R_t = R_0(1+At+Bt^2), \quad t > 0^\circ\,\mathrm{C}; \tag{34}$$

$$R_t = R_0\{1+At+Bt^2+C(t-100)t^3\}, \quad t < 0^\circ\,\mathrm{C}. \tag{35}$$

These forms have certain advantages particularly when the '100° C' calibration point is significantly different from 100·000° C. The constants may be related to those in the Callendar equations:

$$A = \alpha(1+\delta/100) \quad \text{and} \quad \alpha = A+100B$$

$$B = -\alpha\delta/100^2 \qquad \delta = -100^2B/(A+100B)$$

$$C = -\alpha\beta/100^4 \qquad \beta = -100^4C/(A+100B).$$

Convenient methods of calculating t from R_t have been described by Schwab and Smith (1945), Werner and Frazer (1952), and Hales and Herington (1965). The simplest of these, due to Werner and Frazer, tabulates at 1-deg intervals from -190° to 600° C, the calculated values of R/R_0 for an imaginary thermometer having $\alpha = 0{\cdot}0040$, $\delta = 1{\cdot}493$, and $\beta = 0{\cdot}1090$. Then knowing the real α-value for a particular thermometer, an experimental value of R/R_0 can be easily converted into a value of t; small corrections due to difference in δ and β are made with the aid of subsidiary tables given by the authors. Inaccuracies due to linear interpolation should not exceed a millidegree.

The methods of construction which have been found to give a high degree of reproducibility are those in which annealed platinum wire is supported in a protective capsule in a condition of least mechanical constraint and is then re-annealed. In three successful types of thermometers, fine platinum wire 0·005–0·01 in diam. of total resistance about 25Ω at room temperature, is wound into a fine helix and then either (i) freely suspended in a Pyrex U-tube (Barber 1950, 1955), (ii) supported in a bifilar fashion on opposite sides of a twisted silica ribbon (Russian method described by Stimson 1955), or (iii) bifilar wound in a second larger helix which is loosely supported on a notched mica cross (Meyers 1932). With four leads attached the thermometer is then sealed off in a silica, Pyrex, or platinum capsule under a small pressure of oxygen. For low-temperature use, some helium gas must be mixed with the oxygen to ensure good heat dissipation at low temperatures.

More recently Meyers has used straight platinum wire bifilar wound on a closely notched mica cross (e.g. H. F. Stimson in *Temperature*, Vol. 2, p. 141) and a limited number of his thermometers have been available commercially. The Leeds and Northrup Company (Philadelphia) have made encapsulated platinum thermometers of the Meyers coiled-coil type until recently and now produce the later Meyers model. Tinsley Limited (London) have made thermometers of the Barber design. Rosemount Engineering Company of Minneapolis (or Research and Engineering Controls Limited in Britain) have made a reproducible thermometer from straight platinum wire threaded through fine holes in a ceramic cylinder, which in turn is held in a gold cylinder and encapsulated. Platinum thermometers of less elaborate construction, which do not necessarily have the high standards of reproducibility of those just mentioned but which may be very useful as control or sensing elements, are made by various firms including Rosemount Engineering Company, Sensing Devices Inc. of California, Minco Products Inc. of Minneapolis, Degussa, and Hartmann and Braun A.G., of West Germany.

Procedure at low temperatures

As mentioned earlier in the chapter, there is yet no internationally recognized scale below the oxygen point except that covered by the vapour pressure of helium below 5° K. The electrical resistance of platinum (and of sundry other metals and semiconductors) has the necessary reproducibility and sensitivity over much of the temperature range but cannot be represented by a convenient analytic function. Therefore calibration at two or three points cannot specify it and it becomes necessary to calibrate each thermometer at a national laboratory against their standards. These have previously been calibrated against a gas thermometer and constitute a form of national scale.

For those who make or buy resistance thermometers and do not have adequate calibration facilities or recourse to a calibration laboratory, life is more difficult. The best course seems to be the following. Use the best available platinum and ensure that the thermometer is well annealed and not subject to marked hysteresis on thermal cycling. Measure its ice-point resistance, R_{273}. If it is to be used in the region from 90° K to room temperature, the constant α should be determined either by measuring R_{373} at the steam point or more simply by checking its resistance R_T in a well-stirred hot-water bath against an accurate substandard or standard mercury-in-glass thermometer; then assuming a value of about 1·492 for δ, the appropriate α may be calculated from (33). For subsequent use below the ice point a value for β can be assumed; β usually lies in the range 0·109–0·112, so that a trial value of 0·111 should meet the demands of required accuracy. Whether in fact these demands are met by the assumed values of β can be tested by measuring R at the oxygen point and in the intermediate region at the carbon-dioxide sublimation point (see Scott 1941; Barber 1966; Berry 1962, for methods of using oxygen and carbon-dioxide points).

If the thermometer is to be used below 90° K then measure the ice-point and helium-point resistances R_{273} and R_4 and any intermediate values (e.g. R_{90}, R_{54}, or R_{20}) that are possible and compare $(R_T - R_4)/(R_{273} - R_4)$ with a tabulated function for a 'standard' thermometer whose characteristics are similar. If Matthiessen's rule for the additivity of a temperature-independent impurity resistance, R_{r}, and an impurity-independent 'thermal' resistance, R_{i}, were valid then a function of the type

$$Z = (R_{T_1} - R_{T_2})/(R_{T_0} - R_{T_2})$$

would be identical for all resistors made of a particular metallic element. Unfortunately this additivity hypothesis is not exactly valid, and the departures are greatest between 10° and 30° K where the impurity and 'thermal' resistances are comparable in magnitude. Therefore the Z-function should only be used with care and awareness of its weaknesses. Cragoe at the National Bureau of Standards first proposed and tested its usefulness over the restricted range from 54° to 90° K where it seemed fairly successful. Later examination of Z-values for wider ranges 20° to 90° K, etc., showed deviations of many hundredths of a degree. Therefore various efforts have been made to produce corrections, for example by Hoge (1950*b*), Los and Morrison (1951), Van Dijk (1952), Lowenthal, Kemp, and Harper (1958), Corruccini (1960). An illuminating review of resistance thermometry by Barber (1960) discusses this question as also do articles by a number of these authors in *Temperature*, Vol. 3, Pt. 1.

Since most research laboratories have liquid helium on hand rather than liquid hydrogen, it is most practical to consider a Z-function based on this, viz:

$$Z = \frac{R_T - R_4}{R_{273} - R_4} \quad \text{or} \quad Z = \frac{R_T - R_4}{R_{90} - R_4}.$$

The Z-values for a number of platinum thermometers of high purity ($W_4 < 5 \times 10^{-4}$ or $\alpha > 0{\cdot}0039255$) are tabulated here in Table B (Appendix) over the range 10° to 150° K. Some values are also included for an 'ideal' thermometer having zero impurity resistance (Berry 1963).

The difference between the values of Z in the table for say Meyers's No. 459 and Berry's 'ideal' platinum amount to *ca.* 0·06 degK at 15° K, 0·07 at 25° K, and 0·03 at 60° K. These seem intolerable to the person who is searching for millidegree precision. However in many research investigations deviations of a few hundredths of a degree from a true thermo-dynamic temperature may be unimportant provided that the scale is internally smooth. From this point of view, the values marked CCT 64 may be the most suitable in Table B. They are calculated from values of $W_T = R_T/R_{273}$ which are themselves weighted means expressed in terms of a mean temperature scale (see Orlova *et al.* 1966, also § 3 and Fig. 59, pp. 104–5); they resulted from recommendations of the 1964 meeting of the Consultative Committee on Thermometry.

Progress in purifying metals by zone-refining and other heat-treatments may yet lead to platinum wire and platinum thermometers becoming available which have resistance ratios $W_4 \leqslant 10^{-4}$ and values of $a \geqslant 0{\cdot}003928$. In this case it is likely that one table of Z-values could represent the behaviour of many thermometers with deviations of only a few millidegrees in the worst temperature region.

Other metallic elements as thermometers

Another very useful form of resistance thermometer is that in which an insulated strain-free wire of a suitable metallic element (necessarily isotropic in its thermal expansion) is wound onto a former of the same element, so that differential contraction on cooling is small. Dauphinee and Preston-Thomas (1954) described the use as a thermometer of commercial enamelled copper wire (46 B. & S. gauge), wound and held by baked Formel varnish on a copper calorimeter. They reported a high degree of reproducibility and found that for a number of different thermometers made from the same commercial wire, the deviation from one another in terms of $(R_T - R_{4\cdot2})/(R_{273} - R_{4\cdot2})$ represented considerably less than 0·1 degK. Rose-Innes (1964, p. 101) has also described a simply constructed platinum thermometer in which the wire is wound onto a thin piece of varnished platinum sheet which is rolled up to form a small cylinder.

Other elements suitable for thermometers include gold, palladium, and silver as they have relatively low Debye temperatures (200–300° K), high ductility, and high reproducibility due to their cubic structure. They are chemically inactive and can be obtained in a high state of purity. However to displace platinum from its privileged position they would need to have an appreciably higher sensitivity below 20° K and therefore be available in a much purer form than platinum.

If, on the other hand, metallic elements with a $\theta_D \sim 100°$ K are suitable on other grounds, their resistance will be an approximately linear function of temperature down to about 30° K and their sensitivity as thermometers should be tolerably good down to 4° or 5° K. Unfortunately most elements in this category are unsuitable for some reason or other. The alkali metals—of which rubidium, caesium, and potassium have the lowest characteristic temperatures—are chemically very active and must generally be cast in glass capillaries, due to difficulties in handling them; therefore, although ρ_i may be relatively large at 5° or 6° K, it is not highly reproducible, due partly to the mechanical constraint imposed on the element.

Thallium has $\theta_D \simeq 90°$ K and can be obtained in a pure state, but is oxidized very quickly and is difficult to handle; gallium melts at room temperature and is highly anisotropic so might not be reproducible on thermal cycling. Two of the best possibilities are bismuth and indium. Indium is readily available in the state of highest 'electrical purity', with a resistance ratio, $W_4 < 10^{-4}$. It is easily extruded into wire but this is very soft and difficult to handle. Due to its tetragonal structure it is anisotropic and may therefore not be sufficiently reproducible for a standard. It has been used as a thermometer by White and Woods (1957; see also White, Woods, and Anglin 1958; James and Yates 1963; Orlova, Astrov, and Medvedeva 1964; Kos, Drolet and Lamarche 1967).

7. Metallic alloys as resistance thermometers

Introduction

In an alloy we have the extreme case of an impure metallic element, in which the reverse situation holds to that considered desirable in our discussion of the previous section: that is, the major part of the electrical resistance at all temperatures is due to scattering by fixed impurities, so that the electron mean free paths are essentially constant and the total resistivity $\rho \simeq \rho_r$. In most alloys, the total resistivity is considerably greater than the intrinsic resistivity of the individual constituent elements and is rather insensitive to temperature. This insensitivity is not unexpected since the scattering effect of the thermal vibrations is insignificant compared with the scattering effect of the random arrangement of atoms of differing atomic radius on the lattice sites. Alloys might all be expected to be extremely insensitive thermometric systems.

Although this is generally true there are notable exceptions (see also discussion by Daunt 1955) which fall into three main categories:

(i) 'anomalous' alloys such as manganin and constantan;

(ii) dilute alloys exhibiting a resistance minimum at low temperatures;

(iii) alloys with a superconducting component or inclusion.

'Anomalous' alloys

In class (i) manganin and constantan and perhaps a number of other alloys as yet neglected, exhibit an electrical resistance which is almost completely temperature insensitive at room

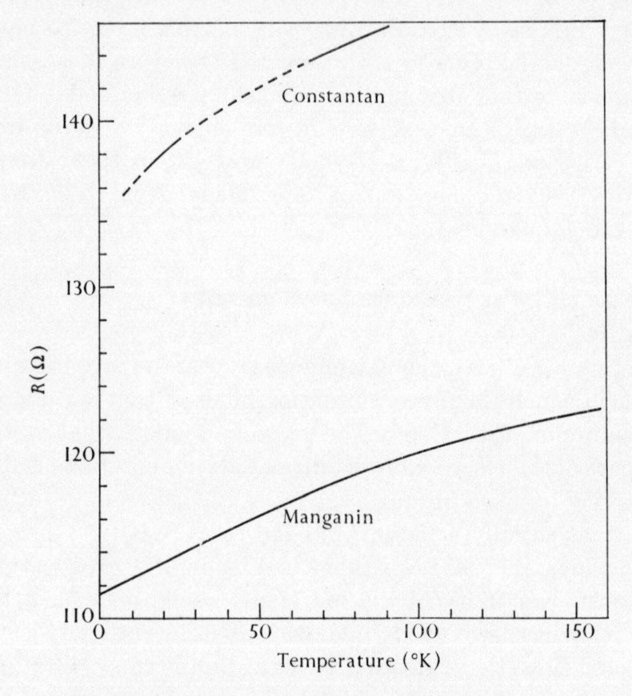

FIG. 66. Variation of resistance with temperature for manganin and constantan.

temperature, but which begins to decrease quite markedly below about 200° K. Kamerlingh Onnes, and Holst (1914) observed the temperature dependence shown in Fig. 66. As a result constantan in particular has been used as a low-temperature resistance thermometer; Parkinson (see, for example, Simon, Parkinson, and Spedding 1951) used it in the range below 20° K, for temperature measurement on a calorimeter.

For some years, the writer used a manganin resistor as a temperature-sensitive element to actuate an electronic temperature controller, in which the manganin resistance in close thermal contact with the experimental chamber of a cryostat (White 1953) formed one arm of an a.c. Wheatstone bridge; in this case the manganin served adequately over the range from 5° to 150° K to control temperature to a few thousandths of a degree. Thermometers of this type generally seem to show some degree of hysteresis when the temperature is changed to a marked extent and then returned to its earlier value. This is probably not an intrinsic property of the alloy, so that annealed wires of manganin or constantan supported in a way which causes no mechanical strain when heated or cooled, rather than in the more common manner of having them wound on a solid former of some dissimilar material, may be quite reproducible in behaviour.

Electrical resistance minimum

There is a minimum in the electrical resistance of many dilute alloys and impure metallic elements at low temperatures, generally below 20° K (review by Van den Berg 1964). It arises from an inelastic scattering of the electrons by magnetic spins localized on the impurity atoms. In some cases the resistance continues to increase as temperature decreases below T_{min} and in others it becomes constant or passes through a maximum and decreases. The temperature of the minimum depends on the impurity and it increases slowly with increase in impurity content.

A few atoms per million of iron in solution in copper or gold has a quite marked effect on the resistance (and on the thermoelectric power). As shown in Fig. 67 (e.g. Dugdale and MacDonald 1957), such dilute alloys can constitute a useful thermometer. In particular dilute copper alloys have T_{min} which is sufficiently high for them to be used from 15° or 20° K down to

2° K; below this their resistance becomes relatively constant. Gold alloys have a smaller value of $T_{\min}$ and can be used to much lower temperatures. As discussed in § IV.9, the anomalous thermopower of these alloys, particularly Au+Fe, serves as a useful thermometer. Since the impurity producing the anomaly is generally present in very small amounts, it can be assumed to

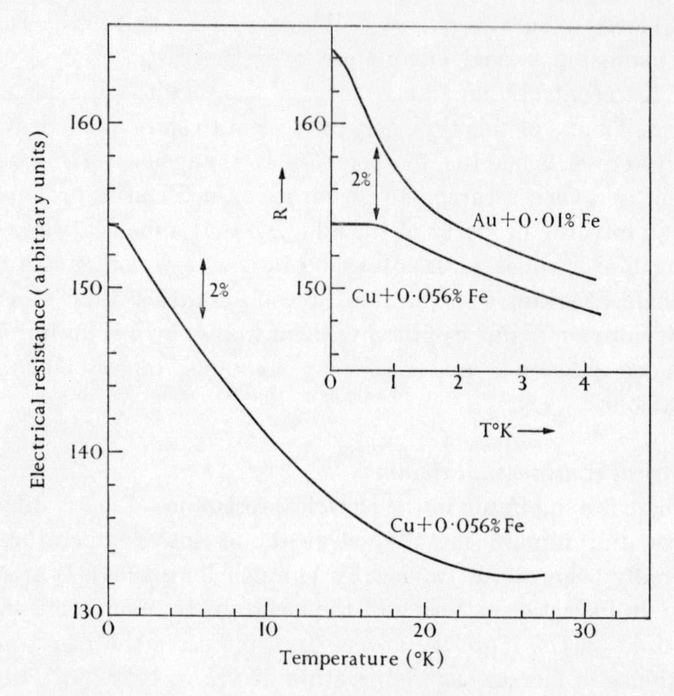

Fig. 67. Electrical resistance of dilute alloys at low temperatures.

be in fairly homogeneous solid solution and therefore reproducible provided that neither strains nor oxidizing atmospheres are present which can change the distribution of the solute atoms.

Another promising thermometer may be made from dilute rhodium–iron alloys which Coles (1964; also Coles, Waszink, and Loram 1964) has shown to have an 'anomalous' fall in resistance at low temperatures. Instead of becoming constant—the normal behaviour—or passing through a minimum, the resistance continues to decrease with decrease in temperatures.

Superconducting inclusions in alloys

The third class (iii) mentioned above are those alloys which contain a superconducting element in relatively small proportion, and which at temperatures below 10° K exhibit a marked decrease in resistance which may extend over 5 or 6 degK. Keesom and Van den Ende (1930), and later Babbitt and Mendelssohn (1935), found that in some samples of phosphor-bronze (92·7 % Cu, 7 % Sn, 0·2 % P + traces of Pb, Be, Se, Ca, Mg) the resistance fell nearly linearly with temperature from about 7° K downward.

It seems that in the samples of phosphor-bronze which had this satisfactory thermometric behaviour, a trace of lead (0·1 %) was present and that segregated inclusions of lead in the form of fine needles become superconducting below 8° K. Van Dijk (1951) and Daunt (1955) have also prepared leaded phosphor-bronze wires which behaved in this manner.

Babbitt and Mendelssohn (1935) also found that a silver-lead alloy containing 5 per cent lead had a very broad superconducting transition making it suitable as a resistance thermometer for the range 3·5°–7° K.

The use of leaded-brass for measuring temperatures in the region below 7° K was first suggested by Mendelssohn. Later samples containing 62 % Cu, 36 % Zn, 0·08 % Ni, and 1·73 % Pb were tested by Parkinson and Quarrington (1954) who found that if suitably annealed the electrical resistance decreased very markedly over a temperature range of 2° or 3° K. Parkinson and Roberts (1955) later reported that by annealing these wires in a temperature gradient, a thermometer with an almost linear response over the range 1·5°–4° K was obtained.

The unfortunate features of this class (iii) of low-temperature thermometers, apart from their narrow range of usefulness, are the slight dependence of resistance on measuring current, their sensitivity to a magnetic field, and the critical nature of the method of preparation. By the very nature of their mode of action these features are to be expected.

8. Semiconducting resistance thermometers

Introduction

Suitable semiconductors make excellent resistance thermometers in the low range of temperatures where metallic elements

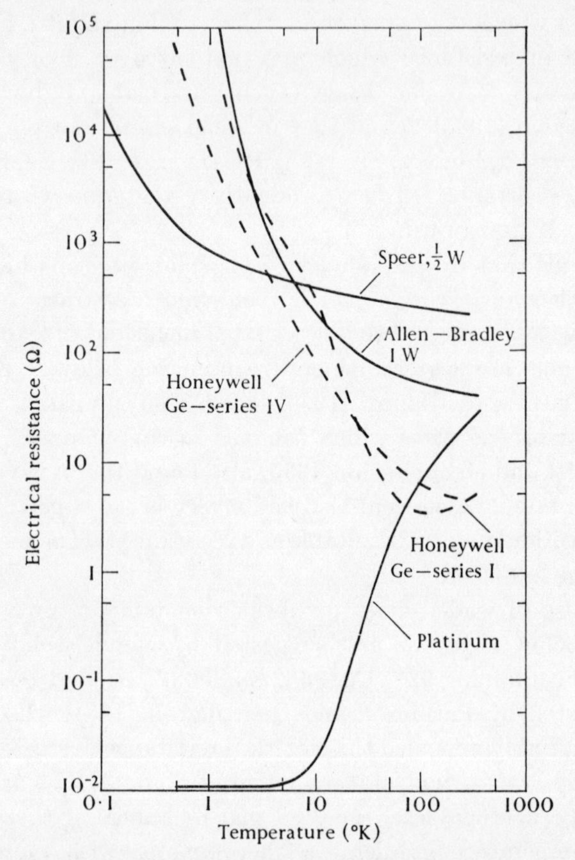

FIG. 68. The electrical resistance of some typical thermometer elements. 'Allen-Bradley' and 'Speer' denote carbon radio-resistors.

have become too insensitive and alloys are not sufficiently reproducible or sensitive (Fig. 68). The first semiconductors to be used as thermometers were the films of carbon-black or 'Aquadag'

(Giauque, Stout, and Clark 1938). Later the commercial radio-resistors (Clement and Quinnell 1950) became and still are widely used. However a high degree of reproducibility was not achieved until the crystals of doped germanium and silicon were tried in the pioneering efforts of Friedberg and Estermann at the Carnegie Institute of Technology (review by Friedberg in *Temperature*, Vol. 2 (1955) p. 355), and Geballe and his collaborators at the Bell Telephone Laboratories.

Thermistors (Martin and Richards 1962) and gallium arsenide p–n junction diodes (Cohen, Snow and Tretola 1963) have also been used successfully as low-temperature thermometers.

Germanium, silicon, etc.

At high temperatures the conductivity is intrinsic, i.e. due to excitation of electrons from the valency band (otherwise full) to the conduction band (otherwise empty). The conductivity is a function of the number of charge carriers and their mobility, and in the intrinsic region it may be shown that

$$\rho \simeq A \exp(\Delta E/2kT),$$

where A is almost temperature independent. E is the energy gap between the valence and conduction bands, having values of 0·72 eV for Ge, 1·12 eV for Si, and 0·34 eV for Te.

At low temperatures, where $kT \ll \Delta E$, the conductivity is due to the presence of impurities which contribute electrons which may be excited across a gap of energy ΔE_D to the conduction band, or accept electrons from the valence band across a gap ΔE_A leaving conduction holes in the valence band. Thus, depending on the dominant type of impurity, the conduction at low temperatures is n-type (due to electron carriers) or p-type (due to hole carriers).

Kunzler, Geballe, and Hull (1957) used arsenic-doped germanium and mounted the elements in the way shown in Fig. 69. The 'bridge' shape is cut from a single crystal and four gold wires are welded to it as current and potential contacts. These wires are in turn welded to four platinum leads. It is carefully mounted to avoid strains and piezoelectric effects. These have been found to be completely reproducible within the limits of measurement—better than 1 millidegree at 4·2° K (see e.g. P. Lindenfeld in *Temperature*, Vol. 3 (1962), Pt. 1, p. 399, and also Kunzler, Geballe, and Hull in the same volume, p. 391).

Following these tests commercial versions were made by Minneapolis Honeywell and by Texas Instruments, Inc. The former used Ga + Sb-doped germanium and encapsulated them in a small copper cylinder with some helium exchange gas, very much like that shown in Fig. 69 (Blakemore, Schultz, and

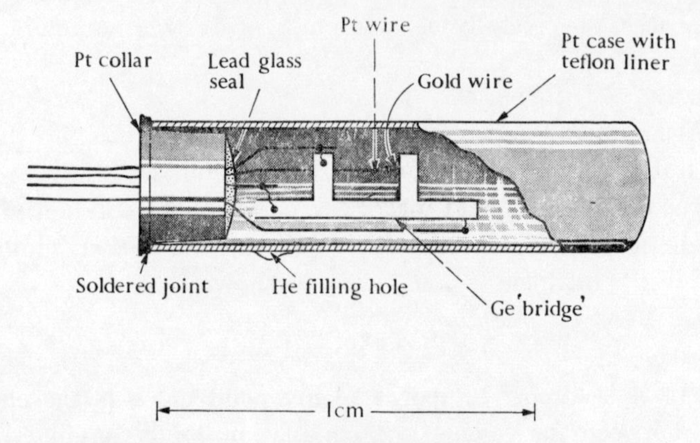

Fig. 69. Encapsulated thermometer (after Kunzler, Geballe, and Hull 1957).

Myers 1962). Tests have shown their different sensors to be quite reproducible. Those made by Texas Instruments have been Ga-doped germanium (Low 1962) mounted in the form of a button or encapsulated in glass as two-terminal devices. Later other manufacturers produced capsule and button types (see pp. 382–3) and in some cases will provide a calibration (at extra cost) over required ranges from $\sim 1°$ to 90° K.

Self-heating effects must be guarded against by checking whether the resistance is sensitive to change in measuring current. With most of the thermometers, 100 μA can be used safely as the current above 10° K and 10 μA used down to 2° or 3° K. Generally power dissipation should be limited to 10^{-7} W at most in the capsule types.

Sensitivity to magnetic field varies with the doping and temperature. Data published for one of the Honeywell series (Blakemore, Schultz, and Myers 1962) indicate that

$$\Delta R/R_0 \approx 2\times 10^{-10}\,\mathrm{H}^2$$

corresponding to a 2 per cent increase in a 10 000-Oe field. Kunzler, Geballe and Hull (*Temperature*, Vol. 3, Pt. 1, p. 391) found increases of 3–6 per cent for different units in a similar field, the effect being slightly larger when the field is transverse.

In this respect and in others silicon thermometers offer some advantages but have not yet been produced commercially. Silicon has a larger Debye temperature and therefore a much lower heat capacity at low temperatures. Slices cut from a boron-doped silicon crystal have been used without encapsulation as thermometers at the Bell Telephone Laboratories for some years for calorimetry. These slices have been cemented onto button-shaped alloy samples ($<\frac{1}{2}$-in diam.) with $\frac{1}{4}$-mil Mylar and cement as insulation between them. In this form they have an extremely small heat capacity and show no measurable hysteresis (Morin and Maita 1963).

Fig. 68 suggests that the resistance-temperature data for a germanium thermometer cannot be easily fitted to any analytic expression. This seems generally true even for those thermometers having rather different doping and $R(T)$ characteristics from those shown in the figure. Over a narrow range of temperatures such as 1°–4° K, a simple relation of the form $A\ln T = B-\ln R$ can be fitted, which leaves an error curve or deviation which is 'sinusoidal' in form and of amplitude 50 millidegrees or more (e.g. Lindenfeld 1961). Over a wider temperature range more complex expressions have been found to fit the data with the same sort of accuracy.

These expressions, including those used more commonly with carbon thermometers (see below), are as follows.

(i) Clement and Quinnell (1952) carbon equation with two constants:

$$\left(\frac{\ln R}{T}\right)^{\frac{1}{2}} = a\ln R+b. \tag{37}$$

(ii) Carbon equation with three constants (e.g. Zimmerman and Hoare 1960):

$$\left(\frac{\ln R}{T}\right)^{\frac{1}{2}} = a\ln R + b + c(\ln R)^2. \tag{38}$$

(iii) Van Dijk, Van Rijn, and Durieux (15th Annual Calorimetry Conference 1960):

$$\ln R = a + \frac{b}{T+\theta}. \tag{39}$$

(iv) Veal and Rayne (1964):

$$\ln T = a(\ln R)^2 + b(\ln R) + c. \tag{40}$$

(v) Ahlers (1964):

$$T = \ln R[a + b\ln R + \{c + d\ln R + e(\ln R)^2\}^{\frac{1}{2}}]^{-2}. \tag{41}$$

(vi) Clairborne, Hardin, and Einspruch (1966):

$$R = eT^{-a}\,e^{-b/T}. \tag{42}$$

(vii) Van der Hoeven and Keesom (1965):

$$\frac{1}{T} = \sum_{n=0}^{3} a_n(\ln R)^n. \tag{43}$$

(viii) Polynomials used for germanium by Cataland and Plumb (1966), Martin (1966), Osborne, Flotow, and Schreiner (1967):

$$\ln R = \sum_{n=0}^{m} a_n(\ln T)^n. \tag{44}$$

From nine to fourteen terms have been used to obtain the best fit over the range 1° to 20° K.

The more complex of these equations, (41), (42), and (44), have been required to represent the resistance of germanium thermometers over the range 1°–20° K, with errors of $\sim$ 0·1 per cent or less. The simpler expressions are generally adequate for carbon thermometers.

Carbon thermometers

These are composed partly of graphite either in the form of a film or composition. The resistance increases markedly at sufficiently low temperatures and in many cases the relation $\ln \rho \propto T^{-1}$ is roughly followed. They are not semiconductors in the strict sense for the following reasons.

Crystalline graphite is highly anisotropic, the resistivity in a natural crystal flake along the basal plane (usually denoted by $\rho_{\perp}$ as this is perpendicular to the hexagonal axis) being about $1 \times 10^{-4}\,\Omega$ cm at room temperature and perpendicular to the basal plane $\rho_{\parallel}$ being approximately $1\,\Omega$ cm. Whereas $\rho_{\perp}$ decreases with decrease in temperature, $\rho_{\parallel}$ increases as the temperature is decreased. This behaviour suggests that in the direction parallel to the basal plane, there is a very small band overlap (cf. bismuth) but in the direction normal to the plane there is a narrow band gap; hence a single crystal should behave as a poor metallic conductor in one direction and a semiconductor in the other.

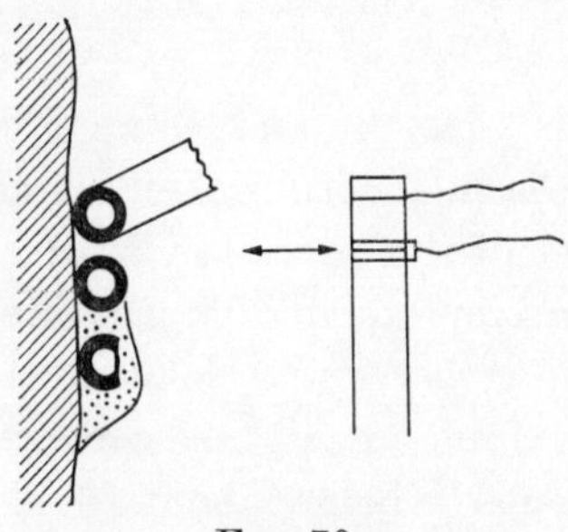

Fig. 70

The early use in California (Giauque, Stout, and Clark 1938) of colloidal carbon films was followed in the ensuing twenty years by many applications of dry carbon blacks or colloidal suspensions of graphite ('Aquadag', Indian ink, etc.) for very low-temperature thermometry. An interesting example is that used by Mendelssohn and Renton (1955) in making measurements of the thermal conductivity of superconductors below 1° K:

Three close turns of 32 SWG enamelled copper wire were wound around the specimen (see Fig. 70) and the ends were twisted so as to tighten the wire down onto the rod. One end of the wire was cut above the twist and the other end used as a lead. Then the enamel on the outside of one of the turns was removed, and a colloidal suspension of graphite in alcohol was painted onto the wire and underlying specimen. After being dried the carbon layer acted as a thermometric substance between the wire and the specimen, the last two acting as end-contacts. . . .

These thermometers have the virtue of high sensitivity at sufficiently low temperature, small heat capacity, and intimate thermal contact with the specimen whose temperature is to be measured. They do, however, show hysteresis effects when warmed up to room temperature and cooled again, and therefore must be recalibrated at each low-temperature experiment.

A further devlopment in 1947 was the application of commercially prepared carbon resistances by Fairbank and Lane (1947). They cut out resistors from the I.R.C. carbon-coated plastic card resistances and used them as second sound detectors, i.e. to detect the temperature waves propagated in liquid helium II; they found that the resistance $R \propto 1/T$ for these thermometers.

Shortly after this Clement and Quinnell (1950, 1952) first reported on the successful results of their search for a commercial carbon thermometer of the normal cylindrical type, which has a high sensitivity and reproducibility at low temperatures. They found that the radio-type resistors manufactured by the Allen–Bradley Company exhibit a resistance which varies rapidly with temperature, particularly below about 20° K (see Fig. 71) and may be expressed to within ± 0·5 per cent by the semi-empirical expression

$$\ln R + K/\ln R = A + B/T.$$

which is similar to (38) discussed above. As may be seen from Fig. 71 (for the 1-W size) such resistors as the 22- or 56-Ω are particularly suitable for use from about 20° down to nearly 1° K.

More recently Clement *et al.* (1953) reported on the use of 2·7- and 10-Ω Allen-Bradley resistors in the region below 1° K, finding them suitable down to about 0·3° K (see Fig. 72).

Most Allen-Bradley resistors reach too high a resistance to be useful below 1° K. Those made by the Speer Carbon Company of Pennsylvania have much lower sensitivity (cf. Fig. 68) and have been used well below 0·1° K. Their behaviour down to $T \sim 0{\cdot}02°$ K has been reported by Black, Roach, and Wheatley (1964) and Edelstein and Mess (1965).

A useful feature of carbon thermometers is their relative insensitivity to magnetic fields. Clement and Quinnell (1952) first showed that $\Delta R/R_0$ varied approximately as H^2; at 4·2° K

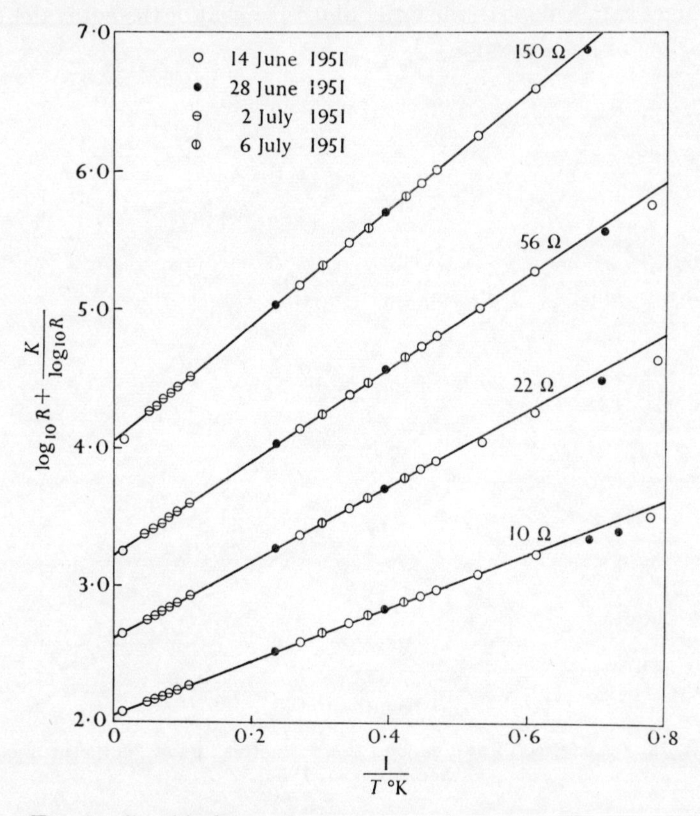

FIG. 71. Graphical test of equation (37) for four carbon resistors (after Clement and Quinnell 1952).

for a 120-Ω resistor it amounted to only one or two parts in a thousand in a field of 10 000 Oe. At lower temperatures this sensitivity to field increases.

Many investigations have been made on the reproducibility of carbon thermometers (e.g. Lindenfeld in *Temperature*, Vol. 3

(1962), Pt. 1, p. 399). Because of their composite nature it is expected that they should show far more hysteresis than single crystals like germanium do on thermal cycling. This seems true particularly for resistors immersed directly to liquid helium during their cycling. But for those which are protected from mechanical and thermal shock and are not directly in the liquid, changes can be kept to the order of 0·1 per cent or the equivalent of a few millidegrees at 4° K.

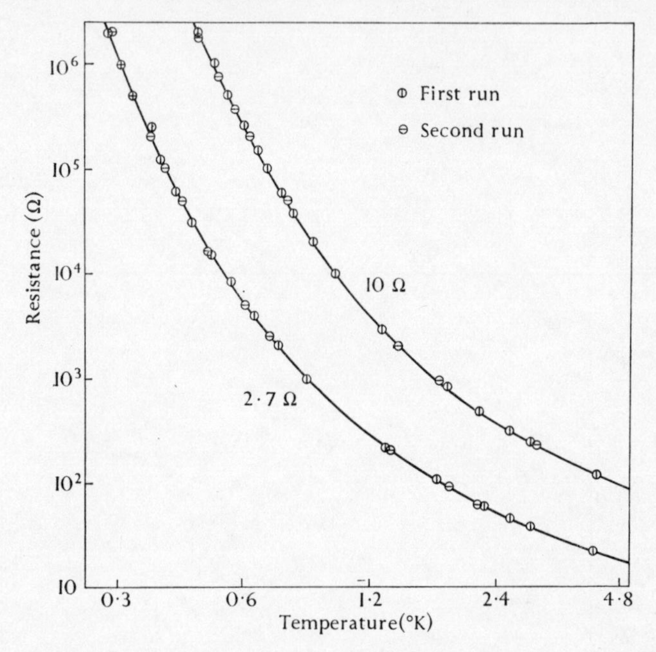

FIG. 72. Resistance of two carbon thermometers at low temperatures (Clement *et al.* 1953).

Self-heating depends on the method of mounting. It can be reduced by grinding off some of the insulation on the resistors and thermally coupling the carbon and/or its copper leads as thoroughly as possible to a heat sink. The relation between the dependence of the temperature rise in the thermometer with power input, dP/dT, and its thermal conductivity was investigated by Berman (1954); he established that in the 100-Ω ($\frac{1}{2}$-W)

Allen–Bradley resistor,

$$dP/dT = 3{\cdot}9\times 10^{-5}T^{1{\cdot}6}\,\mathrm{W/deg},$$

so that using 1 μA current heating is negligible at 4° K; at 1·59° K the mean temperature rise is 7×10^{-4} degK, and at 1·2° K the rise is $1{\cdot}3\times 10^{-2}$ degK. In the 20-Ω size he found dP/dT to be about half as great as for the 100-Ω size.

Apart from their use below 1° K, these commercial carbon resistors have had considerable application in calorimetry measurements below 20° K, where they provide a very sensitive thermometer suitable for measuring small temperature changes. In these situations they are normally calibrated by determining the constants in (37) or (38) by reference to the vapour pressure of liquid helium or liquid hydrogen. Where subsequent interpolation is not considered sufficiently accurate, i.e. between 4·2° and 14° K, $R(T)$ may be checked against a gas thermometer and an error function showing the difference $\Delta T = T(R)_{Eq} - T_g$ can be obtained. Morrison, Patterson, and Dugdale (1955) have given a clear account of such a calibration procedure for a 10-Ω (½-W) Allen–Bradley resistor used in an adiabatic calorimeter. They employed the three-constant equation (38) and found an error function which has a normal maximum amplitude of about 0·02 degK in the range between 20° and 3° K. When the cryostat was warmed to temperatures above that of liquid nitrogen, they found small changes in the constants (notably in A) and in the error function; however, the sensitivity dR/dT seems completely reproducible within the limits of error of measurement.

Rayne (1956) has also described a rather similar calibration procedure for a 10-Ω (½-W) resistor in his calorimeter; working over the range 4·2°–1·2° K, he used the two-constant equation

and found $$\{(\log_{10} R)/T\}^{\frac{1}{2}} \simeq a + b\log_{10} R,$$

where $$a = -0{\cdot}746,\quad b = 0{\cdot}770.$$

Over a rather wider temperature range, this two-constant formula may prove too inaccurate unless a deviation curve is obtained. Keesom and Pearlman (1956) have given a good method of obtaining this: plot $(\ln R/T)^{\frac{1}{2}}$ against $\ln R$ to find the best estimate of the constant b, and then use this fixed value of b to calculate a from the equation as a function of T. The smoothed $a(T)$ curve is then used in interpolation.

The resistance of the semiconducting thermometer can be measured with a potentiometer, digital voltmeter, or a.c. bridge. The latter is to be preferred at the lowest temperatures as the higher sensitivity means that the power dissipation can be kept at a low level of 10^{-10} W or less. A suitable a.c. bridge has been described by Blake, Chase, and Maxwell (1958). In some cases a pair of resistors are used to determine small temperature differences and either an a.c. bridge (Fairbank and Wilks 1955) or a d.c. bridge (Mendelssohn and Renton 1955) may be suitable. The advent of digital voltmeters with a sensitivity of a microvolt has greatly simplified resistance thermometry. These may be frequency counters or potentiometric but all give a direct reading which can be recorded automatically.

9. Thermocouple thermometers

Introduction

In any metal in which there exists a temperature gradient we should expect to find a variation of the electron-energy distribution function with temperature. At normal or low temperatures a small fraction of the electrons have energies in a narrow region of width kT at the top of the Fermi–Dirac distribution. As the temperature is raised, this narrow region expands, resulting in slightly more electrons having higher energies than hitherto. Thus in the metallic bar (Fig. 73 (a)) electrons travel up the bar under a thermal gradient carrying excess thermal energy. To preserve the requirement of no net current flow (since this is an open circuit electrically), a return flow of electrons carrying their normal charge '$-e$' but no excess thermal energy, takes place. This return flow is produced by an electrical driving force, a potential difference ΔE, called the absolute thermoelectric force of the metal. The absolute thermoelectric power is then defined as

$$S = \frac{\Delta E}{\Delta T}.$$

In practice it is usually convenient and necessary to measure the difference in absolute thermoelectric power between two metals. Thus in the situation of Fig. 73 (b) two dissimilar metals A and B have their junctions at temperatures T_1 and T_2. By measuring the thermoelectric voltage E_{12}, we have a thermometric indicator available for establishing the difference between T_1 and T_2. If A or B are not particularly suitable metals for constructing long strain-free connecting leads to the measuring instruments,

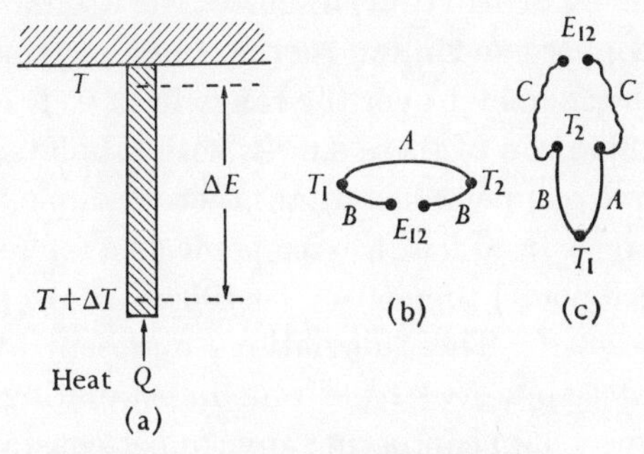

Fig. 73. Thermoelectric force.

then two junctions at a common temperature T_2 between A and a suitable third metal C, and between B and C (Fig. 73 (c)) are made; the potential difference E_{12} at the ends of the leads C is still a direct measure of the thermoelectric potential difference between junctions AB at T_1 and AB at T_2. In practice T_1 is usually an unknown temperature, T_2 is an ice pot (or in some low-temperature applications a bath of liquid helium or liquid nitrogen), and the leads C are of copper.

It appears obvious that in a complete circuit of one homogeneous metal, irrespective of the presence of temperature gradients, no net thermoelectric current flows nor will a thermoelectric e.m.f. be observed on breaking the circuit at a point provided the two adjacent ends are at the same temperature. It follows that the reproducibility and general accuracy of thermoelectric thermometry depend largely on the homogeneity —both chemical and physical—of the metals in the circuit. Any chemical inhomogeneities or physical strains which are present may generate a spurious thermal e.m.f. if exposed to a temperature gradient.

Useful accounts of the history, application, and limitations of thermoelectric thermometry are given in *Temperature*, Vol. 1 (Reinhold, New York, 1941) by W. F. Roeser (p. 180), J. G. Aston (p. 219), Roeser and H. T. Wensel (p. 284), and R. B. Scott (p. 206).

Thermocouple materials

The article by Scott (1941) discusses the calibration of a large number of copper-constantan thermocouples against a platinum resistance thermometer over the range from 0° to $-200°$ C, and the intercomparison of these e.m.f.s with tabulated values for a representative copper-constantan thermocouple. The normal variation ranges from 0 at the ice point (the reference temperature is the ice point) to $\pm 50\,\mu$V ($\pm 100\,\mu$V in the most extreme cases) at $-200°$ C. These deviations represent about ± 1 per cent, so that errors of ± 1 per cent in estimating temperature might be expected in comparing any copper-constantan thermocouple with a standard table.

However, by using a formula of the type

$$E = at + bt^2 + ct^3 \quad \text{for } -200°\,\text{C} < t < 0°\,\text{C} \tag{45}$$

and determining the constants a, b, c by measuring E at three different temperatures, the values of t subsequently obtained by measuring $E(t)$ at intermediate temperatures should not be in error by more than about $2\,\mu$V, representing less than 0·1 deg C. The three calibration temperatures may be the oxygen point, the carbon-dioxide point, and a point in the region from $-30°$ to $-50°$ C where the thermocouple may be calibrated in a stirred alcohol bath against an available resistance thermometer or liquid-in-glass thermometer. A slightly better fit is obtained by calibrating every 50 deg C and producing a graphical tabulated error function to represent the difference between E (experimental) and $E(t)$ calculated from (45).

It is an interesting sidelight on the possible accuracy of thermocouples to note that Giauque and collaborators (see Giauque, Buffington, and Schulze 1927) used six copper-constantan thermocouples—originally calibrated from 15° to 283° K by reference to a hydrogen gas thermometer—as low-temperature secondary standards for many years. On checking their calibration after some years, they were found to be correct within limits of $\pm 0{\cdot}05$ deg K. Thus by observing the necessary precaution of selecting the original wire carefully and not subjecting the thermocouples to mechanical straining, thermocouples can provide a practical secondary temperature scale suitable for most low-temperature applications down to below 10° K. They have the merits of being extremely localized in so far as the temperature-sensitive junction is concerned, of having a very small heat capacity, and they generate negligible heat in the process of measurement.

As Giauque *et al.* (1927) pointed out, serious errors which may arise due to having inhomogeneous lengths of wire in a temperature gradient largely come from wires in alloy form rather than from wires of a pure metallic element; in making a copper-constantan thermocouple, the copper (if annealed) produces insignificant spurious e.m.f.s but the constantan should be tested for chemical inhomogeneity; Giauque used the simple procedure of connecting the ends of each length of constantan being tested, to a potentiometer and passing the wire slowly through a glass U-tube immersed in liquid nitrogen. He then selected samples which did not give e.m.f.s over any regions of their length, greater than, say, 1 μV. For the samples of constantan tested by Giauque, the average thermoelectric e.m.f. developed in this test varied from 0·24 μV in 25 B. & S. wire to 1·84 μV in 40 B. & S. wire, maximum deflexions varying from 0·7 to 3·1 μV respectively. In his thermocouples, the effect of inhomogeneities was partly cancelled by using five constantan wires in parallel, each of 30 B. & S. silk-covered wire.

Values of the thermoelectric e.m.f.s down to about 80° K for representative copper-constantan, iron-constantan, and chromel-alumel thermocouples are available in National Bureau of Standards publications, e.g. *Reference tables for thermocouples* (1955), *N.B.S. Circular* 561, U.S. Government Printing Office, Washington, D.C.

Further data for constantan versus copper, gold + 2·1 per cent cobalt versus copper, and normal silver versus copper have been determined by Powell, Bunch, and Corruccini (1961) down to 1° K, and are reproduced in abbreviated form here in Table C (Appendix).† The wires used were a gold-cobalt and normal silver from Sigmund-Cohn Corp, New York, constantan from Thermoelectric Incorporated, New Jersey, and copper from Leeds and Northrup Company, Pennsylvania. Powell *et al.* remark that deviations among constantan wires from a single batch are small, less than 0·1 per cent above 10° K; they differ from those in N.B.S. Circular 561 above 80° K by less than 1 per cent. Variations are much greater among gold-cobalt samples due to the unstable nature of the alloy and the sensitivity to the amount of cobalt in solid solution: their thermopowers may vary by 1 per cent for different batches and by many per cent for wires from different sources of supply. But if calibrated *in situ* they provide a useful sensing element down to liquid-helium temperatures.

The extensive researches of Borelius *et al.* (1930) first revealed that many dilute alloys might be used for sensitive low-temperature thermocouples, particularly the monovalent metals copper, silver, or gold with small amounts of a magnetic solute such as iron or manganese or cobalt. Later it was shown that a few atoms per million of some of these solutes could carry a local magnetic moment in copper, gold, etc., and produce an anomaly at low temperatures both in the electrical resistance (see § 7, p. 134) and in the thermoelectric power. Dauphinee, MacDonald, and Pearson (1953) suggested the use of some of the copper-rich alloys as thermocouples. Later gold + *ca.* 0·03 per cent iron was used very successfully as a thermometer by Berman and Huntley (1963; also Berman, Brock, and Huntley 1964) at

† These data together with tables for chromel versus alumel and iron versus constantan from 1 to 300° K also appear in Powell, Caywood, and Bunch in *Temperature*, Vol. 3 (1962), Pt. 2, p. 65.

temperatures down to 1° K (see Fig. 74). Such dilute alloys are metallurgically stable provided that they are not heated in an oxidizing atmosphere.

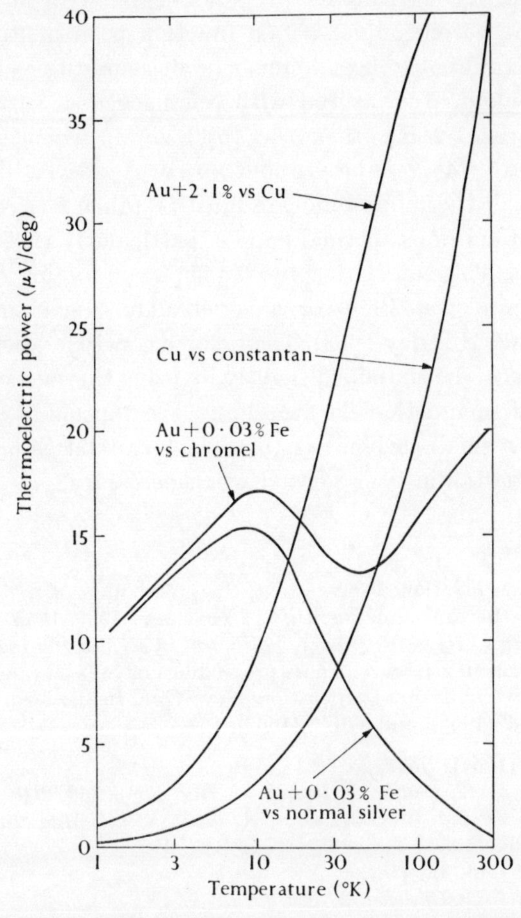

Fig. 74. Thermoelectric power of some thermocouples which are used at low temperatures. Data from Powell *et al.* 1961 and Berman *et al.* 1964.

When working at low temperatures there are often considerable advantages in having the reference junction immersed in liquid nitrogen or in liquid helium. Depending on the experimental arrangement, this may decrease considerably the spurious effects arising from inhomogeneities when wires of an alloy are brought out from a low-temperature cryostat to room temperature and into an ice pot. Suppose we are measuring a temperature of $T \simeq 22°$ K to $\pm 0{\cdot}2$ degK with a copper-constantan thermocouple; if one function is in melting ice a total e.m.f. of about $6000\,\mu$V must be measured to $\pm 1\,\mu$V. On the other hand, if measured with reference to a surrounding helium bath ($4{\cdot}2°$ K), then the total e.m.f. which must be measured to $\pm 1\,\mu$V is only about $60\,\mu$V.

With small thermopowers care must be taken to avoid or to account for spurious thermal e.m.f.s, particularly those arising in the leads. This can be done with a superconducting reversing switch (Templeton 1955) or a superconducting shorting-bar (Berman and Huntley 1963). The reversing switch described by Templeton can be simplified slightly by using thermal switching rather than magnetic-field switching: the tantalum wires are wound onto small carbon resistors which can act as heaters to drive the tantalum coils normal when necessary.

10. References

As has been mentioned in the text, the proceedings of the symposia arranged by the American Institute of Physics in 1939, 1954, and 1961 on *Temperature. Its measurement and control in science and industry* are valuable works of reference. These proceedings have been published by the Reinhold Publishing Corporation, New York, in the form of Vol. 1 (1941), Vol. 2 (1955), and Vol. 3 (1962).

Ahlers, G. (1964). *Phys. Rev.* **135**, A10.
Armstrong, G. T. (1954). *J. Res. natn. Bur. Stand.* **53**, 263.
Babbitt, J. D. and Mendelssohn, K. (1935). *Phil. Mag.* **20**, 1025.
Barber, C. R. (1950). *J. scient. Instrum.* **27**, 47.
—— (1955). Ibid. **32**, 416.
—— (1960). *Prog. Cryogen.* **2**, 149.
—— (1962). *Br. J. appl. Phys.* **13**, 235.
—— (1966). Ibid. **17**, 391.
—— and Horsford, A. (1963). Ibid. **14**, 920.
—— —— (1965). *Metrologia* **1**, 12.
Berman, R. (1954). *Rev. scient. Instrum.* **25**, 94.
—— Brock, J. C. F., and Huntley, D. J. (1964). *Cryogenics* **4**, 233.
—— and Huntley, D. J. (1963). Ibid. **3**, 70.
Berry, R. J. (1962). *Can. J. Phys.* **40**, 859.
—— (1963). Ibid. **41**, 946.

BLACK, W. C., ROACH, W. R., and WHEATLEY, J. C. (1964). *Rev. scient. Instrum.* **35**, 587.

BLAKE, C., CHASE, C. E., and MAXWELL, E. (1958). Ibid. **29**, 715.

BLAKEMORE, J. S., SCHULTZ, J. W., and MYERS, J. G. (1962). Ibid. **33**, 545.

BORELIUS, G., KEESOM, W. H., JOHANSSON, C. H., and LINDE, J. O. (1930). *Leiden Commun.* 206a, 206b.

BRICKWEDDE, F. G., VAN DIJK, H., DURIEUX, M., CLEMENT, J. R., and LOGAN, J. K. (1960). *J. Res. natn. Bur. Stand.* **64**A, 1.

CATALAND, G. and PLUMB, H. H. (1966). Ibid. **70**A, 243.

CLAIRBORNE, L. T., HARDIN, W. R., and EINSPRUCH, N. G. (1966). *Soumal. Tiedeakat. Toim.* A**6**, No. 210, p. 49, also *Rev. scient. Instrum.* **37**, 1422.

CLEMENT, J. R. and QUINNELL, E. H. (1950). *Phys. Rev.* **79**, 1028.

—— (1952). *Rev. scient. Instrum.* **23**, 213.

—— STEELE, M. C., HEIN, R. A., and DOLECEK, R. L. (1953). Ibid. **24**, 545.

COHEN, B. G., SNOW, W. B., and TRETOLA, A. R. (1963). Ibid. **34**, 1091.

COLES, B. R. (1964). *Phys. Lett.* **8**, 243.

—— WASZINK, J. H., and LORAM, J. (1964). *Proc. Int. Conf. Magnetism*, p. 165. Inst. of Physics and Phys. Soc., London.

CORRUCCINI, R. J. (1960). *Rev. scient. Instrum.* **31**, 637.

DAUNT, J. G. (1955). *Temperature*, Vol. 2, p. 327. Reinhold, New York.

DAUPHINEE, T. M., MACDONALD, D. K. C., and PEARSON, W. B. (1953). *J. scient. Instrum.* **30**, 399.

—— and PRESTON-THOMAS, H. (1954). *Rev. scient. Instrum.* **25**, 884.

DUGDALE, J. S. and MACDONALD, D. K. C. (1957). *Can. J. Phys.* **35**, 271.

EDELSTEIN, A. S. and MESS, K. W. (1965). *Physica* **31**, 1707.

FAIRBANK, H. A. and LANE, C. T. (1947). *Rev. scient. Instrum.* **18**, 525.

—— and WILKS, J. (1955). *Proc. R. Soc.* A**231**, 545.

FURUKAWA, G. T. and REILLY, M. L. (1965). *J. Res. natn. Bur. Stand.* **69**A, 5.

GIAUQUE, W. F., BUFFINGTON, R. M., and SCHULZE, W. A. (1927). *J. Amer. chem. Soc.* **49**, 2343.

—— STOUT, J. W., and CLARK, C. W. (1938). Ibid. **60**, 1053.

GRILLY, E. R. (1962). *Cryogenics* **2**, 226.

HALES, J. L. and HERINGTON, E. F. G. (1965). *J. scient. Instrum.* **42**, 203.

HALL, J. A. (1955). *Temperature*, Vol. 2, p. 115. Reinhold, New York.

—— (1956). *Br. J. appl. Phys.* **7**, 233.

HOARE, F. E. and ZIMMERMAN, J. (1959). *Rev. scient. Instrum.* **30**, 184.

HOGE, H. T. (1941). *Temperature*, Vol. 1, p. 141. Reinhold, New York.

—— (1950*a*). *J. Res. natn. Bur. Stand.* **44**, 321.

—— (1950*b*). *Rev. scient. Instrum.* **21**, 815.

—— and ARNOLD, R. D. (1951). *J. Res. natn. Bur. Stand.* **47**, 63.

—— and BRICKWEDDE, F. G. (1939). Ibid. **22**, 351.

HULM, J. K. (1950). *Proc. R. Soc.* A**204**, 98.

JAMES, B. W. and YATES, B. (1963). *J. scient. Instrum.* **40**, 193.

KAMERLINGH ONNES, H. and HOLST, G. (1914). *Leiden Commun.* **142a.**
KEESOM, P. H. and PEARLMAN, N. (1956). *Handb. Phys.* **14**, 282.
KEESOM, W. H. (1942). *Helium.* Elsevier, Amsterdam.
—— and BIJL, A. (1937). *Physica* **4**, 305.
—— and VAN DEN ENDE, J. N. (1930). *Leiden Commun.* 203c.
KILPATRICK, J. E., KELLER, W. E., and HAMMEL, F. F. (1955). *Phys. Rev.* **97**, 9.
KOS, J. F., DROLET, M., and LAMARCHE, J. L. G. (1967). *Can. J. Phys.* **45**, 2787.
KUNZLER, J. E., GEBALLE, T. H., and HULL, G. W. (1957). *Rev. scient. Instrum.* **28**, 96.
LINDENFELD, P. (1961). Ibid. **32**, 9.
LOS, J. M. and MORRISON, J. A. (1951). *Can. J. Phys.* **29**, 142.
LOW, F. J. (1962). *Adv. cryogen. Engng* **7**, 514.
LOWENTHAL, G. C., KEMP, W. R. G., and HARPER, A. F. A. (1958). *Bull. int. Inst. Refrig.* Annexe 1958–1, p. 167.
MARTIN, D. L. (1966). *Phys. Rev.* **141**, 576.
MARTIN, P. E. and RICHARDS, H. (1962). *Adv. cryogen. Engng* **7**, 522.
MENDELSSOHN, K. and PONTIUS, R. B. (1937). *Phil. Mag.* **24**, 777.
—— and RENTON, C. A. (1955). *Proc. R. Soc.* A**230**, 157.
MEYERS, C. H. (1932). *J. Res. natn. Bur. Stand.* **9**, 807.
MONTGOMERY, H. and PELLS, G. P. (1963). *Br. J. appl. Phys.* **14**, 525.
MORIN, F. J. and MAITA, J. P. (1963). *Phys. Rev.* **129**, 1115.
MORRISON, J. A., PATTERSON, D., and DUGDALE, J. S. (1955). *Can. J. Chem.* **33**, 375.
MOUSSA, M. R., MUIJLWIJK, R., and VAN DIJK, H. (1966). *Physica* **32**, 900.
MUELLER, E. F. (1941). *Temperature*, Vol. 1, p. 162. Reinhold, New York.
MUIJLWIJK, R., MOUSSA, M. R., and VAN DIJK, H. (1966). *Physica* **32**, 805.
ORLOVA, M. P., ASTROV, D. N., and MEDVEDEVA, L. A. (1964). *Cryogenics* **4**, 95.
—— SHAREVSKAYA, D. I., ASTROV, D. N., KRUTIKOVA, I. G., BARBER, C. R., and HAYES, J. G. (1966). *Metrologia* **2**, 6.
OSBORNE, D. W., FLOTOW, H. E., and SCHREINER, F. (1967). *Rev. scient. Instrum.* **38**, 159.
PARKINSON, D. H. and QUARRINGTON, J. E. (1954). *Proc. phys. Soc.* B**67**, 644.
—— and ROBERTS, L. M. (1955). Ibid. B**68**, 386.
PLUMB, H. H. and CATALAND, G. (1965). *J. Res. natn. Bur. Stand.* **69**A, 375.
—— —— (1965). *Science* **150**, 155.
POWELL, R. L., BUNCH, M. D., and CORRUCCINI, R. J. (1961). *Cryogenics* **1**, 139.
RAYNE, J. A. (1956). *Aust. J. Phys.* **9**, 189.
ROBERTS, T. R., SHERMAN, R. H., SYDORIAK, S. G., and BRICKWEDDE, F. G. (1964). *Prog. low Temp. Phys.* **4**, 480.
—— and SYDORIAK, S. G. (1956). *Phys. Rev.* **102**, 304.

RODER, H. M. (1965). *J. Res. natn. Bur. Stand.* **69**A, 527.

ROSE-INNES, A. C. (1964). *Low temperature techniques.* English Universities Press, London.

RUHEMANN, M. and B. (1937). *Low temperature physics.* Cambridge University Press.

SCHWAB, F. W. and SMITH, E. R. (1945). *J. Res. natn. Bur. Stand.* **34**, 360.

SCOTT, R. B. (1941). *Temperature*, Vol. 1, p. 206. Reinhold, New York.

—— (1955). *Ibid.* Vol. 2, p. 179. Reinhold, New York.

SHERMAN, R. H., SYDORIAK, S. G., and ROBERTS, T. R. (1964). *J. Res. natn. Bur. Stand.* **68**A, 579.

SIMON, F. E., PARKINSON, D. H., and SPEDDING, F. (1951). *Proc. R. Soc.* A**207**, 137.

STIMSON, H. F. (1955). *Temperature*, Vol. 2, p. 141. Reinhold, New York.

SWIM, R. T. (1960). *Adv. cryogen. Engng* **5**, 498.

TEMPLETON, I. M. (1955). *J. scient. Instrum.* **32**, 172.

VAN DEN BERG, G. J. (1964). *Prog. low Temp. Phys.* **4**, 194.

VAN DER HOEVEN, B. J. C. and KEESOM, P. H. (1965). *Phys. Rev.* **137**, 103.

VAN DIJK, H. (1951). *Proc. int. Conf. low Temp. Phys.*, p. 49 (ed. R. BOWERS), Clarendon Lab., Oxford.

—— (1952). *Maintenance of standards symposium*, p. 51. H.M.S.O., London.

—— (1960). *Prog. Cryogen.* **2**, 120.

VEAL, B. W. and RAYNE, J. A. (1964). *Phys. Rev.* **135**, A442.

WERNER, F. D. and FRAZER, A. C. (1952). *Rev. scient. Instrum.* **23**, 163.

WHITE, G. K. (1953). *Proc. phys. Soc.* A**66**, 559.

—— and WOODS, S. B. (1957). *Rev. scient. Instrum.* **28**, 638.

—— —— and ANGLIN, F. (1958). Ibid. **29**, 181.

WOODCOCK, A. H. (1938). *Can. J. Res.* A**16**, 133.

WOOLLEY, H. W., SCOTT, R. B., and BRICKWEDDE, F. G. (1948). *J. Res. natn. Bur. Stand.* **41**, 379.

ZIMMERMAN, J. E. and HOARE, F. E. (1960). *Physics Chem. Solids* **17**, 52.

CHAPTER V

COOLING WITH HELIUM-3

1. Introduction

THE lighter isotope of helium, ^{3}He, is obtained as a decay product of tritium and therefore is only available in relatively small quantities and is expensive. It has been sold through the Mound Laboratory of the Monsanto Research Corporation at Miamisburg, Ohio, at a price of 15 cents per cm^3 of gas. For research purposes it is liquefied in quantities of the order of 1 cm^3 of liquid by contact with liquid ^{4}He and *not* in a normal circulation liquefier of the type discussed in Chapter I.

The boiling point of ^{3}He is 3·19° K and the critical temperature is 3·33° K at a critical pressure of 1·15 atm. The latent heat and specific heat of the liquid are very sensitive to temperature but between 0·5° and 1·0° K they are approximately 0·25 cal cm^{-3} and 0·025 cal cm^{-3} degK^{-1} respectively (see review by Brewer 1964, for details of physical properties). The density near 1° K is *ca.* 0·08 g cm^{-3}.

There is no indication of an ordering transformation or superfluid film down to a temperature of a few millidegrees. Below 1° K the vapour pressure becomes more than two orders of magnitude greater than that of ^{4}He. A thermally shielded bath of liquid may be boiled under pressures of 10^{-3} mm Hg or less which correspond to temperatures of less than 0·3° K (see vapour pressures in Table A in Appendix). As a result, ^{3}He has become widely used in the past few years to provide an additional stage of cooling which encompasses the range from *ca.* 0·25° to 1° K; this range was accessible hitherto only by using adiabatic demagnetization of a paramagnetic salt. Clearly a quietly boiling bath of liquid ^{3}He, however small, is a more convenient cooling medium than a salt crystal with its attendant problems of magnetic field, thermal contact, temperature measurement, etc. (see Chapter IX). Many of the first ^{3}He cryostats were discussed in a review by Taconis (1961) and some of their essential features are outlined in § 2.

Another important method of cooling with ^{3}He is by adiabatic dilution in liquid ^{4}He. This technique of refrigeration was proposed by London, Clarke, and Mendoza (1962). The first experimental models made at Leiden (Das, de Bruyn Ouboter, and Taconis 1965) and Manchester did not cool below *ca.* 0·2° K but later efforts at Manchester (Hall, Ford, and Thompson 1966), Moscow (Neganov, Borisov, and Liburg 1966), and by the Oxford Instrument Company (M. F. Wood 1966, private communication) showed that temperatures of 0·06° K or less could be maintained. Temperatures of a few millidegrees may yet be achieved by this method as the successful operation is sensitive to the detailed geometry, as discussed below in § 3.

2. ^{3}He cryostats (evaporation cooling)

General principles

Fig. 75 shows two conventional cryostats in schematic form in each of which there is a small ^{3}He chamber attached to the experimental space. The pumping tube usually acts also as the

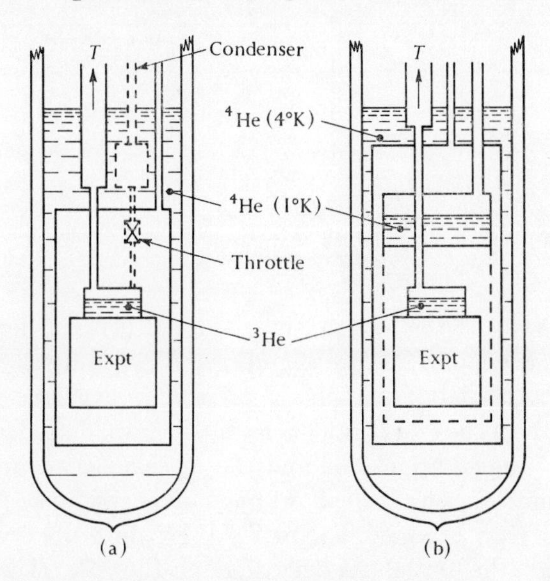

FIG. 75. Schematic diagrams of two alternative cryostats, containing liquid ^{3}He chambers. In (a) 'dashed' lines show how continuous circulation may be achieved.

condenser for initially producing the liquid ^{3}He and therefore must be in thermal contact with a pumped ^{4}He bath along part of its length. This ^{4}He bath may be either the main dewar as in (a), or a separate chamber as in (b).

There *can* be a separate condensing line, shown dotted in Fig. 75 (a), which then allows continuous circulation of the ^{3}He. This is necessary if the heat loads are large or experiments continue for a long time. If the gas is pumped by a system such as illustrated by Fig. 76, the high-pressure side of the pump is

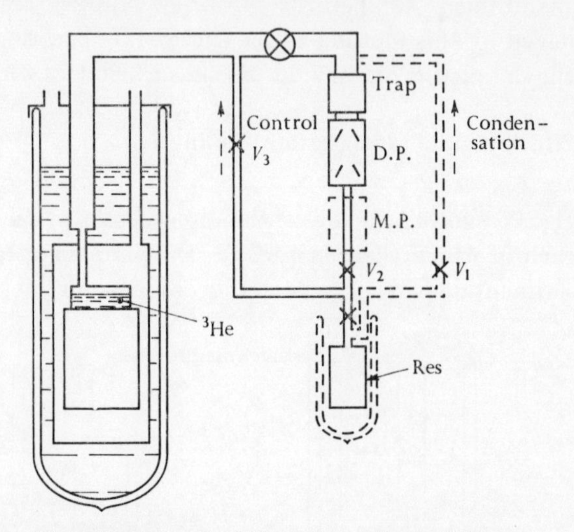

Fig. 76. Flow diagram of a ^{3}He system. 'Dashed' lines indicate optional items.

connected back to the condenser (Fig. 75 (a)) and liquid ^{3}He slowly returns into the ^{3}He chamber at a rate controlled by the throttle; this latter may be a valve, fine capillary tube, or porous plug. The other optional features shown dotted in Fig. 76 are the 'condensation' line and the mechanical pump (M.P.). The 'condensation' line allows gas to be pumped *from* the reservoir (Res.) through valves V_1, V_2, V_3 into the cryostat to accelerate the initial condensation of the ^{3}He charge. The mechanical pump is not necessary if a diffusion pump is used which operates against a relatively high backing pressure. For example, the 'Speedivac' type 2M4 made by Edwards and Company, Sussex, will operate against a maximum backing pressure of 35 mm Hg; this requires that either (a) the reservoir volume must be large enough that its pressure is below this, or (b) the backing side must be connected to a condenser immersed in the pumped ^{4}He bath at a temperature of less than 2° K.

If a mechanical pump is included, it must be a model that is well sealed against gas leakage, such as that produced by the Welch Company, Chicago.

Another alternative is to use a sorption pump, that is to have an adsorber of charcoal or similar material contained in a vessel which is cooled by the ^{4}He bath. The adsorbent is exposed to the gas evaporating from the ^{3}He chamber. This system pioneered largely by the Russians can produce large pumping speeds without vibration but is more difficult to control.

Examples

Some of the first ^{3}He cryostats and most of the more recent ones have used the single condensation or 'single-shot' procedure rather than continuous circulation. The former is certainly more simple to make although the latter offers a greater cooling capacity for a given quantity of ^{3}He. This capacity was a more important factor when the gas cost over a dollar per cm^3 than presently when the price has already fallen to 15 cents per cm^3.

Among these earlier cryostats, those of Roberts and Sydoriak (1955), Seidel and Keesom (1958), and de Bruyn Ouboter *et al.* (1960) used the 'single-shot' while Zinoveva and Peshkov (1958), Reich and Garwin (1959), Rauch (1961), Ambler and Dove (1961), all had continuous circulation of ^{3}He.

Fig. 77 (Seidel and Keesom 1958) shows a calorimeter which is cooled by ^{3}He condensed in the space d. The specimen i can be lowered by the nylon thread f onto the copper plate k. The ^{3}He vapour is pumped away through a 4-mm diam. tube c which enlarges before reaching the oil-diffusion pump. The refrigerating capacity near to the lower limit of $T \approx 0{\cdot}3°$ K is 50 ergs/s^{-1}. The temperature scale is established by means of a paramagnetic salt, m, or ferric ammonium sulphate.

Another single-shot cryostat used for thermal conductivity measurements down to 0·3° K at Cornell University is in Fig. 78 (Seward 1955). Note that the ^{3}He chamber which is machined from copper has internal ribs to increase the area of heat transfer between liquid and metal. The salt used to calibrate the germanium thermometer (Ge) is a sphere of cerium magnesium

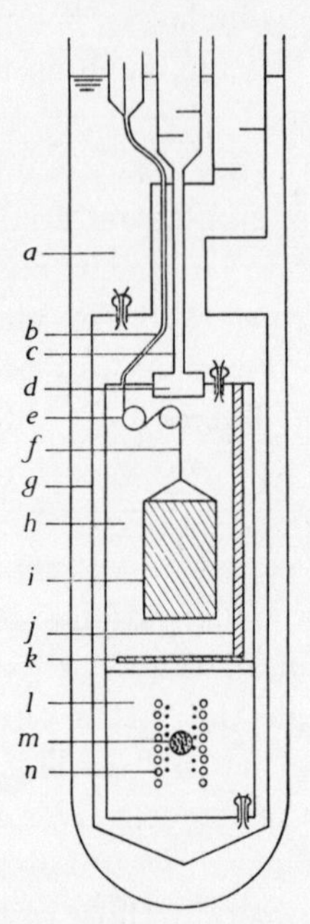

FIG. 77. A calorimeter cooled by liquid ^{3}He in the chamber *d* (after Seidel and Keesom 1958).

nitrate (CMN). The 'crystal holder' is a C-shaped piece of copper which thermally connects the platform holding the crystal to the ^{3}He coolant.

Fig. 79 shows a cryostat (Mate *et al.* 1965) in which the ^{3}He is pumped by adsorption on charcoal. *K* is a cage of copper gauze filled with 40 g of activated charcoal which can be raised or lowered into the cold region by a nylon line from the pulleys *L*. The ^{3}He chamber *A* is machined from solid copper and has a honeycomb structure to improve thermal equilibrium. An alternative construction is to put a tightly rolled cylinder of copper shim into the chamber and solder the bottom down. The sorption pump has very high speed and so gives a lower temperature than rotary or diffusion pumps. Mate *et al.* reached 0·28° K with a heat load of about 500 ergs/s^{-1}. Further details of the

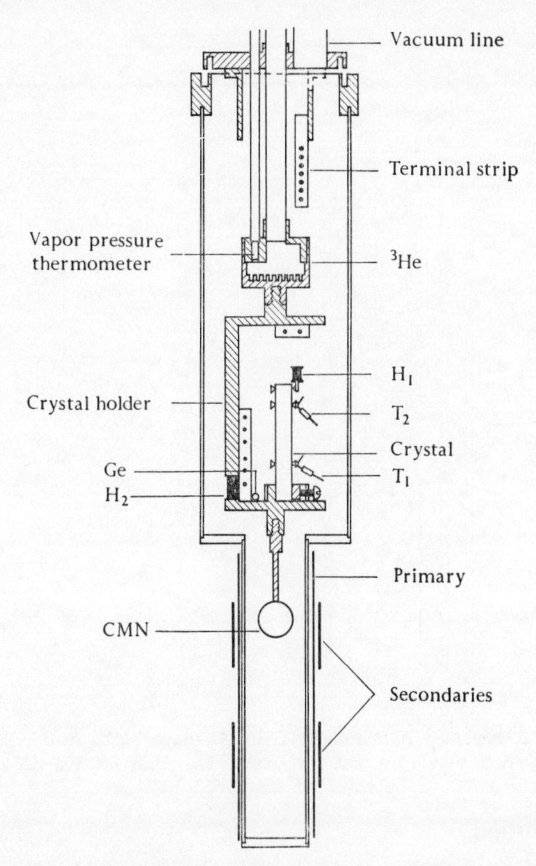

FIG. 78. A thermal conductivity cryostat cooled by liquid ^{3}He (after Seward 1965).

use of sorption pumps for ^{3}He and ^{4}He may be found in papers by Eselson, Lazarev, and Shvets (1962, 1963) and Eselson, Shvets, and Bereznyak (1962).

Among the increasing number of descriptions of ^{3}He cryostats is a useful account by Walton (1966) of a simple 'single-shot' cryostat with carefully chosen sizes of pumping-tube which reaches about 0·21° K.

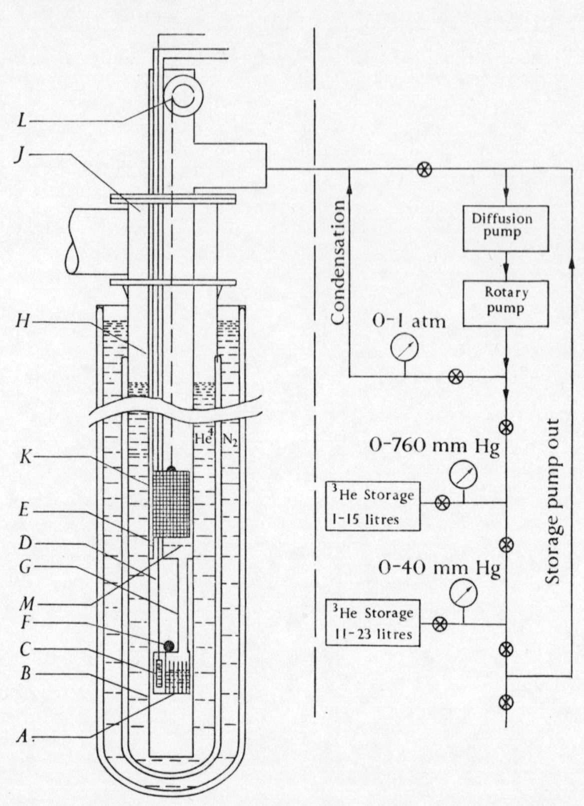

FIG. 79. A cryostat in which ^{3}He (in *A*) evaporates under reduced pressure, produced by the charcoal sorption pump (*K*) (after Mate *et al.* 1965).

3. ^{3}He dilution

At the Low Temperature Physics Conference held in Oxford in 1951, H. London (p. 157 of the *Proceedings*) suggested that 'a valuable cooling process would . . . make use of the entropy of mixing of ^{3}He and ^{4}He . . . at very low temperatures the ^{3}He can 'expand'' by being diluted with ^{4}He. The latter has negligible entropy below 1° K, while the entropy of the former is still appreciable . . . . Thus by a reversible adiabatic dilution . . . the temperature would fall and at (say) 0·01° K the specific heat would still be appreciable.' Some years later the discovery of phase separation in liquid helium mixtures below 0·8° K and experiments on the osmotic pressure of ^{3}He in liquid ^{4}He resulted in London, Clarke, and Mendoza (1962) making more detailed proposals for designs of a dilution refrigerator. Following these, refrigerators were made by Das *et al* (1965), Hall *et al.*

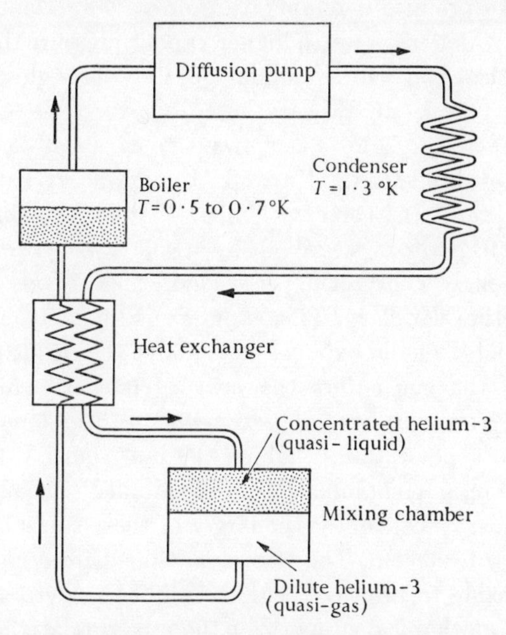

FIG. 80. Flow diagram showing the elements of a dilution refrigerator based on a drawing in *Machine Design*, 30 June 1966, kindly given to me by M. F. Wood.

(1966), Neganov *et al.* (1966). Some of their first models were not completely successful but later versions have reached temperatures of *ca.* 0·07° K in the mixing chamber.

Fig. 80 is a simplified flow diagram which illustrates the operation of the refrigerator. ^{3}He gas at a pressure of *ca.* 30 mm Hg (from the backing side of the diffusion pump) enters the cryostat and condenses in a coil which is immersed in pumped liquid ^{4}He (at 1·3° K). The ^{3}He then goes successively through a flow-throttling capillary and a heat exchanger into the top of the mixing chamber. Phase separation and the *cooling* occur in this chamber. The lighter ^{3}He-rich phase is in the top layer and some ^{3}He diffuses across the phase boundary, the adiabatic dilution causing a cooling. It then diffuses through the quasi-stationary liquid ^{4}He, via the heat exchanger to the boiler or still. This can be electrically heated in a controlled fashion to produce a temperature of *ca.* 0·6° K and evaporate the ^{3}He, which has a much higher vapour pressure than the ^{4}He. And so then the ^{3}He as vapour goes up through the mercury diffusion pump—an Edwards 2M4 booster type—and around the cycle again. Fig. 81 shows the essential features of a refrigerator built for the Harwell Atomic Energy Establishment by the Oxford Instrument Company. It is very similar to the 'final' refrigerator described by Hall *et al.* The maximum flow-rate is *ca.* 1 cm^3 n.t.p. per second which travels through the flow-limiting capillary. This capillary is about 0·5 m in length and is cooled in the exhaust tube above the still (S). From this capillary the gas enters the heat exchanger, which consists firstly of a straight coaxial section about 15 cm long and then a Hampson-type section in which the 0·035 cm i.d. tube spirals up to the mixing chamber (M) in the annular space between the two stainless-steel tubes; the larger of these tubes has an inner diameter of 0·66 cm. The still is machined from copper and is ribbed inside to improve heat transfer. The pyrolitic graphite provides mechanical support for the mixer, relative to the still, and acts as a thermal link in the initial cooling-down. At very low temperatures, it is an effective thermal insulator (see discussion on heat switches in § 7.8 and Shore *et al.* 1960).

There is a small electric heater attached to the still so that the circulation rate of ^{3}He can be controlled. Hall *et al.* reported that a temperature of 0·065° ± 0·005° K was reached at the mixer and that the near 0·1° K the cooling capacity amounted to 40 ergs/s. During operation the still is near 0·5° K and the bottom of the exchanger is 0·2° K. The first commercial model produced by the Oxford Instrument Company has also reached 0·07° K (M. F. Wood, private communication). Neganov *et al.* (1966) appear to have reached temperatures as low as 0·03° K with their refrigerator (see footnote in their paper: May 1966) which is similar in principle to the Hall and Harwell models; Vilches and Wheatley (1967) have reached 0·02° K.

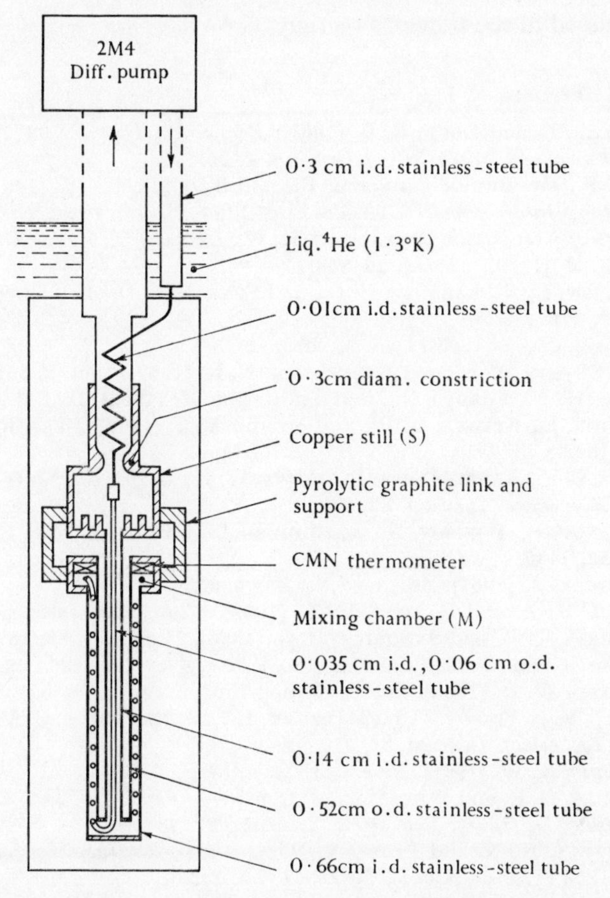

FIG. 81. Schematic diagram of a dilution refrigerator, based on drawings kindly sent by M. F. Wood of the Oxford Instrument Company.

It is clear from the experience of Das *et al.* and Hall *et al.* that the design of dilution refrigerators can be varied considerably but that in the present state of knowledge of their operation such changes are fraught with uncertainty: there are possible convective instabilities in the column of the dilute phase. However, it is a most useful tool which enables experiments to be done more easily in the temperature range below 0·3° K, a range which is inaccessible to the simple evaporation cooler discussed in the previous section.

4. References

AMBLER, E. and DOVE, R. B. (1961). *Rev. scient. Instrum.* **32**, 737.

BREWER, D. F. (1964). *Prog. Cryogen.* **4**, 25.

DAS, P., DE BRUYN OUBOTER, R., and TACONIS, K. W. (1965). *Low temperature physics* LT9, Pt. B, p. 1253. Plenum Press, New York.

DE BRUYN OUBOTER, R., TACONIS, K. W., LE PAIR, C., and BEENAKKER, J. M. (1960). *Physica* **26**, 853.

ESELSON, B. N., LAZAREV, B. G., and SHVETS, A. D. (1962). *Cryogenics* **2**, 279.

—— —— —— (1963). Ibid. **3**, 207.

—— SHVETS, A. D., and BEREZNYAK, N. G. (1962). Ibid. **2**, 361.

HALL, H. E., FORD, P. S., and THOMPSON, K. (1966). Ibid. **6**, 80.

LONDON, H., CLARKE, G. R., and MENDOZA, E. (1962). *Phys. Rev.* **128**, 1992.

MATE, C. F., HARRIS-LOWE, R., DAVIS, W. L., and DAUNT, J. G. (1965). *Rev. scient. Instrum.* **36**, 369.

NEGANOV, B., BORISOV, N., and LIBURG, M. (1966). *Zh. éksp. teor. Fiz.* **50**, 1445.

RAUCH, C. J. (1961). *Adv. cryogen. Engng* **6**, 345.

REICH, H. A. and GARWIN, R. L. (1959). *Rev. scient. Instrum.* **30**, 7.

ROBERTS, T. R. and SYDORIAK, S. G. (1955). *Phys. Rev.* **98**, 1672.

SEIDEL, G. and KEESOM, P. H. (1958). *Rev. scient. Instrum.* **29**, 606.

SEWARD, W. D. (1965). Thesis, Cornell University.

SHORE, F. J., SAILOR, V. L., MARSHAK, H., and REYNOLDS, C. A. (1960). *Rev. scient. Instrum.* **31**, 970.

TACONIS, K. W. (1961). *Prog. low Temp. Phys.* **3**, 153.

VILCHES, O. E. and WHEATLEY, J. C. (1967). *Phys. Lett.* **24A**, 440.

WALTON, D. (1966). *Rev. scient. Instrum.* **37**, 734.

ZINOVEVA, K. N. and PESHKOV, N. P. (1958). *Zh. éksp. teor. Fiz.* **32**, 1256 (Transl.: *Soviet Phys. JETP* **5**, 1025).

PART II

THE RESEARCH CRYOSTAT

CHAPTER VI

INTRODUCTION TO CRYOSTAT DESIGN

1. General considerations

PART II of this book deals largely with the topic of cryostat design and is intended to detail the principles involved and some of the means by which particular physical measurements at low temperatures are made. It is clearly impossible to cover all the eventualities that do and will arise in the plans of the low-temperature experimenter, but it is hoped that the discussion of general principles, heat transfer, methods of temperature control, etc., together with those chapters on temperature measurement and physical data, may help the experimentalist.

It is difficult to conceive of any single cryostat which is suitable for all possible physical measurements on a substance at all temperatures in the range we call 'low'. Even if possible, the complexity of such an apparatus would be considerable, to say the least. As a result it is necessary to formulate clearly the primary experimental problem to be tackled, and try to solve it as simply as possible, always consistent with the fact that a little extra difficulty in construction may allow a much wider range of usefulness or increased accuracy. Whether the primary aim is to measure the electrical resistance of an alloy, the X-ray lattice spacing in a solid or its heat capacity, a decision regarding the range of temperature over which measurements are needed is important in so far as the design is concerned.

With the aid of liquid refrigerants commonly available in a cryogenic laboratory, viz. liquid oxygen, nitrogen, and helium (and possibly liquid hydrogen), the following temperature ranges are immediately available by controlling the pressure at which the liquefied gas boils:

(*a*) 1°–4·2° K (liquid helium);

(*b*) 54°–90° K (liquid oxygen);

(*c*) 63°–77° K (liquid nitrogen);

(and (*d*) 10°–20·4° K (by pumping liquid hydrogen to temperatures ranging from its normal boiling point to well below the triple point)).

Provided that vacuum pumps of the required speed are available, temperatures in these ranges can be attained within the liquefied gases boiling in a dewar and therefore in any experimental chamber immersed in the liquid. With a manostat connected on the pumping tube, they can be maintained with fluctuations of less than one part in a thousand; over much of each range, temperature fluctuations as small as one part in five thousand can be maintained for a long period of time. This accuracy, of course, occurs in the temperature at the surface of the boiling liquid as calculated from the vapour pressure and does not mean that much larger temperature inhomogeneities cannot exist in the liquid itself, and particularly in the solid if pumped below the triple point.

If temperatures other than those within the accessible vapour-pressure range of the available liquefied gases are required, other methods must be used. These other methods generally maintain temperatures within the experimental space that are above the temperature of the surrounding bath; then a balance between the loss of heat (by radiation, conduction, etc.) from the experimental chamber and the supply of heat from a manually or automatically controlled source to the chamber must be produced. The degree of ease or difficulty with which this balance, and therefore relative constancy of temperature, is produced depends ultimately on the type of physical investigation being pursued; for example it depends on whether a steady drift of 1 degK/min or an irregular oscillation of 2 or 3 degK in temperature is unimportant, or whether on the other hand the temperature must be maintained to $\pm 0{\cdot}005$ degK for an hour or more. This degree of difficulty not only depends on the type of experiment but also on the approximate temperature at which it is being carried out: at temperatures below say 20° K, heat capacities are small and the time required to attain thermal equilibrium is usually short, so that the response to the manual control of an electrical heater is quite rapid; but at 150° K equilibrium times are much longer and the difficulty of finding manually a particular heat input necessary to produce ultimate temperature equilibrium is more tedious.

Another important and related factor is whether adiabatic conditions are required. To illustrate this, consider a measurement of heat capacity or heat conductivity. In these measurements stray temperature drifts and spurious heat inputs to the specimen under investigation must be eliminated as completely as possible. The specimen or its immediate container must be in a high vacuum enclosure and the surroundings must be temperature controlled to varying degrees of constancy. On the other hand, a measurement of the Hall effect, electrical resistance, or magnetic susceptibility may be made in a short time with the specimen kept in equilibrium with its surroundings by the presence of exchange gas so that a slight change in temperature will not affect the validity of the experimental value obtained, although it may make the precise temperature at which the value is obtained somewhat uncertain. Therefore before laying plans for a cryostat these questions must be answered:

(i) Will results in the range from 1°–4° K and 55°–90° K give all the information needed, or are measurements, perhaps, only required at 90° K, 77° K, and 4·2° K?

(ii) If the answers to these first questions are no, then how constant must the temperature be maintained and for how long at say 8° K, 40° K, and 115° K?

(iii) Can the experimental space be kept filled with a low-pressure gas which will assist in preserving temperature equilibrium within this space?

(iv) What form of access to the experiment is required, i.e. are ten or twenty fine copper wires as electrical leads all that are necessary, or do we need special windows for an infra-red or X-ray beam? Is some mechanical movement necessary in the experimental space, e.g. a mechanical switch or a suspension to a microbalance?

Having answered these questions—most of them with very obvious answers in a particular investigation—the problem of designing and making a suitable apparatus has to be faced. One of the major problems in this, apart from the obvious ones of available materials, workshop facilities, and a necessary mechanical lifetime, is that of heat transfer.

In any apparatus designed to operate at temperatures considerably different from its surroundings, and perhaps designed to measure properties which are strongly influenced by the inflow or outflow of heat and/or change in temperature, this factor is a vital one. As a result a large part of the success or failure of an apparatus to give accurate results can often be traced to the care taken in reducing stray heat inflows due to (i) conduction through low-pressure gas when the vacuum is not sufficiently good, (ii) radiation, (iii) conduction of heat along the tubes or electrical leads, or (iv) Joule heating or eddy current heating.

Chapter VII deals with the calculation of heat inflows that may be expected and compares the magnitude of the heat transferred by radiation between surfaces of various emissivities with that conducted through low-pressure gas or conducted down electrical lead wires and supporting tubes. However, there seems to be merit in first showing some schematic diagrams, and in some cases more detailed illustrations of cryostats which have been and are being used to measure physical properties, before considering the problems of heat transfer and temperature control in more detail. This may help the reader to understand some of the practical methods of tackling these problems, and he may refer back to them again after reading other chapters.

2. Cryostats for specific-heat measurements

Most specific-heat measurements at low temperatures have been made with the adiabatic calorimeter—originally introduced by Nernst and Eucken in 1910—and the ensuing discussion is restricted to examples of this type. In the simple form illustrated schematically in Fig. 82 (a) the specimen S, with heater H and thermometer T (combined as one in some cases) attached, is suspended by nylon or cotton threads in an evacuated enclosure X, which in turn is surrounded by a liquefied gas A. The specimen may be in the form of a solid block of metal with T and H cemented to its surface or embedded in it; or the specimen may be in a solid, powder, or liquid form contained in a thin-walled calorimeter vessel of known heat capacity in which case T and H are usually attached to the calorimeter; a small pressure of helium exchange gas may then be used to ensure temperature

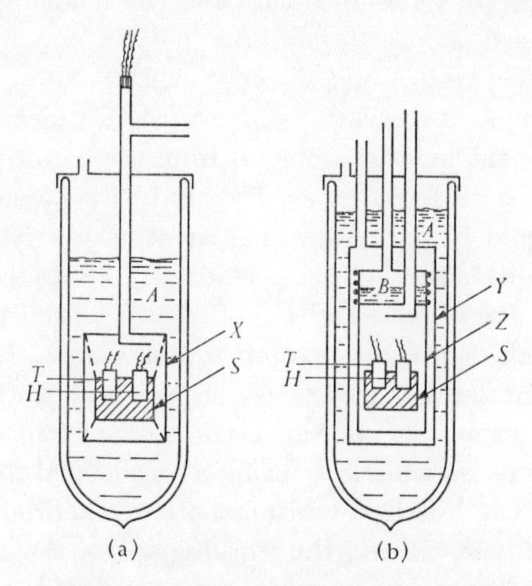

FIG. 82. Types of cryostat suitable for specific-heat determinations.

equilibrium within the calorimeter. If the vacuum is sufficiently good and conduction down the leads is small, such a simple apparatus can adequately provide specific heat data over a quite wide range of temperature, provided that the temperatures of the wall X and the specimen S are not widely different, otherwise radiation and lead conduction become large. As a result this simplest form may be used at temperatures below 20° K, and at temperatures within the range of say 55°–90° K covered by liquid oxygen, provided that $T_S - T_X$ is kept small. At temperatures above 20° K, an adiabatic shield in the space between S and X usually becomes necessary.

Silvidi and Daunt (1950) and Pearlman and Keesom (1952) have described the use of comparatively simple cryostats of this type (see also review by Keesom and Pearlman 1956).

The second cryostat (Fig. 82 (b)) is more complex than 82 (a) in that it has an intermediate wall Z which is attached to the copper chamber B, and can therefore be controlled at temperatures different from that of the outer liquid bath A; either by reducing the pressure over a liquid in B or electrically heating the chamber B, its temperature and the temperature of Z can be controlled so as to preserve only a small temperature difference between Z and the specimen S.

An example of a cryostat of type (b), in which liquid can be drawn into the inner chamber B from the dewar bath itself, is illustrated in Fig. 83 (after Rayne 1956; Rayne and Kemp 1956). Liquid in the dewar A is at atmospheric pressure and can be admitted through the needle valve E into the chamber F. The vessel D is of brass and F is of copper, both being supported by thin-walled (0·1 and 0·2 mm wall thickness) German silver tubes from the top plate. In this cryostat both vessels are evacuated separately and the use of helium exchange gas at low temperatures is avoided by using a mechanical thermal switch similar to that described by Ramanathan and Srinivasan (1955). Quoting Rayne, 'within the working space, the specimen J is suspended by nylon threads from the disk K, which is thermally anchored to the bath F by a flexible copper strap L. Movement of the disk is made by a stainless steel tube M, connected to a bellows N which can be actuated by the screw mechanism P ...'. By lowering the disk and pressing the specimen against the flat bottom of the can, J may be cooled to the temperature of F. The avoidance of the use of exchange gas, particularly at liquid-helium temperatures, reduces the 'starting-up' time considerably and ensures that no heat leak through residual exchange gas can occur. Rayne used a manganin heater wound on a copper former Q, screwed into the top of the specimen, and a 10-Ω Allen–Bradley carbon thermometer in a copper sleeve R for temperature measurement.

A cryostat in which a gas thermometer is attached to the calorimeter and is used for temperature measurement over the range 4°–15° K has been described by Aven, Craig, and Wallace (1956). Their calorimeter–thermometer unit is sus-

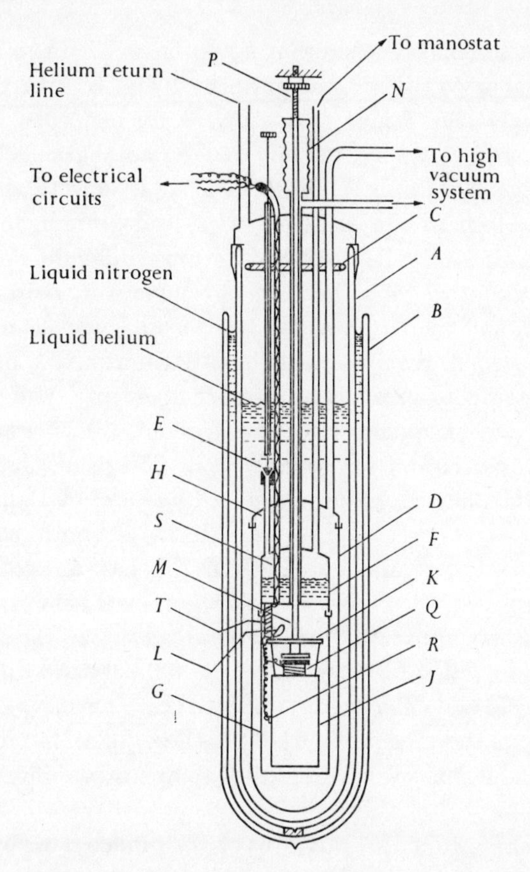

Fig. 83. Schematic diagram of cryostat (after Rayne 1956).

pended below a copper plate on which an electrical heater is wound, and a radiation shield is attached to the copper plate so as to surround the calorimeter. This is all enclosed in an evacuated metal can, surrounded by a dewar of liquid helium. Adiabatic conditions are produced by controlling the shield temperature in response to a Au+Co : Ag+Au thermocouple which measures the temperature difference between shield and calorimeter. In terms of the models in Fig. 82, the apparatus of Aven *et al.* might be considered as type (a) with a shield interposed between S and X or as type (b) without the small liquid chamber B.

The most serious criticism that might be levelled at a specific heat apparatus of the general pattern of Fig. 82 (b) is that at comparatively high temperatures where, for example, T_A may be 90·1° K, and T_B approximately 150° K, adiabatic conditions are not sufficiently well established. That is, due to radiation transfer and to heat flow along electrical leads, serious temperature inequalities may be produced between different points on the shield Z and thereby cause small temperature drifts in the calorimeter, sufficient to adversely affect high-precision calorimetry. Two good examples of adiabatic calorimeters designed for use at temperatures up to room temperature, and to give values of specific heat correct to within about $\pm 0{\cdot}2$ per cent, are those of Morrison and his collaborators (Morrison, Patterson, and Dugdale 1955; Morrison and Los 1950) and of Dauphinee, MacDonald, and Preston-Thomas (1954). In both cases an apparatus rather similar to type 82 (b) has an additional temperature-controlled copper shield interposed between Z and S. In Morrison's apparatus the electrical heating of the shield is manually controlled in response to the temperature difference between shield and calorimeter, indicated by a sensitive chromel P-constantan thermocouple; using a three-junction couple, a signal of about 180 μV per degree of temperature difference is realized.

In the semi-automatic apparatus of Dauphinee *et al.*, of which the principal parts are shown in Fig. 84, Au + Co : Ag + Au differential thermocouples are used to actuate an electronic temperature controller. Great care is taken to ensure temperature equilibrium over the adiabatic shield, and to a lesser extent on the outer shield, as their method of measurement consists of continuous heating of the calorimeter and calculation of heat capacity from the continuously recorded temperature–time chart. In such a continuous process in which constant electrical heating produces a temperature rise of about 5 degK/h, no corrections for spurious temperature drift can be made; hence the temperature drift must be negligible. In more conventional specific heat determinations, the heat input to the calorimeter is essentially discontinuous—occupying perhaps 10–30 s—and temperature drift rates are carefully ascertained before and after the brief heating period. Martin (1960, 1962) has modified this calorimeter by adding an extra adiabatic shield.

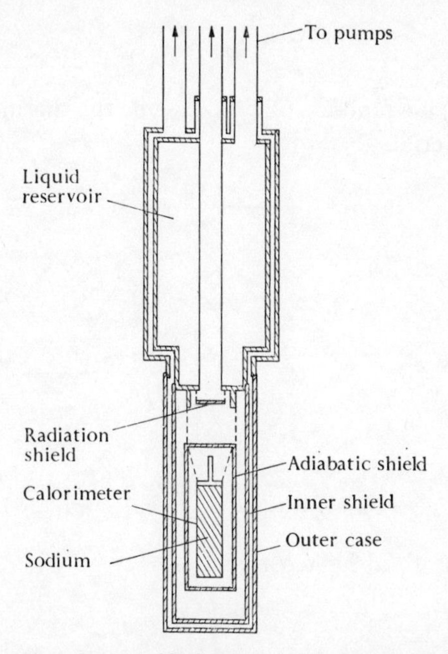

FIG. 84. A cryostat for specific-heat measurements. Heaters and thermocouples are not shown (after Dauphinee, MacDonald, and Preston-Thomas 1954).

Further details of calorimeters can be found in the following papers: Westrum, Hatcher, and Osborne (1953); Seidel and Keesom (1958, see also Fig. 77 in § 5·2); Phillips (1959); Martin (1961); Lounasmaa and Guenther (1962); Martin, Zych, and Heer (1964); O'Neal and Phillips (1965).

Excellent discussions of calorimetry below 20 °K have been written by Hill, Martin, and Osborne (1967) and by Hill (1959).

They compare various forms of cryostat and discuss the problems of thermometry, thermal contact, etc. Further details of heat switches which are useful in calorimetry are given below (§ 7.8).

3. Cryostats for measurement of thermal conductivity and thermoelectric power

In the solid rod shown in Fig. 85, the axial heat flow $\dot{Q}$ is given by

$$\lambda(T) = \frac{\dot{Q}}{A}\frac{\delta l}{\delta T},$$

where $\lambda(T)$ is a function of T expressing the thermal conductivity of the material.

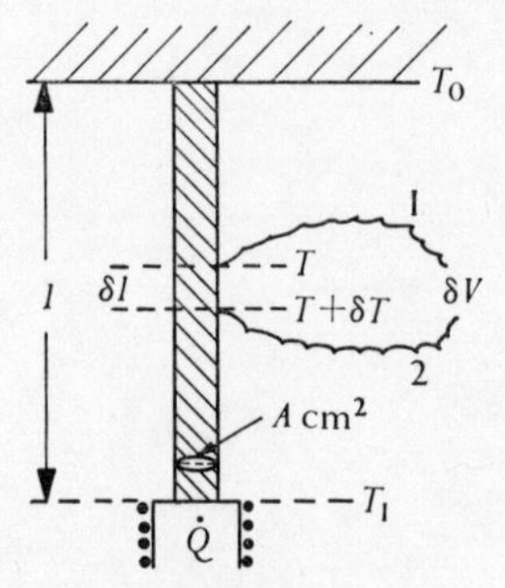

FIG. 85. Heat flow through a solid rod.

Thus the total heat conduction

$$\dot{Q} = \frac{A}{l}\int_{T_0}^{T_1} \lambda(T)dT.$$

If the rod is metallic (or a semiconductor), a thermoelectric potential difference is produced along the rod and the thermoelectric power with respect to the metal of which the potential leads 1 and 2 are made is $S = \delta V/\delta T$; the thermoelectric force, denoted by V or E, is

$$E = \int_{T_0}^{T_1} S(T)dT.$$

In principle and in practice the thermoelectric power $S(T)$ may be determined either (i) by measuring δV and δT, where $\delta T \ll T$

and hence it may be assumed $S(T) \simeq S(T+\delta T)$; then, varying T_1 and T_0, a series of values for $S(T)$ may be obtained; or (ii) by keeping T_0 fixed and measuring the total thermoelectric force $E(T_1 - T_0)$ as T_1 is changed; $S(T)$ is then obtained by differentiating the curve $E(T)$.

Similarly, in the case of thermal conductivity, an integrated conductivity is obtained by measuring $\dot{Q}$ and T_1 for different values of $\dot{Q}$, and differentiating the resultant function $\dot{Q}(T_1)$ to obtain $\lambda(T)$. Alternatively, by keeping $T_1 - T_0$ small, values of δT and $\dot{Q}$ are obtained at various temperatures and a graph of $\lambda(T)$ obtained directly from these.

Since the major interest in thermoelectricity has been practical thermometric interest in the thermoelectric e.m.f. E produced by a thermocouple junction with respect to a second junction fixed at a reference temperature (often 0° C but sometimes 4·2° or 77° K), the integrated quantity E is usually measured. This quantity has greater practical value and since it is normally a simple monotonic function of T, S can be derived by differentiation if required. In recent years, however, there has been an increased fundamental interest in thermoelectric power and many differential measurements giving S directly have been made (see, for example, Frederikse 1953; and Jan, Pearson, and Templeton 1955). In these latter measurements the procedure is almost identical with the measurement of thermal conductivity except that a potential difference δE ($= \delta V$) must be obtained corresponding to δT and the heat flow $\dot{Q}$ need not be known.

In thermal conductivity measurements, the interest has been a largely fundamental interest in $\lambda(T)$ and so the differential measurement giving $\lambda(T)$ directly has been used. Since $\lambda(T)$ is frequently not a simple monotonically increasing function of T but a rather complicated function, and also since δT can be measured with a fairly high degree of accuracy with a differential thermometer system, this yields more reliable results than the 'integral' method.

In certain cases (see Wilkinson and Wilks 1949) where, for technical reasons, values of the total heat $\dot{Q}$ conducted by a rod of a material when the ends of the rod are at quite different temperatures (e.g. 4·2° and 20° K) are needed, the 'integral' method has been used; in these cases values of $\dot{Q}l/A$ for certain specific values of T_0 and T_1 have been required, and not any detailed knowledge of $\lambda(T)$.

We shall confine our discussion below to cryostats in which values of $\lambda(T)$ can be obtained, and if desired $S(T)$ can be measured at the same time. In some of the early low-temperature measurements on heat conduction at Leiden, one end of the specimen was soldered directly to a 'heat sink', e.g. copper plate of which the opposite surface was in contact with liquid helium or hydrogen, and the heater was wound directly onto or adjacent to a thermometer attached by a solder connexion to the other end of the specimen. The accuracy of this method depends on the contact thermal resistance of the solder junctions being small in comparison with that of the specimen, which is frequently not the case.

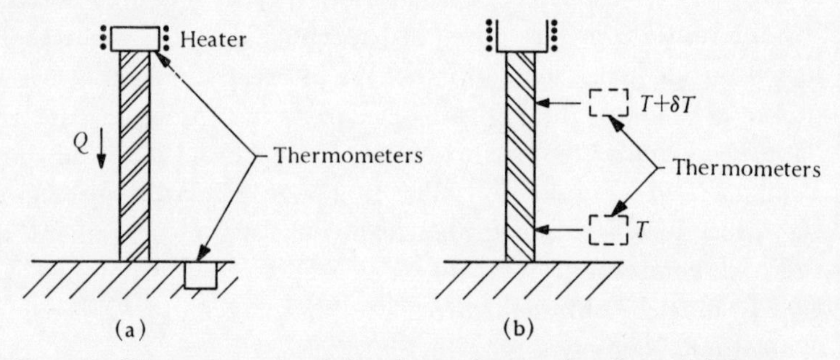

FIG. 86. Thermal-conductivity determination by the axial flow method; (b) represents the more reliable *potentiometric* form.

In an interesting paper on the heat conductivity of steels de Nobel (1951) has described the use of the type of assembly shown in Fig. 86 (a) and also the subsequent adoption of the 'potentiometric' form of measurement (Fig. 86 (b)), which is largely used today. This latter form, provided that the heat flow along the thermometer connexions is negligible when equilibrium is established, represents a thermal potentiometer; measurement of T and $T+\delta T$ at the thermometers is a true indication of the respective temperatures at the points on the specimens where the thermometer 'probes' are attached. Helium-filled gas thermometers which are connected by fine capillaries to a differential manometer—mercury or oil—were used first by de Haas and Rademakers (1940) and later by Hulm (1950), Berman (1951), Olsen (1952), White (1953), and others; because they are useful over a wide temperature range and are not affected by magnetic fields they continue to be used in some cryostats, particular for

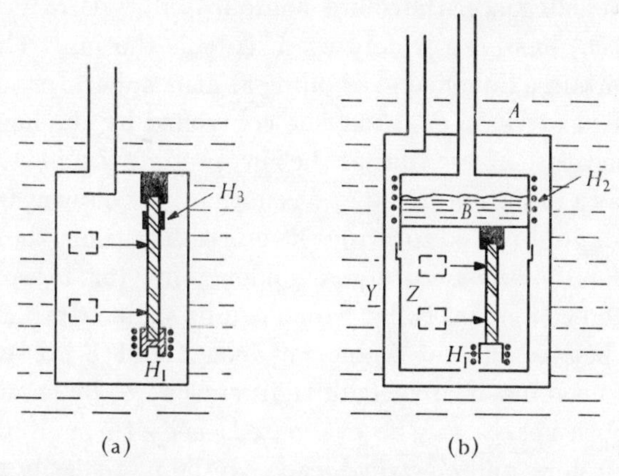

FIG. 87. Types of cryostat suitable for determination of thermal conductivity and thermoelectric power.

in situ calibration. However differential measurements are made usually now with semiconducting resistance thermometers of carbon or germanium below *ca.* 50° K or with suitable thermocouples such as Au + 2·1 per cent Co versus copper or Au + 0·03 per cent Fe versus chromel (§ 4.9) from 1° or 2° K up to room temperature.

Leaving the details of temperature measurement (discussed in Chapter IV), we turn to the type of experimental cryostat suited to measurements of $\lambda(T)$. The simple form of Fig. 87 (a) shows the specimen S thermally anchored to an evacuated metal chamber which is immersed in a liquid refrigerant A; S has a heater H_1 attached to its lower end. This cryostat is very suitable for measurements in the range of temperature covered by the liquid A and measurements may often be extended to slightly higher temperatures by having an additional heater H_3 to raise the temperature of the rod *S* above T_A. Such a relatively simple type of cryostat has been used very successfully in the liquid-helium and liquid-hydrogen temperature regions by de Haas and Rademakers (1940), Hulm (1950), and others; at these temperatures radiation heat transfer is not a serious problem.

For working at temperatures outside the range covered by liquefied gases, the more complicated type of cryostat in Fig. 87 (b) has been widely used. In this the inner chamber *B* may contain a liquefied gas boiling at atmospheric or some other controlled pressure or it may be controlled by the heater H_2 at a temperature above that of the surrounding liquid *A*. The wall *Z* acts as a radiation shield and is normally of copper, frequently silver- or gold-plated to reduce its emissivity. Since the specimen is thermally anchored by a solder joint (or in some cases cemented to a metal collar which can be soldered) to *B*, and the spaces between *S* and *Z*, and between *Z* and *Y* are evacuated, *Z* itself need not be a vaccum-tight enclosure; by leaving a hole in *Z*, both spaces can be evacuated easily through a common pumping line, and electrical leads can be more easily taken out from H_1 (for example) and then thermally anchored both around chamber *B* and to a copper post attached to *Y*.

Fig. 88 from Berman (1951) shows a thermal conductivity cryostat similar in principle to 82 (b), but in which the inner liquid chamber is the expansion vessel of a Simon liquefier. As is discussed in Chapter VIII, such an expansion vessel can be temperature controlled below 4·2° K by pumping on liquid helium, between 10° and 4·2° K by controlled expansion, and from 10° to 33° K by admitting liquid hydrogen into *B* and letting it boil at a controlled pressure.

Details are given below (with Figs. 89, 90) of a cryostat similar to that described briefly by White and Woods (1955), and based on that of White (1953). Such cryostats have been found useful not only for measurements of thermal conductivity, thermoelectric power, and electrical resistance of specimens in rod form, but also for the electrical resistance of fine wires (using exchange gas in the inner container which is then made vacuum tight), and has been adapted for use in measuring mechanical properties over the temperature range from 2° K to room temperature; one has been in operation for ten years and has

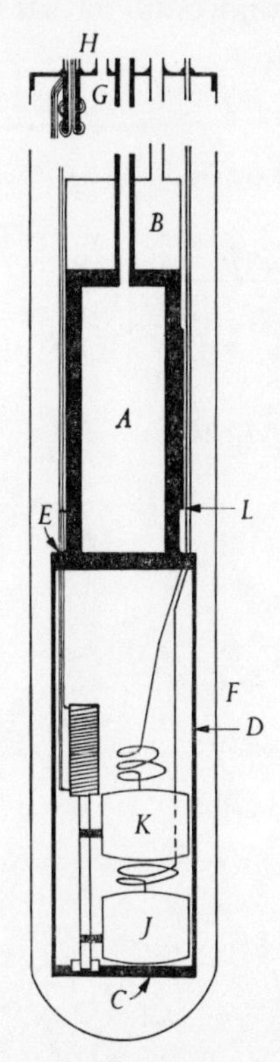

FIG. 88. A thermal conductivity cryostat associated with a Simon expansion liquefier (after Berman 1951).

given no trouble, the only vacuum leaks having been occasional ones due to leaking glass stop-cocks or careless soldering of the Wood's alloy joint which seals the outer vacuum jacket.

With a voltage signal from a Au + Co : Ag + Au thermocouple (see Chapter IV) a chopper-amplifier can be used to regulate the power fed to the electrical heater attached to the inner experimental chamber, so that its temperature may be controlled (see White and Woods 1955; Dauphinee and Woods 1955) to a constancy of about 0·001 degK at temperatures outside the range of available liquids. In experiments not involving such precise temperature control the heater (H_2 in Fig. 84) could be manually controlled. H_2 has normally been a 200-Ω resistor of constantan or manganin wound onto a baked Formel varnish on

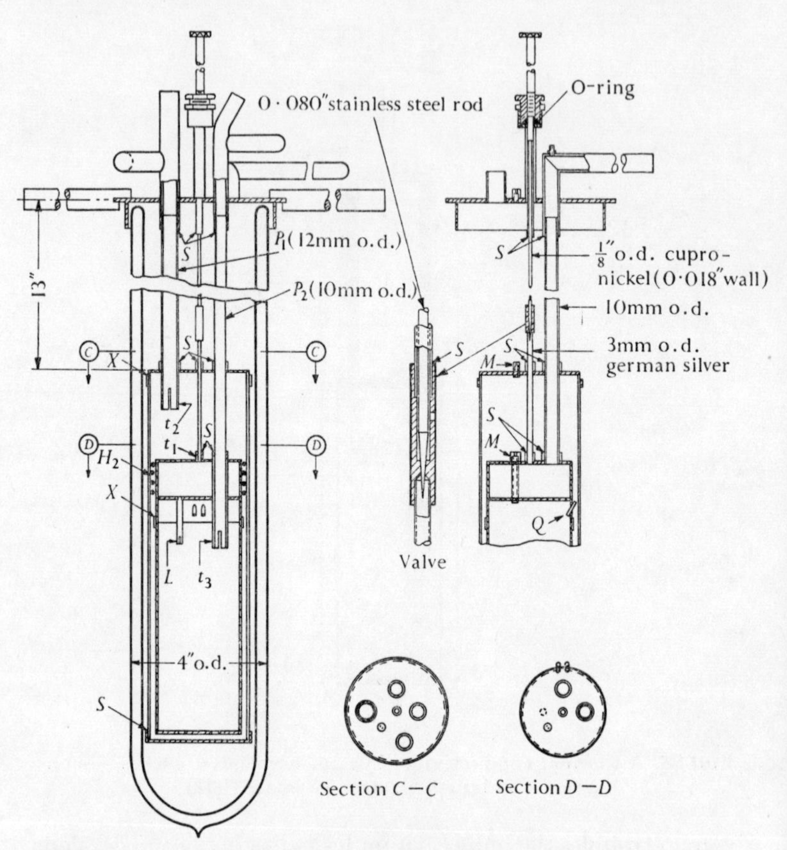

Fig. 89. A temperature-controlled cryostat used at the National Research Council (Ottawa).

the copper chamber and then revarnished and baked; however, a 200-Ω carbon resistor (I.R.C. resistor) in a copper sleeve soldered to the inner chamber seems to work adequately.

The Au + Co : Ag + Au junction of the control thermocouple is soldered to the copper bush at t_1; the Au + Co : Cu (40 B. & S.

enamelled copper wire) and Ag + Au : Co junctions are thermally anchored to the re-entrant copper bush t_2, with cigarette paper and nail polish as insulation and cement respectively.

The outer can is of 3-in diameter brass tube ($\frac{1}{16}$-in wall thickness) and is suspended by the German silver pumping tubes (0·2-mm wall thickness) from the main top plate. These German silver tubes are soft soldered (marked S) through bushes of brass or copper which are hard soldered into the plates. The two joints

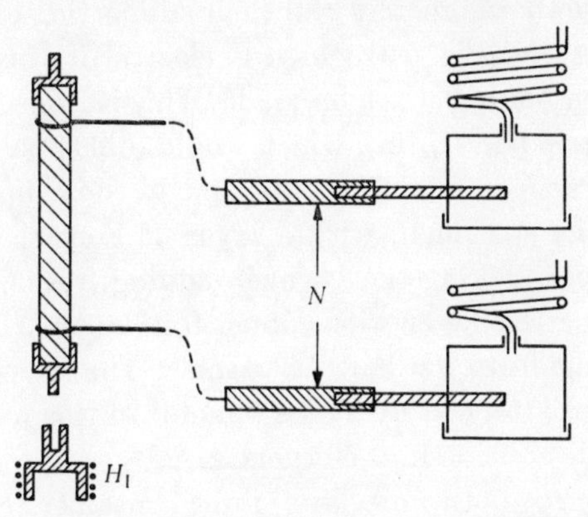

FIG. 90. Method of mounting thermal conductivity specimens.

marked X are Wood's metal joints by means of which the outer brass can and inner copper shield (gold-plated) can be easily removed and replaced after the mounting of specimens. The valve for inlet of liquid from the dewar into the small copper inner chamber is a stainless-steel needle of 10° total taper seating in a brass shoulder and is operated by turning the knurled head above the main top plate.

For measurements of the conductivity of a sample in rod form, the rod is fitted with small copper end-pieces (shown in Fig. 90) which are soldered to or, if soldering is not possible, are cemented to the rod. One copper end-piece is then fixed by Wood's metal into the 0·040-in hole in the copper pillar L (seen in Fig. 89) and the heater H_1 is similarly fixed to the other end-piece on the specimen. For potential leads, thermal and electrical, two short lengths of copper wire (about 0·030-in diameter) are tightly wrapped around the specimen, then soldered or cemented to it. The copper gas-thermometer bulbs (Fig. 90) are gold-plated to prevent tarnishing and hence to keep the radiation corrections small and constant; they are connected by lengths of 0·5 mm o.d. (0·3 mm i.d.) German silver capillary tubing to the external manometer system, the two capillaries passing through small holes in the bushes marked *M* on Fig. 89. In many experiments it is necessary to prevent a partial electrical short-circuit through these thermometers and their supporting capillaries, so that a linkage *N* (Fig. 90) which is electrically insulating but thermally conducting must be made. This is done by covering the copper strip ($\frac{1}{16} \times \frac{1}{64}$ in), which would otherwise connect the thermometers via a Wood's metal joint to the copper potential leads from the specimen, with a layer of Formel varnish and cigarette paper, baking the varnish, adding fresh varnish, and then wrapping this layer with copper foil (0·003–0·004 in thick) and baking again to harden the varnish. The copper potential leads are then attached by Wood's metal to this foil, when the specimen is in position in the cryostat.

As mentioned, both inner and outer containers (Fig. 89) are evacuated during thermal conductivity measurements through a common pumping line P_1. In this case two holes at the bottom of the inner shield are left open, the inner shield is merely attached at *X* by a heavy vacuum grease, the second pumping tube P_2 is sealed at its top end and all electrical leads are brought down the pumping tube P_1. This is convenient since the leads may be thermally anchored at the temperature of the bath by wrapping them around the copper bush t_2 (holding them by cotton thread and nail polish), and then bringing them into the inner experimental space through the holes at *Q* and anchoring them again to the copper bush t_3.

However, when the electrical resistance of fine wire specimens is being measured, it is essential to have exchange gas in the inner space to preserve thermal equilibrium. In this case the inner shield is made vacuum-tight and electrical leads are brought in through pumping tube P_2 and thermally anchored at t_3; so that these leads may not conduct too much heat to the inner chamber, they are thermally anchored at the temperature of the liquid in the surrounding dewar via a copper pillar inserted into the pumping tube P_2 (see Fig. 91).

In either case the electrical leads emerging from the top of the pumping tubes are taken through short lengths of ceramic tubing which rest in a brass bush (not shown); a vacuum wax Apiezon W, is used to seal the top of the tube where the leads emerge.

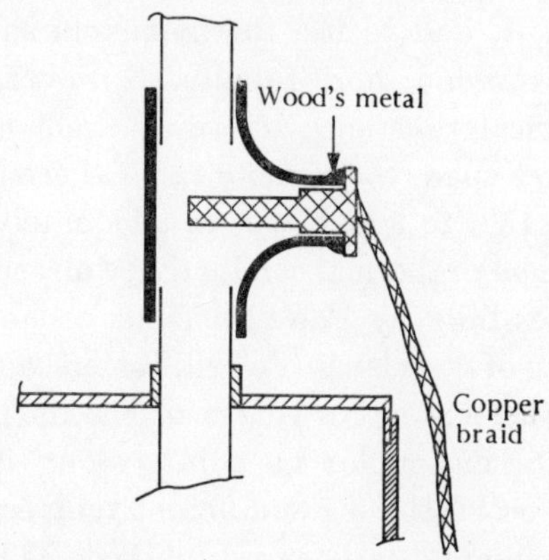

FIG. 91. A modification to the cryostat (Fig. 91) which allows electrical leads to be thermally anchored at the bath temperature.

The operation of such a cryostat is as follows.

(*a*) Below 4·2° K: liquid helium is at atmospheric pressure in the surrounding dewar and boils under controlled pressure in the inner chamber.

(*b*) 4·2°–54° K: with liquid helium in the surrounding dewar, electronically controlled heating of the inner vessel maintains the required temperature.

(*c*) 55°–90° K: liquid oxygen is at atmospheric pressure in the surrounding dewar, boils under controlled pressure in the inner chamber. Similarly, with liquid nitrogen the range 63°–78° K can be controlled.

(*d*) Above 90° K: with liquid nitrogen or oxygen in dewar, the inner vessel is electronically temperature controlled.

Note that in measurements of thermal conductivity or thermo-electric power, temperature drifts which are comparable with the difference δT being measured, over a period of time of the order of the equilibrium time of the specimen and thermometer assembly, must be eliminated; with specimens which are poor heat conductors, this 'equilibrium' time may vary from minutes ($\sim 10°$ K) to hours ($\sim 100°$ K.) As a result few measurements of these quantities have been carried out above 150° K in cryostats of this type, as it is usually less time-consuming and more accurate to use specimens of very different shape factor, i.e. of much smaller l/A, and to use thermometers such as thermocouples which respond more rapidly. However, for measurements of electrical resistance in which exchange gas can be used in the inner space to promote rapid thermal equilibrium, the range from 150° to 300° K can be adequately covered.

Another method of thermal conductivity measurement is that used very successfully by Powell and his collaborators at the National Bureau of Standards (Powell, Rogers, and Coffin 1957). They use a rather extensive length of specimen in rod form, attach eight thermocouples at points along its length and automatically record the eight different temperatures when a temperature gradient has been established. This yields seven values of thermal conductivity, each at a slightly different temperature, from which in practice a weighted mean is obtained.

A good example of a ^{3}He cryostat in which heat conductivity can be measured down to *ca.* 0·3° K is that of Seward (1965) illustrated in Fig. 78 of § 5.2. Other conductivity cryostats include the following: Holland and Rubin (1962); Guénault (1961); Colvin and Arajs (1964); Rowell (1960); Berman, Bounds, and Rogers (1965); Jericho (1965); Klein and Caldwell (1966). The methods of measuring thermal conductivity are also reviewed by White (1967).

4. Cryostats for electrical resistance or Hall-effect measurements

From the cryogenic viewpoint electrical resistance and such quantities as magneto-resistance or Hall effect are among the properties whose measurement presents no great difficulties. For one thing such determinations require the entry of only a limited number of electrical leads and are unaffected by exchange of thermal energy with the surroundings; the specimen may be kept in thermal equilibrium with a thermometer either by direct contact, by exchange gas, or by immersion of specimen and thermometer in a common refrigerating bath. Equally important is the fact that slow drifts in temperature of the specimen and its surroundings do not directly affect the determination as they do in calorimetric or thermal conduction measurements, except in so far as they introduce a small uncertainty as to the precise temperature of the specimen at the particular time that the measurement is made.

It is relatively difficult to measure the specific heat or heat conductivity of a solid at low temperatures, even when a comparatively crude estimate is required. But by merely attaching current and potential leads to a wire or rod and immersing it in a dewar of liquid helium, hydrogen, oxygen, etc., a reliable value of its electrical resistance may be obtained provided that there is a suitable electrical measuring circuit.

Of course, for precise calibration of electrical-resistance thermometers as secondary standards many precautions must be observed in the temperature determination as we discussed in Chapter IV. Frequently the resistance thermometer in capsule form is placed in good thermal contact with a copper block, which also houses a vapour-pressure bulb and gas thermometer.

More generally, values of electrical resistance of a metal in wire or rod form are determined at certain fixed temperatures or over limited ranges of temperature by immersion in a series of liquefied gas refrigerants—boiling at atmospheric pressure or under controlled pressure.

For values over the whole range from about 1° K to room temperature, a cryostat of the type described in some detail in § 2 (Fig. 89) may be used. In this cryostat several specimens, in the form of wires or rods, are suspended in a strain-free manner from the copper pillar in the inner experimental vessel. To each specimen current and potential leads are attached, and the inner vessel is filled with a small pressure of helium exchange gas

($\sim$ 1 mm Hg) to preserve temperature equilibrium. By controlling the pressure over the evaporating liquid in the small inner chamber or by controlled electrical heating of this chamber, temperatures over the range 1° K to room temperature can be attained; temperatures are measured by the gas thermometers in the inner space, by the vapour pressure of the evaporating liquid, or with a platinum resistance thermometer in the inner space. At temperatures above, say, 170° K, a convenient alternative to the use of liquid oxygen (or nitrogen) in the dewar and electrical heating of the inner chamber, is the use of a well-stirred bath of alcohol or other low-freezing-point liquid, cooled by an immersed copper coil with a slow stream of liquid nitrogen passing through it.

Commonly available liquids which are used for such a low-temperature bath are methyl alcohol (melting point at −98° C), ethyl alcohol (melting point at about −116° C), and *iso*pentane (melting point at −160° C). However, their highly inflammable nature may make it preferable to use a non-flammable mixture of organic liquids, if such baths are to be used frequently. As suggested by Scott (1941; see also Kanolt 1926, for recommended non-flammable mixtures), a non-flammable eutectic mixture of carbon tetrachloride and chloroform may be used down to −75° C, and a five-component system (14·5% chloroform, 25·3% methylene chloride, 33·4% ethyl bromide, 10·4% trans-dichloroethylene, and 16·4% trichloroethylene) may be used to below −140° C.

5. Cryostats for investigating mechanical properties

There is considerable interest in low-temperature studies of such metallurgical problems as the mechanism of flow stress, brittle fracture, and the hardness and tensile strength of materials.

From the point of view of designing a suitable cryostat for measurement of stress–strain relations, no very new cryogenic problem is presented as the material may be surrounded by exchange gas to preserve thermal equilibrium and its temperature may therefore be easily measured by a suitable

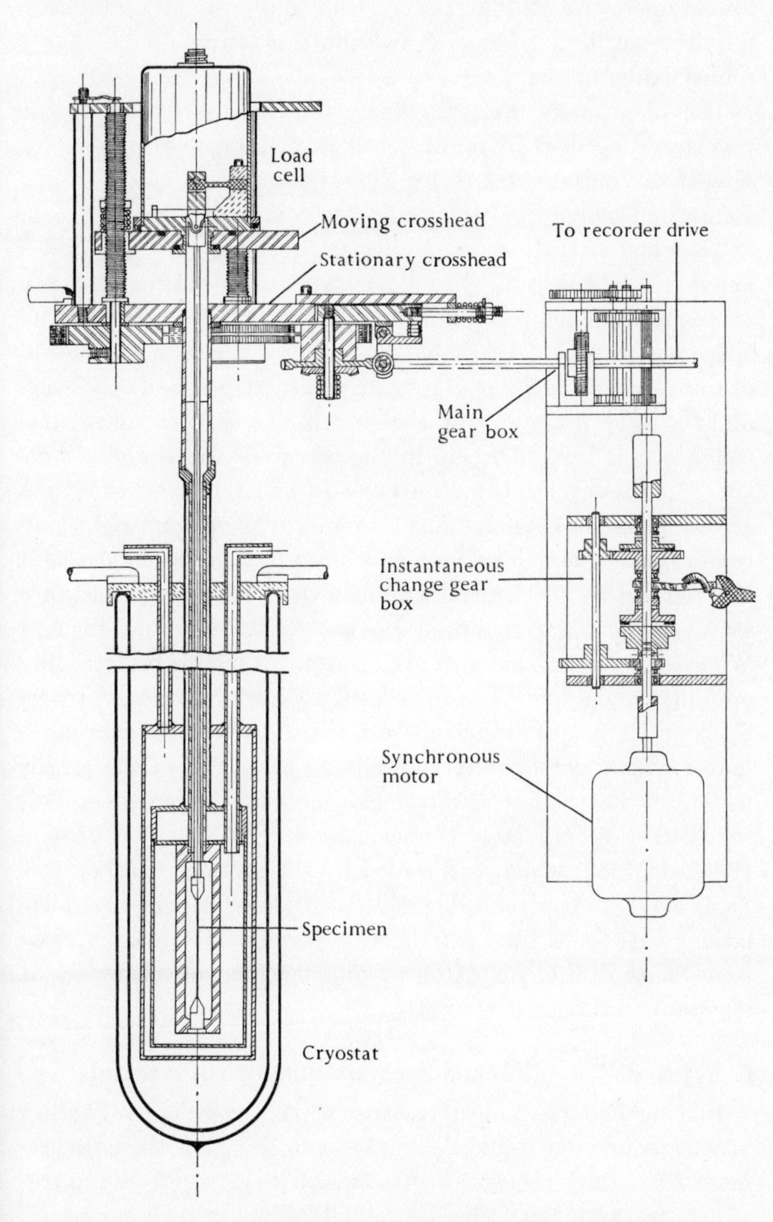

Fig. 92. Cryostat of Fig. 84 adapted for the measurement of flow-stress of materials.

low-temperature thermometer; the only slightly different requirement is a means of transmitting stress to the body under examination.

The diagram of Fig. 92 illustrates the adaptation of the cryostat described in some detail in § 2 (Fig. 89) above, for flow-stress measurements by Z. S. Basinski of the National Research Council in Ottawa (private communication).

The chief alterations to the design of the cryostat are that a heavy wall ($\frac{5}{8}$-in o.d., $\frac{1}{4}$-in i.d.) stainless-steel tube is taken centrally from the top plate down into the inner experimental space and to this is rigidly fixed a stainless-steel 'cage'; one end of the specimen is clamped by a grip at the bottom of this 'cage' and the other is clamped to the bottom of a $\frac{3}{16}$-in stainless-steel rod which is located freely in the centre of the stainless-steel tube. At the top the tube is attached to a fixed crosshead; on a screw-driven moving crosshead is mounted a vacuum-tight load cell in which a 24S aluminium cantilever beam is used to register the tension in the central stainless-steel rod; four resistance strain gauges are mounted on the cantilever and form a Wheatstone bridge network, the output from this bridge being amplified and applied to a Speedomax recorder. The chart of the Speedomax recorder is driven from the gearbox and synchronous motor which also operate the moving crosshead of the tensile tester, so that a direct load–elongation curve is recorded. This tensile-testing cryostat of Basinski and the earlier version (Basinski 1957) which was cooled by the Swenson method (see § 8.5) have proved suitable for determining stress–strain relations between 2° and 300° K. Rosenberg (1959) has reviewed some of the problems of measuring mechanical properties at low temperatures.

6. Cryostats for optical and X-ray examination of materials

One particular avenue of research which the wider availability of refrigerants like liquid helium has encouraged is the measurement of optical properties of solids, e.g. infra-red and ultraviolet transmission; closely linked from the viewpoint of cryostat design are X-ray diffraction studies, whether they be an investigation of phase change or measurement of expansion coefficient from the measured lattice spacing. In all these cases, solids must be cooled, maintained at specified low temperatures, and subjected to electromagnetic radiation, which must therefore have free access to and exit from the experimental region of the cryostat.

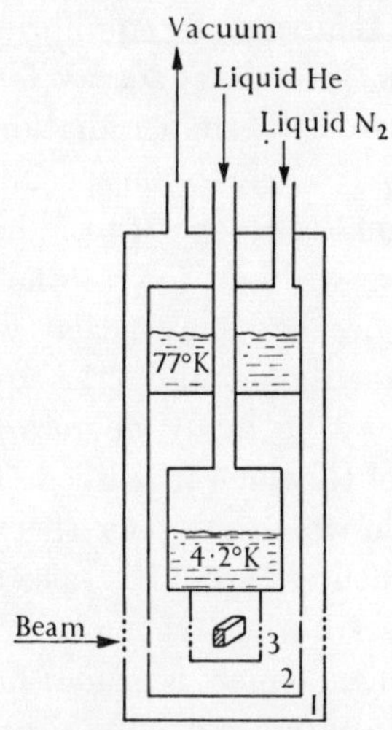

FIG. 93. Schematic diagram of the type of cryostat which may be used for optical or X-ray examination of materials.

Usually temperature control need not be precise in experiments of this type as fluctuations of one or two degrees in temperature are unimportant. However, the necessity for providing unrestricted passage to a beam of infra-red radiation or X-radiation poses its own special problems.

A suitable type of cryostat is shown schematically in Fig. 93. The radiation beam in its passage from source to specimen must pass through three barriers: (1) the outer wall of the vacuum jacket which is at room temperature; hence the required type of window, e.g. beryllium or cellophane for X-radiation, silica for ultra-violet, or rock salt for infra-red radiation, may be cemented with Araldite or sealed with a flange and rubber O-ring; (2) the liquid-nitrogen-cooled radiation shield—usually of polished copper—which need not be vacuum-tight, as the regions on both sides of (2) may be pumped through a common line, and therefore holes or slits may be freely cut in to act as windows for the beam; (3) the final wall, which may or may not require a vacuum-sealed transmission window depending on whether the specimen must be surrounded by exchange gas or can be supported in good thermal contact with the liquid-helium chamber to maintain temperature equilibrium.

Since the windows providing access for the beam must necessarily transmit some thermal radiation from surfaces at room temperature and at 77° K to the specimen, the latter must be in quite close thermal contact with the helium vessel if it is to be at a temperature within a few tenths of a degree of the helium bath. Experience has shown that with most pressure contacts and many cemented contacts the equilibrium temperature of a specimen may be many degrees above that of the low-temperature vessel to which it is fixed. This is particularly so in the case of solids which are poor thermal conductors, as their surface temperatures may be raised considerably by absorption of incident radiation. Of course, in the case of a solid metallic X-ray specimen which is solder-bonded to a heavy copper pillar, itself in contact with liquid helium, temperature equilibrium may be preserved quite well.

Optical cells

Two examples of low-temperature optical cells are shown in Figs. 94 and 95. In the first schematic diagram is a cryostat similar to that of Duerig and Mador (1952) in which no exchange gas is used around the specimen; they find that by thermally bonding their crystal to a copper holder with silicone grease (impregnated with silver dust) or a silver conducting paste or indium, the crystal reaches a temperature of about 5°–8° K. A thermocouple of Au + Co : copper is attached to the crystal surface to check its temperature. The 1-litre helium vessel is copper plated and polished to reduce radiation inflow and is suspended by a $\frac{5}{8}$-in o.d. (0·028-in wall) stainless-steel tube; a polished copper radiation shield surrounds the section of the helium chamber and specimen chamber which are below the annular liquid-nitrogen vessel. A sylphon bellows and a rack and pinion arrangement (not shown) allow vertical movement of the specimens in order that they may be aligned with the optical beam.

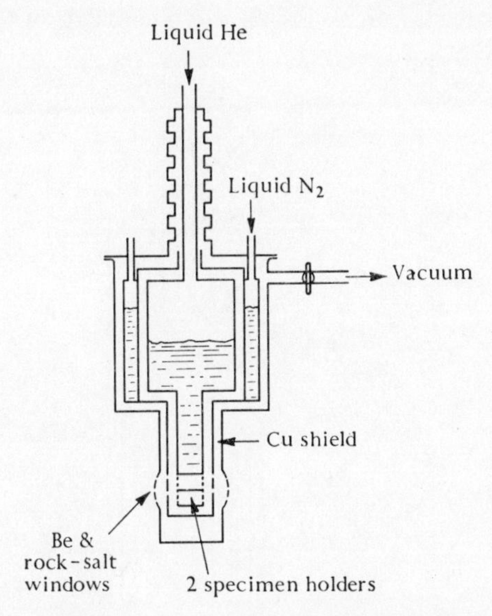

FIG. 94. Schematic diagram of a cryostat, similar to that used by Duerig and Mador (1952) for optical measurements.

In the absorption cell of Roberts (1955) shown in Fig. 95, a rock-salt window is used to seal the specimen compartment so that it can be filled with helium gas and the specimen kept within 1 deg K of the cooling bath at temperatures from 4·2°–330° K. This cryostat is rather similar to that described in 1949 by McMahon, Hainer, and King (1949) who used special dome-shaped silver chloride windows sealed with a spring washer and lock-nut arrangement. Roberts (1954) found that a rather simpler seal, capable of withstanding repeated cooling, could be made by cementing a thin flexible diaphragm of copper (0·003 in thick) to the window with thermo-setting Araldite. The outer edge of the copper shim is soldered to the copper chamber (see Fig. 96). In Roberts's cryostat, the temperature of the gas-filled specimen chamber is measured by a gas thermometer which occupies the annular volume V (Fig. 95) surrounding the radiation beam.

The method of construction used in Roberts's cryostat in which the two liquid chambers are directly above one another has advantages, in so far as the construction and maintenance

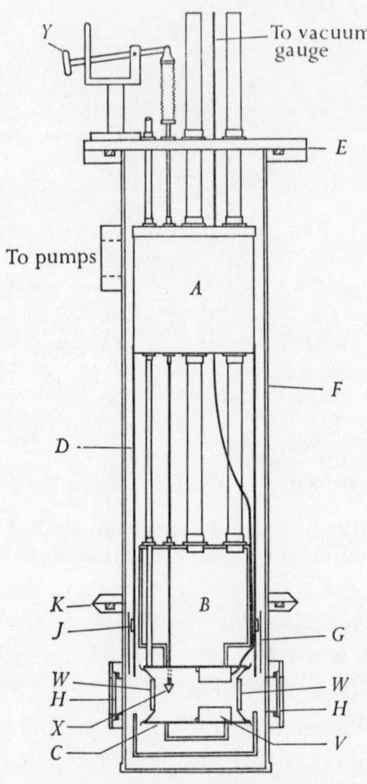

FIG. 95. Low-temperature absorption cell (after Roberts 1955).

of high vacuum-tight joints are concerned. The container B is of copper and A is of brass or copper; brass bushes are hard soldered into these containers and thin-walled cupro-nickel (or German silver) tubes are soft soldered through these. The radiation shield D should be of polished copper sufficiently thick-walled (say $\sim 0{\cdot}050$ in) to ensure adequate heat conduction along it and be well soldered to chamber A; then temperatures within a few degrees of the liquid nitrogen in A should be maintained over the entire shield. A simple and effective alternative to polishing or plating the containers and shields is to wrap them with thin aluminium foil which has an emissivity of about 0·04.

An interesting development in optical windows was described by Warschauer and Paul (1956), who tested a thin polyethylene window and found it remained vacuum-tight after repeated

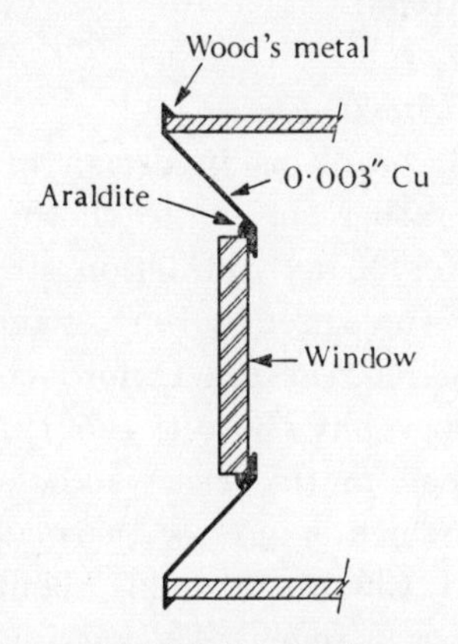

FIG. 96. An optical window for low-temperature use (after Roberts 1954).

cooling to 4·2° K and warming to room temperature. They milled a window aperture from brass or copper stock about $\frac{1}{8}$ in thick, leaving a flat area about $\frac{1}{4}$ in wide surrounding the opening. The polyethylene film, 0·005 in thick, was stuck to this supporting area with a thin film of silicone grease. A greased flat retaining plate was then clamped down tightly onto the polyethylene by four screws. Under a 1-atm pressure difference the film window assumed a domed shape but remained vacuum-tight.

Some other optical cryostats with special features have been described by Taylor, Smith, and Johnston (1951); Geiger (1955); Brebner and Mooser (1962); Schoen and Broida (1962). Fitchen (1963) has described a high-pressure cell for optical measurements.

Richards (1964) has given an interesting account of Fourier-transform spectroscopy in the far infra-red which discusses the use of light pipes for bringing infra-red radiation into a cryostat. Instead of windows in the tail of the cryostat, a simple brass tube of $\sim$ 1 cm diam. is very effective in funnelling long wave-length radiation from the top of the cryostat down to the bottom where an absorption sample and bolometer-detector are placed (see also short review of infra-red spectroscopy at low temperatures by Parkinson 1963).

X-ray and neutron diffractometers

As we mentioned, problems in design of low-temperature X-ray cameras are rather similar to those arising in optical cells. The cryostat for X-ray diffraction studies of the alkali metals at low temperatures used by Barrett and his collaborators at Chicago was described at the Oxford conference on low-temperature crystallography (Barrett 1956) and is illustrated in Fig. 97. The specimen in this case is cooled by conduction through the bar (9) which is in close thermal contact with the bottom of chamber 1. Chamber 1 holds about 2 litres of liquid helium (or hydrogen or nitrogen). The specimen can be changed by opening the O-ring joint between (3) and (4) and the conical joint (7).

Other diffractometer-cryostats have been described by Figgins, Jones, and Riley (1956); Mauer and Bolz (1961); Chopra (1962); Robertson (1960).

When a neutron beam is used for diffraction studies, the materials exposed to the beam have to be selected with care. Generally the window and radiation shields are made of aluminium. A good example of such a cryostat is that of Abrahams (1960) shown in Fig. 98. It is spun from a type-316 stainless steel and Heli-arc welded; the helium chamber (9) terminates in a copper block (13) into which the crystal holder (14), made of a Zr : 2Ti alloy, is screwed. This holder is surrounded by a spun-aluminium thimble (11) which is attached to the gold-plated copper radiation shield. The removable base (16) with its window (17) are also of aluminium. In the figure, (8) is liquid nitrogen and (10) are Teflon spacers. Similar cryostats have been made by Hofman Laboratories Inc. of New Jersey.

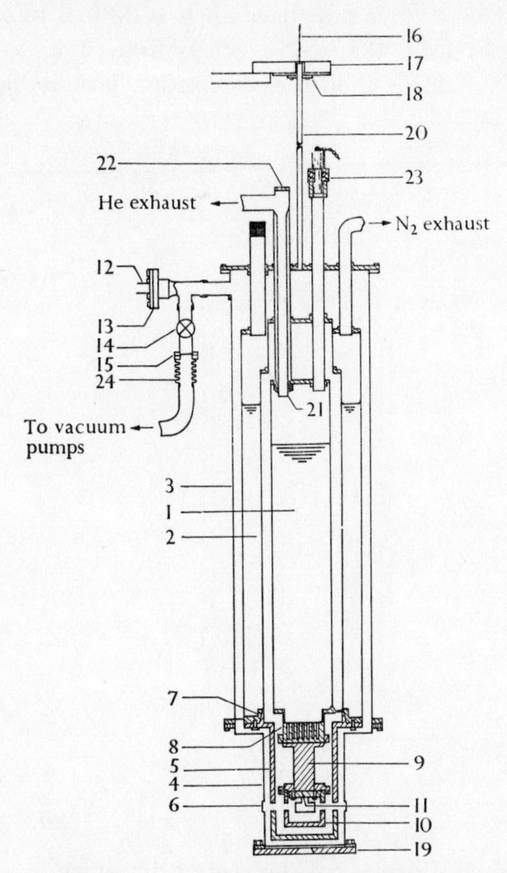

FIG. 97. A cryostat for X-ray diffraction studies (after Barrett 1956).

Temperature control

Finally, returning to the general form of cryostat sketched in Fig. 93 at the beginning of this section, there arises the problem of controlling specimen temperatures outside the easily accessible ranges 1°–4° K, 55°–90° K, and 10°–20° K. One way of solving this problem is to separate the specimen cavity from the lower liquid reservoir by a metal post of carefully chosen material and dimensions. Then with the aid of an electrical heater attached to the cavity or on the post adjacent to the cavity, it can be raised to temperatures above that of the liquid boiling in the reservoir. This method has obvious drawbacks if it is desired to operate at temperatures over the entire range from 2° K to room temperature; no single choice of connecting post is likely to

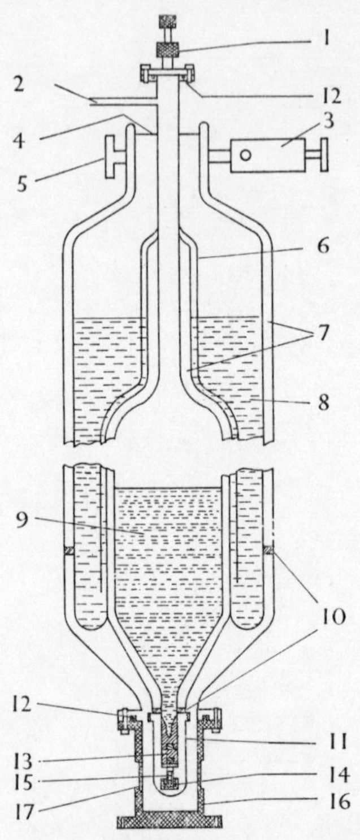

Fig. 98. Cryostat used for neutron diffraction (after Abrahams 1960).

allow temperature control at both 5° and 40° K with liquid helium in the lower reservoir, or at both 60° and 150° K with liquid oxygen, and yet preserve a relatively low rate of evaporation of the refrigerant in each case.

One solution is to use a gas-filled tube as a thermal switch between the sample and the liquid chamber, rather than simply have a metal part of fixed dimensions and fixed conductivity. By varying the pressure of helium-exchange gas in the tube, the thermal linkage can be altered easily. An electrical heater on the bottom of the tube or beside the sample can then be used to give fine control of the temperature (see also Chapter VIII).

Alternatively liquid and cold gas from the liquid chamber can be made to flow slowly through a coil soldered to the copper sample-holder (Fig. 99). This is a variant of the Swenson form of temperature control (see § 8.5 for more details): the flow is produced by either an overpressure above the liquid or by

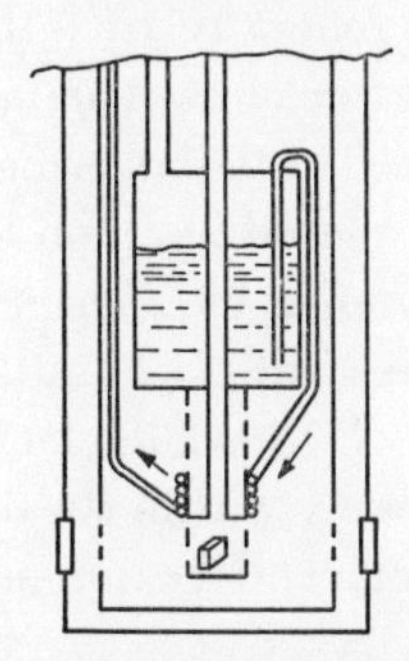

FIG. 99. Possible method of temperature control for a low-temperature camera.

sucking the liquid through the tube. The evaporated gas is then led to a needle valve at room temperature, through which a mechanical vacuum pump extracts it; by using a fine control needle valve, the rate of flow can be varied sufficiently to control the chamber over a wide temperature range. The cryostat used by Brebner and Mooser (1962) operates from 4° to 300° K with a coil of this sort on the crystal-holder. That of Mauer and Bolz (1961) also operates from 4° to 300° K but has a stainless steel tube between the main liquid chamber and the crystal-holder with a valve that can let liquid helium into this tube and cool the crystal to 4° K; with the tube empty it can be heated electrically well above bath temperature.

7. Magnetic susceptibility cryostats

The many methods of measuring magnetic susceptibilities can be divided into two broad classes—static and induction methods. The latter involve a measurement or comparison of mutual or self-inductances. Using an a.c. inductance bridge or a ballistic throw method, they have been widely used for the measurement of susceptibilities of paramagnetic salts at temperatures near 1° K and below but have not been greatly used for absolute measurements at higher temperatures (see, however, inductance bridge of Erickson, Roberts, and Dabbs 1954, for use in measuring susceptibilities below 77° K and the development of an improved inductance method by McKim and Wolf 1957).

Of the static methods generally used, those involving a weight determination and therefore classed as balance methods appear to be the most precise. The balance methods depend on the force exerted on a body placed in a non-homogeneous magnetic field; this force is measured by suspending the body from a sensitive balance. In the so-called Faraday method of measurement, a sample is used which is sufficiently small that the field H does not vary appreciably within the dimensions of the body, but is nevertheless sufficiently inhomogeneous to exert a measurable force given by

$$\mathbf{F} = \mathbf{M} \wedge \operatorname{grad} \mathbf{H},$$

where $\mathbf{M}$ is the magnetic moment. Hence, if χ_{m} is the mass susceptibility and m the mass, then

$$F_y = m\chi_{\mathrm{m}} H_x \frac{\partial H_x}{\partial y}$$

is the vertical force exerted by an inhomogeneous field of strength H_x in a horizontal direction and gradient $\partial H_x/\partial y$ in the vertical direction.

An alternative balance method is that of Gouy in which the specimen is in the form of a long cylindrical rod. If the lower end is in a homogeneous field H and the upper end in a much weaker field H_0, then the force acting on a sample of cross-sectional area A is

$$F = \frac{\chi_{\mathrm{v}} - \chi_{\mathrm{g}}}{2}(H^2 - H_0^2),$$

where χ_{v} is the volume susceptibility of the sample and χ_{g} is the susceptibility of the gas medium.

Many of the methods of measuring magnetization or susceptibility have been discussed in texts by Bates (1948) and McGuire (1959). Descriptions of individual methods include those of Hutchinson and Reekie (1946); Bowers and Long (1955); Bozorth, Williams, and Walsh (1956); Foner (1959); Hedgcock and Muir (1960); Condon and Marcus (1964).

Returning to the cryogenic aspects of the problems associated with susceptibility measurement by a balance method, we note two distinctive features:

(*a*) the sample must be freely suspended with no mechanical constraint from a microbalance;

(*b*) vibrations of the sample or balance should be avoided as much as possible.

The result is a narrow vertical tube in which the sample is suspended; the tube and balance (or part of them) are filled with exchange gas to keep the sample in temperature equilibrium with the nearby tube wall. The tube is then immersed in a dewar which, to ensure maximum field strength, is narrow in the region that is in the pole gap of the magnet. The requirement of no constraint prevents any thermometer being attached to the sample under investigation unless perhaps a thermocouple is attached to the sample and the two thermocouple leads form the suspension.

A common type of susceptibility cryostat is that illustrated schematically in Fig. 100 (cf. Kamerlingh Onnes and Perrier, 1913; McGuire and Lane, 1949). The dewar containing liquid helium or liquid nitrogen may be pumped to control the temperature of the liquid bath if care is taken to avoid vibration from the mechanical pump being transmitted to the dewar and balance.

The chief problems arise when control of temperatures outside those covered by a liquefied gas is required. One method of controlling temperature at say 30° or 110° K has been used by Hedgcock and Muir (1960) and is illustrated in Fig. 101. The lower section of the sample tube is of copper of sufficient wall thickness to ensure a relatively small temperature gradient along the length of the specimen. This copper tube has an electrical heater and thermometers (carbon thermometer and a copper resistance thermometer) attached to it, and a heavy copper wire (six strands of no. 24 B. & S.) acts as a partial

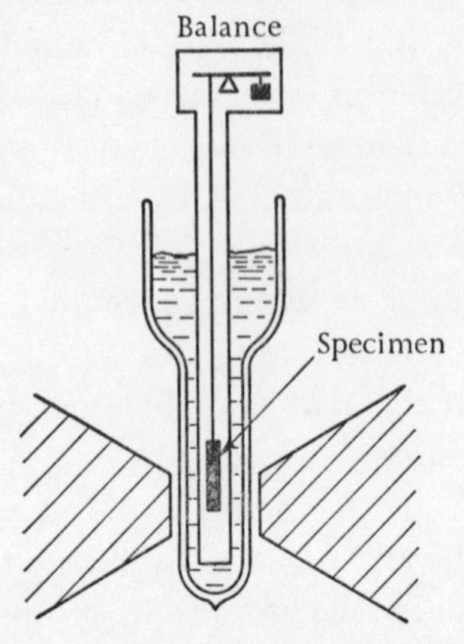

FIG. 100. Schematic diagram of a cryostat for susceptibility determinations.

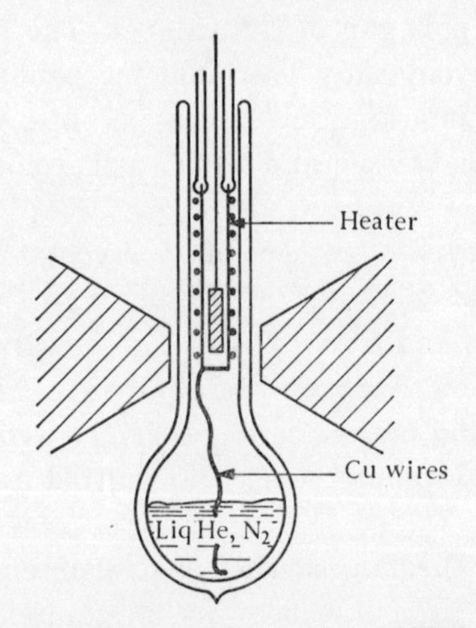

FIG. 101. A similar cryostat to that of Hedgcock and Muir (1960).

thermal link between the copper tube and the liquid in the wide tail of the dewar vessel. By carefully selecting the dimensions of this thermal link and varying the heater power, temperatures above that of the liquid refrigerant can be maintained around the specimen.

An alternative to this might be the use of a 'Swenson' type of control (see Chapter VIII) in which copper tubing about $\frac{1}{8}$ in o.d., 0·015-in wall) is closely wound around the copper specimen tube; one end of this fine tubing extends to the bottom of the dewar while the other leads to a needle valve and mechanical pump. By controlling the setting of the needle valve, liquid is drawn up through the fine copper tube to cool the specimen chamber. Again a thermometer and a resistance heater would be attached to the chamber in the vicinity of the specimen.

When making sensitive balances and cryostats for measuring specimens of relatively small magnetic susceptibility, use materials which themselves have very small susceptibility. The susceptibility of a number of these common materials has been measured between 1° and 4° K by Salinger and Wheatley (1961). They expressed the results in the form

$$\chi_v/\rho = B + C/T,$$

where B and C (in units of 10^{-6} e.m.u. per g) have the following values:

	B	C
Pyrex glass tube (Corning no. 7740)	$1{\cdot}6 \pm 1$	12 ± 3
Nylon	$-0{\cdot}6 \pm 1{\cdot}5$	$0{\cdot}3 \pm 3$
Aquadag	12 ± 2	$0{\cdot}1 \pm 4$
Copper magnet wire (Formex insulated)	$0{\cdot}03 \pm 1$	$0{\cdot}01 \pm 1{\cdot}5$
Mylar	50 ± 7	$1{\cdot}5 \pm 6$
Inconel	$2{\cdot}6 \times 10^5$	$-2{\cdot}3 \times 10^5$
304 stainless steel	108 ± 9	$-5{\cdot}7 \pm 4{\cdot}8$
316 stainless steel	312 ± 25	-105 ± 20

8. Other cryostats

Many other physical measurements are performed at low temperature and they impose their own particular requirements on cryostat design. Some can be done in an experimental space of limited size with only a few electrical connexions to the outside world. Some may dissipate little heat and not need very fine temperature control.

For example, ultrasonic velocity and attenuation measurements can be done in the kinds of cryostats that we have discussed earlier in this chapter. No peculiar cryogenic difficulties arise except perhaps the need to cement transducers onto crystals so they do not reflect ultrasonic waves too much nor crack at low temperatures: a variety of cementing materials have been used to make the seal including rubber solutions, Nonaq stop-cock grease, viscous silicone oils, alcohol, isopetane, and even methane. More details of the methods of measurements are given in reviews by McSkimin (1964), Van Itterbeek (1955), and E. R. Dobbs (Hoare, Jackson, and Kurti 1961, p. 254).

Microwave studies require a resonant cavity in the cryostat and need waveguides to connect this to the microwave source and detectors which are at room temperature. Precision-drawn stainless steel suitable for waveguides is now available from sources such as Superior Tube Company, Norristown, Pa. Naturally there must be a gas seal in the waveguide to prevent air travelling down into the cavity and condensing; pieces of mica sheet are commonly used as a seal at the junction of two pieces of waveguide. Bagguley and Owen (Hoare *et al.* 1961, p. 297) discuss some of the cryogenic problems of microwave measurements.

Irradiation at low temperatures in a nuclear reactor presents special problems. There is a limitation on space and on the materials exposed to the neutron flux, also there is the continual heating from the γ-ray background. In a high-flux reactor continuous refrigeration at a rate of 10 to 100 W may be needed to balance this heating. Therefore a considerable flow of liquid or cold gaseous helium must be circulated around the irradiated sample to keep it at the required low temperature. Usually the cooling tubes and sample chamber are made of aluminium as in the cryostat used by Coltman, Blewitt, and Noggle (1957); Blewitt and Coltman (in Hoare *et al.*, p. 274) have discussed the cryogenic techniques used in irradiation studies.

Other cryostats which may be of interest include the following: bubble chambers discussed in reviews by Barford (1960) and Shaw (1964); Mossbauer studies at low temperatures (e.g.

Wiedemann, Mundt, and Kullmann 1965); optical microscopy at liquid-helium temperatures (e.g. Hull and Rosenberg 1960; Conroy, Gottlieb, and Garbuny 1965); field-emission microscopy (Hardin, Blair, and Einspruch 1965; Attardo, Galligan, and Sadofsky 1966); high-pressure cryostats (Swenson 1960; Levy and Olsen 1965; Lyon, McWhan and Stevens 1967).

9. References

Abrahams, S. C. (1960). *Rev. scient. Instrum.* **31**, 174.

Attardo, M. J., Galligan, J. M., and Sadofsky, J. (1966). *J. scient. Instrum.* **43**, 607.

Aven, M. H., Craig, R. S., and Wallace, W. E. (1956). *Rev. scient. Instrum.* **27**, 623.

Barford, N. C. (1960). *Prog. Cryogen.* **2**, 89.

Barrett, C. S. (1956). *Br. J. appl. Phys.* **7**, 426.

Basinski, Z. S. (1957). *Proc. R. Soc.* A**240**, 229.

Bates, L. F. (1948). *Modern magnetism*, 2nd edn. Cambridge University Press.

Berman, R. (1951). *Proc. R. Soc.* A**208**, 90.

—— Bounds, C. L., and Rogers, S. J. (1965). Ibid. **289**, 66.

Bowers, R. and Long, E. A. (1955). *Rev. scient. Instrum.* **26**, 337.

Bozorth, R. M., Williams, H. J., and Walsh, D. E. (1956). *Phys. Rev.* **103**, 572.

Brebner, J. L. and Mooser, E. (1962). *J. scient. Instrum.* **39**, 69.

Chopra, K. L. (1962). *Cryogenics* **2**, 167.

Coltman, R. R., Blewitt, T. H., and Noggle, T. S. (1957). *Rev. scient. Instrum.* **28**, 375.

Colvin, R. V. and Arajs, S. (1964). *Phys. Rev.* **133**, A1076.

Condon, J. H. and Marcus, J. A. (1964). Ibid. **134**, A446.

Conroy, J., Gottlieb, M., and Garbuny, M. (1965). *Cryogenics* **5**, **348.**

Dauphinee, T. M., MacDonald, D. K. C., and Preston-Thomas, H. (1954). *Proc. R. Soc.* A**221**, 267.

—— and Woods, S. B. (1955). *Rev. scient. Instrum.* **26**, 693.

Duerig, W. H. and Mador, I. L. (1952). Ibid. **23**, 421.

Erickson, R. A., Roberts, L. D., and Dabbs, J. W. T. (1954). Ibid. **25**, 1178.

Estermann, I. and Zimmerman, J. E. (1952). *J. appl. Phys.* **23**, 578.

Figgins, B. F., Jones, G. O., and Riley, D. P. (1956). *Phil. Mag.* **1**, 747.

Fitchen, D. B. (1963). *Rev. scient. Instrum.* **34**, 673.

Foner, S. (1959). Ibid. **30**, 548.

Frederikse, H. P. R. (1953). *Phys. Rev.* **92**, 248.

Geiger, F. E. (1955). *Rev. scient. Instrum.* **26**, 383.

Guénault, A. M. (1961). *Proc. R. Soc.* A**262**, 240.

de Haas, W. J. and Rademakers, A. (1940). *Physica* **7**, 992.

Hardin, W. R., Blair, J. C., and Einspruch, N. G. (1965). *Cryogenics* **5**, 138.

Hedgcock, F. T. and Muir, W. B. (1960). *Rev. scient. Instrum.* **31**, 390.
Hill, R. W. (1959). *Prog. Cryogen.* **1**, 180.
—— Martin, D. L., and Osborne, D. W. (1967). *Experimental thermodynamics*, Vol. 1: *Calorimetry of non-reacting systems*, Chapter 7. Butterworths.
Hoare, F. E., Jackson, L. C., and Kurti, N. (1961). *Experimental cryophysics*. Butterworths, London.
Holland, M. G. and Rubin, L. G. (1962). *Rev. scient. Instrum.* **33**, 923.
Hull, D. and Rosenberg, H. M. (1960). *Cryogenics* **1**, 27.
Hulm, J. K. (1950). *Proc. R. Soc.* A**204**, 98.
Hutchison, T. S. and Reekie, J. (1946). *J. scient. Instrum.* **23**, 209.
Jan, J. P., Pearson, W. B., and Templeton, I. M. (1955). *Conférence de physique des basses températures*, p. 418, Annexe 1955–3, *Suppl. Bull. int. Inst. Refrig.*
Jericho, M. H. (1965). *Phil. Trans. R. Soc.* A**257**, 385.
Kamerlingh Onnes, H. and Perrier, A. (1913). *Leiden Commun.* 139a.
Kanolt, C. W. (1926). *Natn. Bur. Stand. Sci. Paper* 520.
Keesom, P. H. and Pearlman, N. (1956). *Handb. Phys.* **14**, 282.
Klein, M. V. and Caldwell, R. F. (1966). *Rev. scient. Instrum.* **37**, 1291.
Levy, M. and Olsen, J. L. (1965). In *Physics of high pressures and the condensed phase*, Chapter 13 (ed. A. Van Itterbeek). North-Holland Publishing Company, Amsterdam.
Lounasmaa, D. V. and Guenther, R. A. (1962). *Phys. Rev.* **126**, 1357.
Lyon, D. N., McWhan, D. B., and Stevens, A. L. (1967). *Rev. scient. Instrum.* **38**, 1234.
McGuire, T. R. and Lane, C. T. (1949). *Rev. scient. Instrum.* **20**, 489.
—— (1959 in *Methods of experimental physics*, Vol. 6B, p. 171 (ed. K. Lark-Horovitz and V. A. Johnson). Academic Press, New York.
McKim, F. R. and Wolf, W. P. (1957). *J. scient. Instrum.* **34**, 64.
McMahon, H. O., Hainer, R. M., and King, G. W. (1949). *J. opt. Soc. Am.* **39**, 786.
McSkimin, H. J. (1964) in *Physical acoustics*, Vol. 1, Part A, p. 272 (ed. W. P. Mason). Academic Press, New York.
Martin, B. D., Zych, D. A., and Heer, C. V. (1964). *Phys. Rev.* **135**, A671.
Martin, D. L. (1960). *Can. J. Phys.* **38**, 17.
—— (1961). *Proc. R. Soc.* A**263**, 378.
—— (1962). *Can. J. Phys.* **40**, 1166.
Mauer, F. A. and Bolz, L. H. (1961). *J. Res. natn. Bur. Stand.* **65**C, 225.
Morrison, J. A. and Los, J. M. (1950). *Discuss. Faraday Soc.* **8**, 321.
—— Patterson, D., and Dugdale, J. S. (1955). *Can. J. Chem.* **33**, 375.
de Nobel, J. (1951). *Physica* **17**, 551.
Olsen, J. L. (1952). *Proc. phys. Soc.* A**65**, 518.
O'Neal, H. R. and Phillips, N. E. (1965). *Phys. Rev.* **137**, A748.
Parkinson, D. H. (1963). *Cryogenics* **3**, 1.
Pearlman, N. and Keesom, P. H. (1952). *Phys. Rev.* **88**, 398.
Phillips, N. E. (1959). Ibid. **114**, 676.

POWELL, R. L., ROGERS, W. M., and COFFIN, D. O. (1957). *J. Res. natn. Bur. Stand.* **59**, 349.

RAMANATHAN, K. G. and SRINIVASAN, T. M. (1955). *Phil. Mag.* **46**, 338.

RAYNE, J. A. (1956). *Aust. J. Phys.* **9**, 189.

—— and KEMP, W. R. G. (1956). *Phil. Mag.* **1**, 918.

RICHARDS, P. L. (1964). *J. opt. Soc. Am.* **54**, 1474.

ROBERTS, V. (1954). *J. scient. Instrum.* **31**, 251.

—— (1955). Ibid. **32**, 294.

ROBERTSON, J. H. (1960). Ibid. **37**, 41.

ROSENBERG, H. M. (1959). *Prog. Cryogen.* **1**, 121.

ROWELL, P. M. (1960). *Proc. R. Soc.* A**254**, 542.

SALINGER, G. L. and WHEATLEY, J. C. (1961). *Rev. scient. Instrum.* **32**, 872.

SCHOEN, L. J. and BROIDA, H. P. (1962). Ibid. **33**, 470.

SCOTT, R. B. (1941). *Temperature*, p. 206. Reinhold, New York.

SEIDEL, G. and KEESOM, P. H. (1958). *Rev. scient. Instrum.* **29**, 606.

SEWARD, W. D. (1965). Thesis, Cornell University.

SHAW, D. F. (1964). *Cryogenics* **4**, 193.

SILVIDI, A. A. and DAUNT, J. G. (1950). *Phys. Rev.* **77**, 125.

SWENSON, C. A. (1960). *Solid State Physics* **11**, 41.

TAYLOR, W. J., SMITH, A. L., and JOHNSTON, H. L. (1951). *J. opt. Soc. Am.* **41**, 91.

VAN ITTERBEEK, A. (1955). *Prog. low Temp. Phys.* **1**, 355.

WARSCHAUER, D. M. and PAUL, W. (1956). *Rev. scient. Instrum.* **27**, 419.

WESTRUM, E. F., HATCHER, J. B., and OSBORNE, D. W. (1953). *J. chem. Phys.* **21**, 419.

WHITE, G. K. (1953). *Proc. phys. Soc.* A**66**, 559; *Aust. J. Phys.* **6**, 397.

—— (1967) in *Thermal conductivity*, Chapter 2 (ed. R. P. TYE). Academic Press, New York.

—— and WOODS, S. B. (1955). *Can. J. Phys.* **33**, 58.

WIEDEMANN, W., MUNDT, W. A., and KULLMANN, D. (1965). *Cryogenics* **5**, 94.

WILKINSON, K. R. and WILKS, J. (1949). *J. scient. Instrum.* **26**, 19.

CHAPTER VII

HEAT TRANSFER

1. Introduction

A DETERMINATION of the physical properties of materials at temperatures considerably different from the ambient temperature requires some degree of thermal isolation of the material from its surroundings; the degree of isolation must be sufficient to meet the demands of temperature control, temperature measurement, and in the case of low-temperature research, the available refrigerating capacity of the coolant. In a measurement such as the determination of heat capacity by the adiabatic method, the sample must also be thermally isolated from its immediate environment—the cooling medium. These conditions imply that an important factor in the design of a successful cryostat is the ability to predict the degree of thermal isolation or in other words, to calculate the transfer of heat that will take place between the specimen and it surroundings; then materials and methods of construction must be used which will ensure this heat transfer being within or below certain allowable limits.

In general, heat may be transferred by conduction, convection, and radiation. In most low-temperature applications, thermal isolation is assisted by partial evacuation of gas from the interior of the cryostat so that convection is eliminated. Then effective heat transfer takes place by conduction through the residual low-pressure gas, conduction through the solids that interconnect the various parts of the cryostat, and by radiation. In addition such factors as Joule heating in electrical leads, eddy current heating, mechanical vibration, adsorption or desorption of gases may contribute to the heat transfer.

Of the three major processes responsible for heat conduction at low temperatures that due to conduction by solids can be estimated generally with a fair degree of accuracy; low-pressure gas conduction and radiation transfer may be estimated with rather less accuracy, the uncertainty depending on our lack of knowledge of the accommodation coefficient and emissivity respectively. However, it is usually possible to estimate an upper bound for the heat transfer by these processes and therefore ensure that materials, degree of high vacuum, etc., used are sufficient to meet the required demands.

The remainder of this chapter deals in some detail with the method of calculating heat transferred by these processes.

2. Conduction of heat by a gas

An elementary treatment on the basis of kinetic theory indicates that in a gas at normal pressures the thermal conductivity λ and viscosity η are given by (see, for example, Roberts 1940 or Jeans 1948)

$$\lambda = \tfrac{1}{3}mln\bar{v}C_v, \quad \eta = \tfrac{1}{3}mln\bar{v},$$

where m = mass of one molecule,

l = mean free path,

n = number of molecule per cm^3,

$\bar{v}$ = mean velocity,

C_v = specific heat per g.

Thus if ρ is the density,

$$\lambda = \tfrac{1}{3}\rho l\bar{v}C_v, \quad \eta = \tfrac{1}{3}\rho l\bar{v},$$

and

$$\lambda = \eta C_v.$$

Since the mean free path $l \propto 1/p$, both λ and η are seen to be pressure independent at least to a first approximation, and to depend on the temperature through their dependence on the mean velocity $\bar{v}$.

Experiment (and a more detailed theory) indicates that $\lambda = \text{constant}.\eta C_v$ where the constant has a value of 1·5–2·5 for most common gases. As may be seen from Figs. 51 and 52 (Chapter III), λ and η increase monotonically with increasing temperature as T^n where the exponent n has experimental values in the range 0·6–0·9 for hydrogen, helium, nitrogen, and oxygen.

However, as we have pointed out before, the residual gas pressure in a cryostat is nearly always reduced to a point where the mean free path becomes comparable with the dimensions of the system; at room temperature a pressure of $\sim 10^{-4}$ mm Hg is sufficiently low that the mean free path is ~ 100 cm. At such pressures, the average molecule may travel from a hot wall to a cold wall without collision with another gas molecule, and the thermal conductivity becomes a function of the number of molecules present (and also their mean velocity), i.e. at low pressures $\lambda \propto n \propto p$.

For approximately parallel surfaces at temperatures T_1 and T_2, the heat transferred, $\dot{Q}$, by conduction through a gas at low pressure p dyn cm^{-2} is given by (see Kennard 1938):

$$\dot{Q} = \frac{a_0}{4}\frac{\gamma+1}{\gamma-1}\sqrt{\left(\frac{2R}{\pi M}\right)}\,p\frac{T_2-T_1}{\sqrt{T}} \text{ ergs cm}^{-2}\,\text{s}^{-1}, \tag{46}$$

where M is the molecular weight and γ is the ratio of specific heats of the gas, R is the gas constant in ergs mole^{-1} deg^{-1}; a_0 is related to the individual accommodation coefficients a_1 and a_2 and the areas A_1 and A_2 of the two surfaces by

$$a_0 = \frac{a_1 a_2}{a_2+(A_2/A_1)(1-a_2)a_1},$$

so that if $A_1 \simeq A_2$, then

$$a_0 = \frac{a_1 a_2}{a_1+a_2-a_1 a_2};$$

if also $a_1 = a_2 = a$,
$$a_0 = \frac{a}{2-a}$$
$$\simeq \tfrac{1}{2}a \text{ for } a \to 0$$
$$\simeq a \quad \text{for } a \to 1.$$

Accommodation coefficients are normally measured for the case of a wire at temperature T_2 slightly higher than a surrounding gas at temperature T_1. In this case molecules colliding with the wire have an average energy corresponding to T_1 so that T in the denominator of (46) may be identified with T_1. The molecules leave the wire with an energy corresponding to a temperature T_2' intermediate between T_1 and T_2; then

$$(T_2'-T_1) = a(T_2-T_1)$$

defines the accommodation coefficient a.

In many low-temperature applications the molecules may travel from one wall to the other without collision and it is impossible to identify T in (40) with either T_1 or T_2. Also the value of p used in eqn (46) must be considered very carefully, since at low pressures (under molecular or Knudsen conditions) the thermal transpiration or thermo-molecular pressure effect arises and the local pressure and temperature vary together as expressed by

$$p/(T)^{\frac{1}{2}} = \text{constant}.$$

Corruccini (1959) has examined recently this question of gaseous heat conduction at low pressures and low temperatures. He points out that p and T are associated in the derivation of (46), and that p is observed by a vacuum gauge usually at room temperature. Under low-pressure conditions $l \gg d$, the diameter of the tube connecting the gauge to the cryostat chamber, and if the system is in equilibrium, $p/(T)^{\frac{1}{2}}$ is constant. Therefore, if a value of p obtained from the gauge is inserted in (46), the appropriate value for T is the temperature at the pressure gauge.

Equation (46) may be simplified to

$$\dot{Q} = 0{\cdot}243\frac{\gamma+1}{\gamma-1}a_0\frac{T_2-T_1}{\sqrt{(MT)}}p_{\text{mm}}\ \text{W cm}^{-2}, \tag{47}$$

and using $T = 295°\,\text{K}$,

$$\dot{Q} = 0{\cdot}014\frac{\gamma+1}{\gamma-1}a_0\frac{T_2-T_1}{\sqrt{M}}p_{\text{mm}}$$

$$= \text{constant}\ a_0 . p_{\text{mm}}(T_2-T_1)\ \text{W cm}^{-2}, \tag{48}$$

where the constant has approximate values of 0·028, 0·059, and 0·016 for helium, hydrogen, and air respectively. The chief uncertainty in calculating the heat conducted by a low-pressure gas lies in the accommodation coefficient a. As an upper limit $a = 1$, but experimental research has shown that with a clean metallic surface exposed to helium gas, a may be as low as 0·025. Keesom (1942) has given a number of values for a for helium gas obtained by measuring the heat loss from a wire stretched along the axis of a tube. Some of the values found (original references may be found in Keesom, 1942) are given in Table XII below. The experimental figures suggest that for a metal surface in the

TABLE XII

Accommodation coefficient for helium gas

Metal	
Platinum	0·49 (90° K, 153° K); 0·38 (34°–264° C)
Bright platinum	0·44 (50°–150° C)
Blackened platinum	0·91 (50°–150° C)
Clean fresh tungsten	0·025 (79° K); 0·046 (195° K); 0·057 (22° C)
Gas-filled tungsten	0·19 to 0·82
Gas-free nickel	0·048 (90° K); 0·060 (195° K); 0·071 (273° K)
Nickel (gas layer adsorbed)	0·413 (90° K); 0·423 (195° K); 0·360 (273° K)
Glass	0·67 (12° K); 0·38 (77° K); 0·34 (273° K)

usual condition encountered in a cryostat, exposed to helium gas at low pressure, a value

$$a \leqslant 0{\cdot}5$$

should provide a useful upper limit for calculating heat transfer.

Available data for the accommodation coefficient of other gases on various metallic surfaces seem rather meagre. Values quoted in the *International critical tables* (1929) for the accommodation coefficients at about room temperature for He, H_2, air, N_2, O_2, and argon at a platinum surface range from 0·2 to 0·9.

More recent experimental work on accommodation coefficients (for example, Bremner 1950; Eggleton, Tompkins, and Wanford 1952; Schäfer 1952; Thomas and Schofield 1955) deals principally with surfaces of tungsten which are clean and gas free. The results, of fundamental rather than practical cryogenic interest, serve to confirm some of the experimental disagreements and the considerable experimental difficulties that are faced in these determinations. In contrast to Roberts (1930), Thomas and Schofield find very little change in accommodation coefficient with temperature for helium on a clean gas-free tungsten filament; from 80° to 300° K they find a has a value of 0·015–0·017 but increases to ~ 0·2 when the surface is contaminated by gas (oxygen). Eggleton, Tompkins, and Wanford report values of $a \simeq 0{\cdot}056$ after flashing a tungsten wire, but find $a \simeq 0{\cdot}3$ when the surface has adsorbed gas layers on it.

Generally for clean gas-free surfaces, a increases with molecular weight: for example, from Bremner (1950) for flashed tungsten at 90° K, $a = 0{\cdot}041$ for He, 0·081 for Ne, 0·16 for A, 0·09 for H_2, 0·20 for O_2. For heavy polyatomic molecules (e.g. organic gases) it appears that generally $0{\cdot}8 < a < 0{\cdot}9$.

3. Heat transfer through solids

As we mentioned in the last chapter, the heat flow $\dot{Q}$ through a solid of cross-section A cm² under a temperature gradient $\partial T/\partial x$ is given by

$$\dot{Q} = \lambda(T)A\frac{\partial T}{\partial x}. \tag{49}$$

Thus, if the ends of a solid bar of uniform cross-section and length l are at temperatures T_1 and T_2,

$$\dot{Q} = \frac{A}{l}\int_{T_1}^{T_2} \lambda(T)\,dT. \tag{50}$$

$\lambda(T)$ is the temperature-dependent thermal conductivity of the solid, of which some typical examples are shown in the graphs of Fig. 102. In Chapter XII are summarized experimental data (see also Table J in Appendix) of the thermal conductivity of the various typical groups of solids: (i) glasses, (ii) metallic alloys, (iii) pure metallic elements, and (iv) crystalline dielectric solids.

In the case of metallic alloys, e.g. brass, German silver, monel, stainless steel, available data enable us to give a fairly accurate estimate of $\lambda(T)$ at any temperature and therefore of the integrated heat conductivity $\int \lambda(T)\,dT$, provided that the alloy composition is not markedly different from that of alloys already investigated and that we know the physical state, e.g. strained or annealed. In the case of pure metallic elements, however, the conductivity at low temperature is very sensitive to small traces

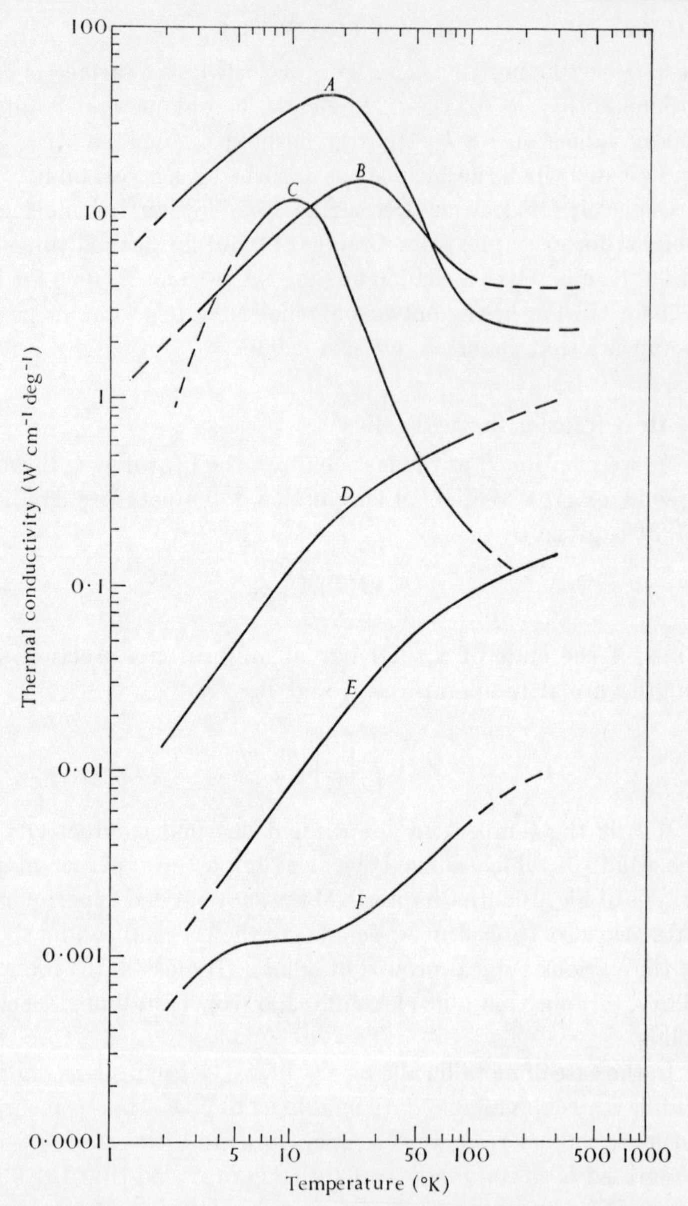

FIG. 102. Thermal conductivity of some solids, from data in Powell and Blanpied (1954). *A*, a metallic element, Al (high purity); *B*, a metallic element, Cu (electrolytic tough pitch); *C*, a dielectric crystal, quartz; *D*, an alloy, annealed brass (70 Cu, 30 Zn); *E*, an alloy, stainless steel; *F*, a glass, vitreous silica.

of chemical impurities and physical defects; but as discussed in Chapter XII, a comparatively simple measurement of the residual electrical resistance (normally at about 4·2° K) enables a good estimate of the heat conductivity to be made.

The knowledge most often needed for practical cryogenic calculations is the effective heat conductance of a solid bar of a few common materials, with certain end temperatures. Commonly encountered pairs of end temperatures may be 300° K (room temperature) and 77° K, 300° and 4·2° K, 77° and 20° K, 77° and 4·2° K, 4·2° and 2° K. In Table XIII are given values for the mean heat conductivity,

$$\bar{\lambda} = (T_1 - T_2)^{-1} \int_{T_1}^{T_2} \lambda(T)\, dT,$$

under these temperature conditions for a number of common materials, e.g. Pyrex glass, stainless steel, inconel (hard-drawn), monel (annealed), German silver, constantan, brass, phosphorus deoxidized copper (representing the material from which many copper items are frequently made: pipe, tube, and some rod and bar), electrolytic tough pitch copper (representing the type of copper frequently used in commercially available spools of wire). The values in the table, with the exception of those on coppers and brass, were calculated from the data compiled by Powell and Blanpied (1954). The values for the copper specimens were based on measurements by Powell, Rogers, and Roder (1957) and those for brass were taken from Kemp, Klemens, Tainsh, and White (1957).

It is well to note that values for $\bar{\lambda}$ for most commercial glasses appear to be close (within 30 per cent) to those given for Pyrex glass in the table. Similarly, values for manganin approximate to the constantan values given. Values for annealed inconel and annealed K-monel will approximate to those for annealed monel, and values for hard-drawn monel and K-monel will be not more than about 20 per cent greater than those given for hard-drawn inconel.

For comparatively pure metallic elements such as commercial copper, small variations in chemical or physical purity may

TABLE XIII

Mean values of thermal conductivity expressed in W/cm degK

	$\bar{\lambda}$ $T_2 = 300°$ K $T_1 = 77°$ K	$\bar{\lambda}$ $T_2 = 300°$ K $T_1 = 20°$ K	$\bar{\lambda}$ $T_2 = 300°$ K $T_1 = 4°$ K	$\bar{\lambda}$ $T_2 = 77°$ K $T_1 = 20°$ K	$\bar{\lambda}$ $T_2 = 77°$ K $T_1 = 4°$ K	$\bar{\lambda}$ $T_2 = 20°$ K $T_1 = 4°$ K	$\bar{\lambda}$ $T_2 = 4°$ K $T_1 = 2°$ K
Pyrex glass	0·0082	0·0071	0·0068	0·0028	0·0025	0·0012	0·0007
Stainless steel†	0·123	0·109	0·103	0·055	0·045	$0{\cdot}009_7$	0·0022
Inconel (*ca.* 72 Ni, 14–17 Cr, 6–10 Fe, 0·1 C) hard-drawn	0·125	0·111	0·106	0·061	0·051	0·012	0·003
Monel (*c.* 66 Ni, 2 Fe, 2 Mn, 30 Cu) annealed	0·207	0·192	0·183	0·133	0·113	0·040	0·007
German silver (47 Cu, 41 Zn, 9 Ni, 2 Pb) as received	0·20	0·19	0·18	0·14	0·12	0·039	0·005
Constantan (60 Cu, 40 Ni) wire as received	0·22	0·21	0·20	0·16	0·14	$0{\cdot}04_6$	0·006
Brass (30 Zn, 70 Cu) as received	0·81	0·70	0·67	0·31	0·26	0·078	0·015
Copper (phosphorus deoxidized) as received	1·91	1·71	1·63	0·95	0·80	0·25	0·07
Copper (electrolytic tough pitch) as received	4·1	5·4	5·7	9·7	9·8	10	4

† These figures for stainless steel are calculated from thermal conductivity data which are representative of the behaviour of types 303, 304, 347, and therefore an appropriate composition could be 18% Cr, 9% Ni, traces of Mn, Nb, Si, Ti totalling 2·3% with remainder Fe.

affect the conductivity considerably, particularly at low temperatures. This is discussed more fully in Chapter XII. Data for $\lambda(T)$ of materials such as nylon, Teflon, Perspex, soft solder, Wood's metal, and a silicon bronze† are also given in tabular form in Table J.

The values in this table make it evident why alloys of the cupro-nickel family (copper-nickel, constantan, German silver) or of the monel and stainless-steel group are so frequently chosen for the tubes in a cryostat where low thermal conductivity is a requirement. Stainless steel and inconel have somewhat lower heat conductivities than the cupro-nickel alloys although the latter are often preferred due to the greater ease with which they may be soft-soldered.

4. Heat transfer by radiation

A perfect black body may be defined as one which absorbs all radiation falling upon it. For such a body the absorptivity a and emissivity ε are unity, and so its reflectivity

$$R = 1-\varepsilon = 1-a$$

is zero.

It may be shown (see, for example, Roberts 1940) that for a black body at a temperature T the total radiant energy emitted per second per unit area is given by

$$E = \sigma T^4,$$

where the constant σ, Stefan's constant, has an experimental value of $5{\cdot}67\times10^{-12}$ W cm^{-2} deg^{-4}. Such radiant energy is distributed over a range of wavelengths, the energy $E(\lambda)\,d\lambda$ emitted over a narrow interval λ, $\lambda+d\lambda$ being a function of λ and T. The function $E(\lambda)$ at any temperature T has a maximum value for $\lambda = \lambda_m$, and it may be shown that

$$\lambda_m T = \text{constant} \quad \text{(Wien's constant)}$$

for which the experimental value is 0·290 cm deg.

† A useful low-conductivity alloy which is non-magnetic at low temperatures; contains 96% Cu, 3% Si, 1% Mn.

Thus for $T = 300°\,\mathrm{K}, \quad \lambda_m = 9{\cdot}67\,\mu\mathrm{m},$

$$T = 200°\,\mathrm{K}, \quad \lambda_m = 14{\cdot}5\,\mu\mathrm{m},$$
$$T = 77°\,\mathrm{K}, \quad \lambda_m = 37{\cdot}7\,\mu\mathrm{m},$$
$$T = 4{\cdot}2°\,\mathrm{K}, \quad \lambda_m = 690\,\mu\mathrm{m}.$$

Most calculations dealing with the radiant heat transfer in low-temperature equipment are not restricted to black bodies but deal with metallic surfaces whose emissivity may be anywhere between 0·01 and 1·0.

Most non-metallic surfaces, of which glass and perhaps baked varnishes are the most important from our viewpoint, do approximate to black bodies in that their emissivities are in the neighbourhood of 0·9. However, for a particular metallic conductor the emissivity or the reflectivity depends on the wavelength of incident radiation and the physical state of the surface.

The classical theory of Drude yields a relation between the reflectivity R, the wavelength λ (microns), and the d.c. electrical resistivity of the metal ρ (Ω cm):

$$1-R = \varepsilon = 36{\cdot}5(\rho/\lambda)^{\frac{1}{2}}.$$

Evidence for the experimental validity of this formula has been discussed by Worthing (1941), Blackman, Egerton, and Truter (1948), Reuter and Sondheimer (1948), and Ramanathan (1952), etc. At relatively high temperatures the experimental agreement is reasonably good in many instances.

At low temperatures, however, the theoretical relation might be expected to fail even with smooth uncontaminated metallic surfaces, as the relaxation time of the electrons (period of the electron mean free path) becomes comparable with the period of vibration of the incident radiation. The discrepancy is still very marked for infra-red radiation at low temperatures, and it is difficult to believe that surface roughness or a surface electrical resistance higher than the bulk resistance is sufficient to explain discrepancies of a factor of 100 or more which occur with annealed electro-polished specimens at helium temperatures. Reuter and Sondheimer (1948; see also review on electron mean free paths in metals by Sondheimer 1952) treated theoretically the problem of the anomalous skin effect, i.e. the effect on surface resistance when electron mean free paths become so long that at high frequencies the applied electric field may change appreciably within the extent of an electron mean free path. They assumed specular reflection of electrons at the metal surface and deduced a formula for the reflectivity or absorptivity in considerably better agreement with low-temperature experiments (e.g. Ramanathan 1952). Dingle (1952, 1953) deduced a slightly simpler expression and by assuming diffuse reflection showed that reasonable agreement with experiment could be obtained. In Table XIV (Dingle 1953*b*) are compared experimental values of the *percentage* absorptivity for 14 μm radiation on electro-polished copper, with theoretical values.

TABLE XIV

Temperature	*Experimental*	*Classical theory (incl. relaxation)*	*Anomalous skin effect (diffuse reflection)*
Room temperature	1·2	0·5	0·8
Liquid-oxygen temperature	0·8	0·09	0·5
Liquid-helium temperature	0·6	0·003	0·4

Returning to our practical problem of calculating the approximate radiant heat transfer between surfaces, it is apparent that despite the partial success of more sophisticated theoretical analyses, it is generally necessary to select experimental data which may be considered most appropriate for the particular surfaces in question. In Table XV below are collected experimental values for the emissivity of a number of commonly used metals; as indicated in the table these values have been obtained on surfaces in various physical conditions, e.g. electro-polished, normally smooth and clean, highly oxidized, etc. By some intelligent guessing, the most suitable value of ε, or a probable upper limit for ε, must then be selected for calculation of heat transfer. Values of ε for these and other materials are listed in the *American Institute of Physics handbook*, pp. 6–153 (McGraw Hill, New York, 1963).

For two plane parallel surfaces each of area A, and emissivities ε_1 and ε_2, and at respective temperatures T_1 and T_2, the heat transfer by radiation per unit time is

$$\dot{Q} = \sigma A(T_1^4 - T_2^4)\frac{\varepsilon_1 \varepsilon_2}{\varepsilon_1 + \varepsilon_2 - \varepsilon_1 \varepsilon_2} \tag{51}$$

whence $\dot{Q} \simeq \sigma A(T_1^4 - T_2^4)$ for $\varepsilon_1 = \varepsilon_2 \simeq 1$.

In the case where $\varepsilon_2 \ll \varepsilon_1$, (51) reduces to

$$\dot{Q} = \sigma A(T_1^4 - T_2^4)\varepsilon_2. \tag{52}$$

Similarly, if $\varepsilon_1 = \varepsilon_2 = \varepsilon$, and $\varepsilon \ll 1$, (51) becomes

$$\dot{Q} = \sigma A(T_1^4 - T_2^4)\varepsilon/2.$$

TABLE XV

Experimental values of emissivity

Material	*Fulk, Reynolds, and Park* (1955), 300° K *radiation on* 78° K *surface*	*McAdams* (1954), *room temperature*	*Ramanathan* (1952), 14μm *radiation on* 2° K *surface*	*Blackman, Egerton, and Truter* (1948), 293° K *radiation on* 90° K *surface*	*Ziegler and Cheung* (1957), 273° K *radiation on* 77° K *surface*
Al, clean polished foil	0·02	0·04	0·011†	0·055	0·043‡
Al, plate	0·03	—	—	—	—
Al, highly oxidized	—	0·31	—	—	—
Brass, clean polished	0·029	0·03	0·018†	0·046	0·10‡
Brass, highly oxidized	—	0·6	—	—	—
Cu, clean polished	0·015–0·019	0·02	0·0062–0·015†	0·019–0·035	—
Cu, highly oxidized	—	0·6	—	—	—
Cr, plate	0·08	0·08	—	0·065	0·084‡
Au, foil	0·010–0·023	0·02–0·03	—	0·026	—
Au, plate	0·026	—	—	—	—
Monel	—	0·2	—	—	0·11‡
Ni, polished	—	0·045	—	—	—
Rh, plate	0·078	—	—	—	—
Ag, plate	0·008	0·02–0·03	—	0·023–0·036	—
Stainless steel	0·048	0·074	—	—	—
Sn, clean foil	0·013	0·06	0·013†	0·038	—
Soft solder	0·03	—	—	—	0·047‡
Glass	—	0·9	—	0·87	—
Wood's metal	—	—	—	—	0·16

† These surfaces were electro-polished (Ramanathan).
‡ These surfaces were neither highly polished nor heavily oxidized, but as encountered in normal practice (Ziegler). Ziegler observed that a thin layer of oil or Apiezon grease on a low emissivity surface raised the emissivity to 0·2 or 0·3. He also found that varnishes such as GEC adhesive No. 7031 and bakelite lacquer gave an emissivity $e \simeq 0{\cdot}87$; similarly, Scotch tape (Sellotape) had an emissivity of about 0·88.

As discussed in Chapter II, a radiation shield is commonly used in cryostats and storage flasks to reduce the heat inflow and liquid evaporation. This shield is often a polished copper cylinder kept at a temperature intermediate between the ambient temperature and that of the experimental space being cooled by liquid air or by evaporating helium gas. It is useful to know how the temperature distribution on this shield varies with time, gas pressure, and emissivity. Dixon, Greig, and Hoare (1964) have measured distributions of temperature on a cylindrical shield made of phosphorus deoxidized copper, attached to a liquid-air chamber. Their shield of 0·16-cm thick copper was 60 cm long and 8 cm in diameter. It approached thermal equilibrium with a time-constant of about 1 h. After 3 h the temperature-difference along the length was *ca.* 15 degK for polished surfaces provided that the vacuum was better than *ca.* 10^{-4} mm Hg; in a poor vacuum of 5×10^{-3}, this amounted to nearly 40 degK. Oxidation of the surfaces increased it by a factor of about 2.

5. Other causes of heat transfer

While the processes of heat transfer discussed in the preceding sections, viz. low-pressure gas conduction, heat conduction by solids, and radiation transfer, are those chiefly encountered, other less common but quite troublesome sources of energy often arise.

Joule heating

Joule heating in connecting leads and in resistance thermometers, giving rise to a heat input $\dot{Q} = I^2R = V^2/R$, is not difficult to estimate but may be overlooked as a source of temperature drift or temperature inhomogeneity in a cryostat. The opposing demands of low thermal conduction along electrical leads and insignificant heat dissipation in these same leads often present a problem. At temperatures below 7° K this has frequently been solved by using wires of a poor thermal conductor such as constantan, and tinning the surface with a thin lead coating; this becomes superconducting below 7° K and the wire remains a poor heat conductor.

As is discussed in a later chapter (XII), most alloys (excluding superconducting alloys) exhibit an electrical resistance which does not decrease appreciably as the temperature falls from room temperature to liquid-helium temperatures, but their heat conductivity usually decreases by a factor of 10–100. On the other hand, pure metallic elements have an electrical resistance which may fall by a factor of 100 or more as the temperature is changed from 300°–4·2° K; over the same range the thermal conductivity of the pure metallic element normally increases as T falls below 100° K, passes through a maximum at a temperature in the vicinity of $\theta_D/20$ (say 10°–20° K), and falls linearly towards zero at the absolute zero. As a result a piece of copper wire may have approximately the same heat conductivity at 4° and 300° K while its electrical resistance is different by a factor of 100; a wire of constantan may have approximately the same electrical resistance at 4° and 300° K but a heat conductivity which has changed by a factor of 100. It may be noted that at both low temperatures ($\sim$ 4° K) and high temperatures ($\sim$ 300° K) the ratio $\rho\lambda/T$ for most metals has a value which approximates to the theoretical Lorenz value of $2{\cdot}45 \times 10^{-8}$ W Ω/deg^2.

With the development of superconducting magnets there is a need to carry relatively large currents (10–100 A) into cryostats and this has made it more necessary to find the optimum size of leads. McFee (1959) has calculated optimum dimensions for copper wires of normal electrical quality (resistivity ratio $\approx$ 100) and Mallon (1962) has made similar calculations for aluminium and sodium. The results for copper are illustrated in Fig. 103; for example, these show that if there is a continuous current through a lead (or leads) of length l and cross-section A going from room temperature into a liquid-helium bath the heat input $\dot{Q}$ is minimized if $lI/A = 5 \times 10^4$ A cm^{-1}. If $I = 20$ A, it follows that l/A should be $2{\cdot}5 \times 10^3$; therefore if $l = 25$ cm, A should be 10^{-2} cm^2 corresponding to either a single wire of No. 17 B and S or say 10 wires of No. 27 B. & S. $\dot{Q}_{\min}$ will be 0·8 W.

These authors point out that heat flow can be reduced considerably by changing the cross-section at one or more intermediate temperatures (say 77° K) and optimizing lead sizes for the different temperature-intervals. In practice the current does not usually flow continuously so that somewhat smaller diameters may be used at the expense of increasing the Joule heating (while current flows) but decreasing direct heat conduction; heat transfer to the evaporating helium gas can also

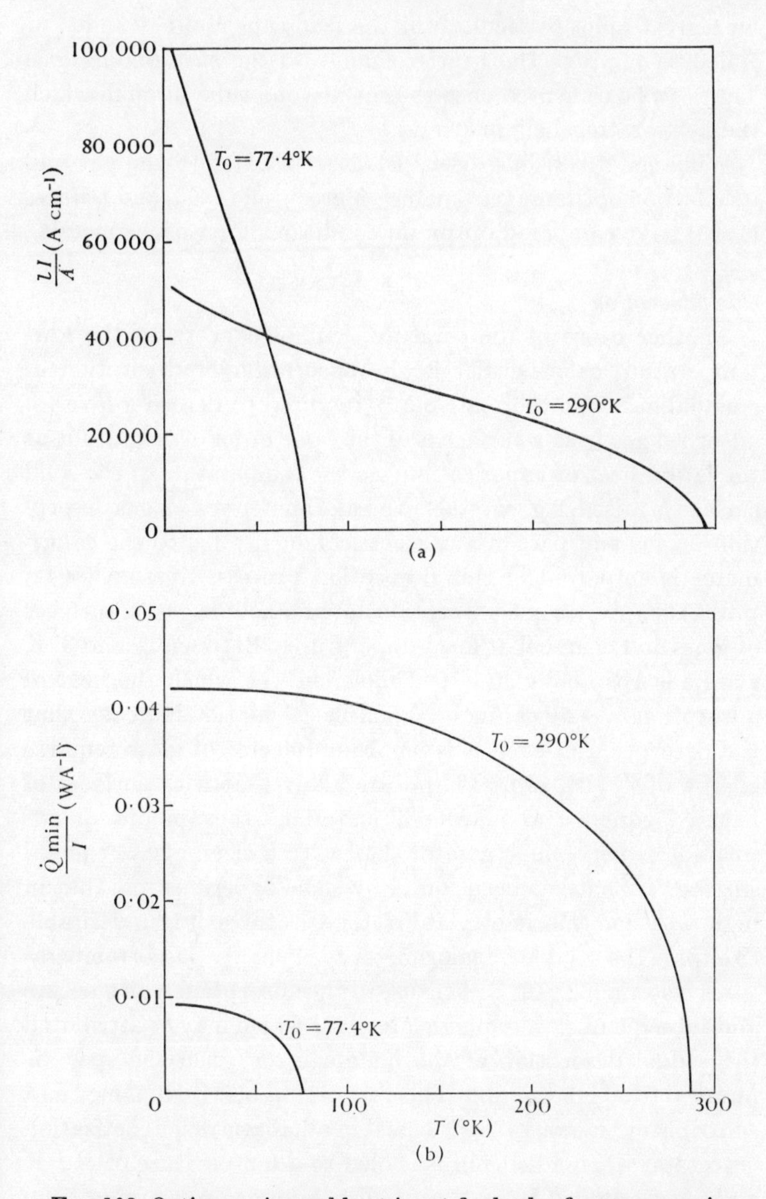

FIG. 103. Optimum size and heat input for leads of copper carrying current I from region at temperature T_0 to a lower temperature T. In (a) l/A is the ratio of length to cross-sectional area and in (b) $\dot{Q}$ is the heat input (after McFee 1959).

be a great help, particularly if the leads are multi-stranded or coiled to improve the heat exchange to the surrounding gas. They can be cemented onto the surface of a tube through which the evaporating helium passes.

Williams (1963) has discussed heat transfer to the gas and its effect on optimum parameters. Mercouroff (1963) and Deiness (1965) have compared optimum conditions for various metals.

Gas adsorption

Another cause of temperature drift of concern to the low-temperature calorimetrist is the adsorption or desorption of residual gas. The thermal energy required to desorb a layer of adsorbed gas from a surface is of the same order of magnitude as the latent heat of vaporization. As the temperature of the solid surface is raised, e.g. crystals in a calorimeter vessel, gas desorption begins and part of any electrical energy fed to the calorimeter is employed in this desorption process. Keesom (1942, p. 127) has given some adsorption data for helium gas on surfaces of glass and charcoal at low temperatures. Between 2° and 3° K glass adsorbs about 30×10^{-10} mole/cm² for which the heat of adsorption is $\sim$ 70 cal/mole. A simple calculation indicates that the desorption of such a layer from 100 cm² of glass requires energy of $\sim$ 1000 ergs. It appears likely that with surfaces of baked lacquers and powdered materials, the amount of adsorption is very much greater than with a clean glass or metal surface. Thus it has been found by some experimenters that in high-precision calorimetry at low temperatures, it is undesirable to expose the sample or calorimeter to exchange gas at temperatures below about 10° K, because of the adsorption which occurs and subsequent slow temperature drifts which may be attributed to gradual desorption of the helium layer when the space is pumped to a high vacuum. This process has also been named as a contributing factor to 'heat leaks' in adiabatic demagnetization cryostats; when a salt pill is cooled to a temperature of $\frac{1}{100}$° K and is surrounded by a 1° K wall, helium gas (remaining from exchange gas used in cooling) slowly desorbs and becomes adsorbed on the colder pill, transferring thermal energy in an amount which may cause a serious temperature rise, owing to the small heat capacities of the system.

The use of mechanical and superconducting heat switches has now become common in calorimeters and other cryostats requiring minimal heat leakages below 5° K. This avoids the need for helium 'exchange' gas and the annoying heat leaks which result from subsequent adsorption and desorption effects.

Helium film creep and thermal oscillations

Two annoying sources of heat can occur with liquid helium. One of these is due to the superfluid film and therefore is confined to liquid He II, that is to the normal isotope ^{4}He below the λ-point (2·17° K). The other arises from spontaneous oscillations which sometimes occur in columns of helium gas in narrow tubes.

When a dewar or chamber contains liquid He II, a mobile film covers the walls of the container up to a height where it is warm enough for it to evaporate. As it evaporates more helium flows up through this film so that the process leads to continuous loss of helium. The amount of this additional evaporation depends on the minimum perimeter of the walls covered by the film and the microscopic nature of the wall. For clean glass the flow rate (below 2° K) is *ca.* 7×10^{-5} cm^3 of liquid per s per cm of perimeter; this rate produces *ca.* 3 cm^3 of gas (n.t.p.) per min so that a mechanical pump of speed of 100 l/min would not be able to reduce the pressure below 3×10^{-5} atm $\approx$ 0·02 mm Hg. Clean metal surfaces have film transfer rates about three times larger than that of clean glass, due probably to microscopic roughness. In practice, glass or metal surfaces are often 'dirty' due to adsorbed layers of air and may show transfer rates ten times that of clean glass (Bowers and Mendelssohn 1950). The result is that in small chambers containing pumped liquid helium it is difficult to reduce the pressure below, say, 0·5 mm of mercury or about 1·2° K unless the film flow is reduced by having a small restriction in the pumping tube below the evaporation point (Ambler and Kurti 1952). This constriction may be 1 mm or even 0·1 mm in diameter but if too small it may be blocked more easily by impurities and make initial pumping-down intolerably slow. A suggested compromise is a 0·5-mm diam. orifice in the form of a piece of glass capillary tube a few millimetres long, sealed into the lower part of the pumping tube with epoxy resin.

Thermal oscillations in a tube (Keesom 1942, p. 174) leading down into a liquid-helium vessel can increase the evaporation rate by hundreds of cubic centimetres of liquid per hour. Their occurrence is rather unpredictable. Once they occur and are recognized they can be cured by altering the tube dimensions or inserting some form of damping (e.g. thread of cloth) in the tube. The necessary (but not always sufficient) conditions for these resonant oscillations to occur seem to be a diameter of less than 1 cm and a large temperature gradient with the bottom end of the tube being below *ca.* 20° K. Occasionally the oscillations themselves can be quite useful—witness the helium level detector produced by Clement and Gaffney, which was described above in § 2.3—although the associated inflow of energy is always a nuisance.

Mechanical vibration

In most cryogenic experiments gas desorption is not a very serious factor, nor is mechanical vibration. Quantitative information on the amount of heat released in an otherwise thermally isolated system by its being in mechanical vibration—a result of mechanical linkage to its surroundings, i.e. external vibrations of pumps, building vibration, etc.—is rather lacking. However, it has been noted frequently by workers in the field of temperature below 1° K, where heat inputs of 100 ergs/min may be considered excessive, that the rate of temperature rise of their salt pills after demagnetization was affected by the operation of pumps. In an interesting series of experiments on thermal contact and insulation below 1° K, Wheatley, Griffing, and Estle (1956) mechanically coupled their cryostat to a motor whose speed could be adjusted to vary the vibration frequency from 0 to 13 per s; they found at the maximum frequency a heat leakage to a pill (mounted on a rather rigid support) of

235 ergs/min, compared with about 10 ergs/min when the motor was switched off. It is apparent that a fairly rigid suspension, e.g. glass rods as opposed to nylon or cotton threads, is less susceptible to external vibrations as its resonant frequency is much higher than the frequency of most normal building or pump vibrations. A salt pill suspended by fine threads is much more likely to be set in resonant vibration by the action of external sources, and heat leaks of the order of 1000 ergs/min have been encountered in some such instances.

Darby *et al.* (1951) reported that in their two-stage demagnetization equipment, in which the two pills were suspended by nylon threads, vacuum pumps produced a serious vibrational heat source; even the action of mercury 'bumping' in a diffusion pump attached to the cryostat caused 'heat leaks' in excess of 300 ergs/min to the upper salt pill. On the other hand, Malaker (1951) used a nylon thread suspension and calculated the heat leak to be only $\sim$ 4 ergs/min at very low temperatures.

An excellent discussion of the problems of thermal insulation at very low temperatures was given some years ago by Cooke and Hull (1942).

Inadequate 'anchoring' of leads, etc.

This common source of heat inflow is really just another case of conduction by solids. Any lead wires, control rods, etc., which offer a flow-path for heat from room temperature to the experimental space need to be connected to a heat sink or guard-ring before reaching the experiment. If this thermal connexion is not adequate then annoying heating effects can arise. In calorimetry, heat conductivity measurements and in most experiments below 1° K, it is most important to guard against these heating effects by careful bonding of the leads to a surface which is at the temperature of the liquid bath. For electrical reasons it is impractical in most cases to solder the leads down: they must be cemented so as to preserve electrical insulation. If the wires are silk- or cotton-covered, a considerable length (e.g. 30 or 40 cm) should be tightly wound onto a copper post (high-conductivity copper) or cylinder which forms the heat sink and then cemented with a varnish such as Glyptal, G.E. 7031, Duco cement, etc. If the wires have only a coating of Formvar, Bicalex, etc., and no silk or cotton, then the heat transfer will be better and a shorter length need be anchored.

Alternatively metallized ceramic materials can be used as heat sinks, one side being soldered to the metal wall which is in contact with the bath and the other side having the lead soldered to it. Tests have been made on metallized discs of alumina, beryllia, sapphire, which show how effective this method is (Dahl 1963) although it may be impractical in cases needing 10 or 20 leads.

Of course the most effective way of connecting the leads to the bath is to take them through a vacuum-seal into the refrigerant liquid but this demands a multi-lead seal which is high-vacuum-tight in liquid helium. If these are metal-to-glass seals with say 20 leads they are usually rather bulky. However compact seals have been made with the leads through nylon bushes sealed onto thin-walled metal tubes with epoxy resin (see § 10.4).

6. Example of heat-transfer calculation

As an illustration of the calculations that may be necessary in cryostat design, consider the magnitudes of the heat inflows from various sources in the following example (see Fig. 104): A polished copper inner chamber of surface area 500 cm^2 is supported by a German silver tube (2·0-cm diam., 0·3-mm wall thickness) inside a tarnished brass vacuum chamber in which the measured pressure (of helium gas) is 10^{-5} mm; the length of the tube separating the two chambers is 6 cm. Twelve electrical leads (8 leads of 38 B. & S. gauge copper and 4 of 32 B. & S. constantan) enter the outer chamber through the pumping tube and are effectively thermally anchored on a copper bush at 77° K; the leads are then taken to the inner chamber, each lead having a length of about 12 cm between its points of attachment on the respective chambers.

We require to calculate the following.

(i) Radiant heat inflow from the brass to the copper chamber.

(ii) Radiant heat inflow down the German silver tube to the inner chamber when no radiation baffle is present.

(iii) Heat conducted down the German silver tube from the outer to the inner chamber.

(iv) Heat conducted down the electrical leads from one chamber to the other.

(v) Heat conducted through the low-pressure helium gas in the inter-chamber space.

(vi) And to examine whether Joule heating due to a current of 5 mA in the electrical leads is serious in view of the other heat inflows.

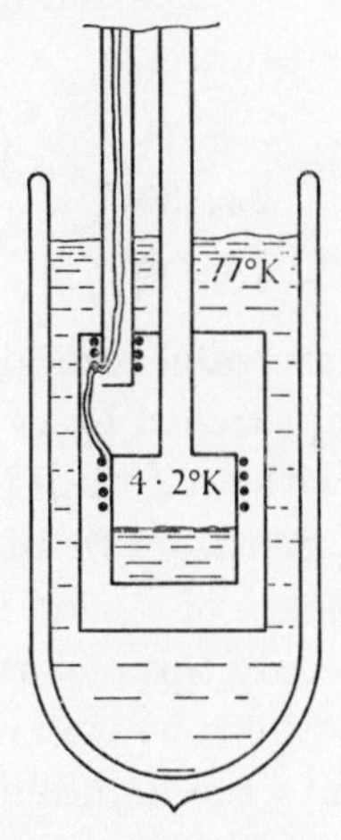

FIG. 104. Diagram of a cryostat.

First,

(i) assuming that for tarnished brass $\varepsilon_1 \to 1$, and for polished copper $\varepsilon_2 \simeq 0{\cdot}03$,

from (52)
$$\dot{Q}_r = \sigma A(T_1^4 - T_2^4)0{\cdot}03$$
$$\simeq 5{\cdot}67 \times 10^{-12} \times 500 \times 0{\cdot}03 \times (77)^4$$
$$= 3 \times 10^{-3}\ \mathrm{W}.$$

(ii) Any calculation of the radiant heat reaching the inner space by 'funnelling' of room temperature radiation down the German silver tube must yield a very crude approximation to the true experimental result, at best. An upper limit $\dot{Q}_f(\max)$ may be calculated easily, assuming 'complete funnelling', i.e. perfect specular reflection from the inner wall of the tube, so that all the radiation $\dot{Q}(=\sigma\pi 1^2(295)^4)$ reaches the inner space.

$$\dot{Q}_f(\text{amx}) = 5{\cdot}67\times10^{-12}\times\pi\times(295)^4$$
$$= 0{\cdot}135 \text{ W}.$$

Alternatively, if we assume almost complete absorption of the radiation which reaches the inner wall of the tube, $\dot{Q}_f(\min)$ is merely the radiation from a small solid angle subtended by the lower exit of the tube at its upper (room temperature) end. If the length of the tube be about 30 cm between 4·2° K end and that part at room temperature

$$\dot{Q}_f(\min) \simeq \frac{\pi.1^1}{2\pi.(30)^2}\times\sigma A(295)^4$$
$$\sim 10^{-4}\,\text{W}.$$

In practice this minimum value for radiant heat inflow via the pumping tube may be approached by painting the inner wall of the German silver tube with an optical black paint, 'Aquadag', etc. More complete isolation from room-temperature radiation is usually obtained by means of a bend in the tube or insertion of a small radiation baffle. By this means the room-temperature radiation is adsorbed by the tube wall near the bend or at the baffle and transmitted into the liquid-nitrogen bath; the inner experimental space then only sees radiation emitted from a relatively small area of surface at 77° K which is transmitted by a direct path or successive internal reflections down the tube.

(iii) For the heat conducted by the German silver tube

$$\dot{Q}_c = \bar{\lambda}\frac{A}{l}\Delta T$$
$$= 0{\cdot}12\times\frac{2\pi\times0{\cdot}03\times73}{6} \quad (\bar{\lambda} = 0{\cdot}12 \text{ from Table XIII})$$
$$= 0{\cdot}275 \text{ W}.$$

(iv) The heat conducted down the electrical leads is given by $\dot{Q}_c^* = \dot{Q}$ (8 copper leads of 0·010-cm diam.) $+\,\dot{Q}$ (4 constantan leads of 0·020-cm diam.); using data from Table XIII, $\bar{\lambda}$ (copper) = 9·8, $\bar{\lambda}$ (constantan) = 0·14, therefore

$$\dot{Q}_c^* = 8\times 9{\cdot}8\times\frac{0{\cdot}786}{12}\times 10^{-4}\times 73+4\times 0{\cdot}14\times\frac{0{\cdot}314}{12}\times 10^{-3}\times 73$$

$$= 6{\cdot}08(0{\cdot}00616+0{\cdot}000176)$$

$$= 0{\cdot}0385\text{ W}.$$

(v) For gas conduction $\dot{Q}_g$ is given by (48), which in the case of helium gas reduces to

$$\dot{Q}_g = 0{\cdot}028 p_{\text{mm}}\, a_0\, \Delta T\text{ W cm}^{-2}.$$

Assuming that the respective surface areas are approximately equal and parallel, and that

$$a_{\text{max}} \simeq 0{\cdot}5 = a_1 = a_2,$$

then $$a_0 = 0{\cdot}5/(2-0{\cdot}5)$$

$$= 0{\cdot}333.$$

Since $\Delta T = 73$ degK, $p = 10^{-5}$ mm,

therefore $$\dot{Q}_g = 0{\cdot}028\times 0{\cdot}333\times 73\times 10^{-5}\text{ W cm}^{-2},$$

and since $$A \simeq 500\text{ cm}^2,$$

$$\dot{Q}_g = 0{\cdot}0034\text{ W}.$$

(vi) (*a*) Eight copper leads of 38 B. & S. wire each have an approximate room-temperature resistance of 0·024 Ω/cm. Hence, assuming for normal commercial wire that $\rho_r/\rho_{0\,°C} \simeq 10^{-2}$, the resistance will be about 0·0046 Ω/cm at 77° K and 0·0003 Ω/cm at 4·2° K.

If the mean resistance is 0·0025 Ω/cm, the total heat produced by 5 mA is

$$(5\times 10^{-3})^2\times 0{\cdot}0025\times 12 = 0{\cdot}75\times 10^{-6}\text{ W per wire.}$$

A rather crude calculation based on a mean thermal conductivity of 10 W/cm degK shows that the temperature rise in the wire will be $< 0{\cdot}01$ degK at any point. If all the Joule heat were transferred to the inner chamber this would only amount to about 6 μW.

(*b*) In the case of constantan (4 wires of 32 B. & S.) the electrical resistance of each wire is about 0·14 Ω/cm and is not particularly sensitive to change in temperature.

Total Joule heat produced is

$$4\times(5\times10^{-3})^2\times12\times0{\cdot}14 = 1{\cdot}68\times10^{-4}\ \mathrm{W}.$$

Again, assuming $\bar{\lambda}\simeq 0{\cdot}14$ W/cm deg, it may be calculated that the temperature rise will not be greater than about 5 degK.

Summarizing, we note that the major source of heat leakage is via the supporting tube, and provided that precautions are taken to prevent room-temperature radiation entering the inner chamber via the tube this leak amounts to about 0·275 W. Radiant heat of 10^{-3} W reaches the inner chamber from the surrounding 77° K chamber and heat leakage through the residual 10^{-5} mm pressure of helium gas is less than 0·004 W; by comparison a 5 mA current through each electrical lead could contribute a maximum heat inflow of about 100 μW.

The total heat leak of about 0·28 W is equivalent to 240 cal/h, which would evaporate nearly 400 cm³ of liquid helium per hour; this assumes that the cold evaporating gas does not play any useful role in cooling the wall of the German silver tube and thereby reduce the heat leak.

7. Heat transfer through pressed contacts

Elsewhere we make passing reference to the problem of heat transfer across the boundary between two solid surfaces in contact. This has a direct application to at least two practical cryogenic problems, namely to the design of mechanical heat switches (for example, as applied to calorimetry by Westrum, Hatcher, and Osborne 1953; Webb and Wilks 1955; Ramanathan and Srinivasan 1955; Rayne 1956) and in the design of insulating supports for low-temperature equipment (as applied to the internal supporting members in large dewar vessels by Birmingham *et al.* 1955).

The first comprehensive investigation of thermal contact has been that of Berman† (1956) who studied the thermal

† In this paper Berman includes a brief account of some earlier work in this field by Jacobs and Starr, Fulton, Zavaritski, etc.

conductance of various solid contacts at liquid-helium and liquid-nitrogen temperatures, using loads of 50–250 lb. His results suggest that:

(i) The measured thermal conductance is always greater than that calculated by the Wiedemann–Franz–Lorenz law from the measured electrical conductance. This discrepancy, often a factor of 100 or more, is as high as 10^5 when contact and measurement are made at 4·2° K. The only possible conclusion seems to be that the majority of the heat is carried across the interface by thermal waves rather than by electrons.

(ii) The thermal conductance varies nearly linearly with the pressure, and hence is not sensitive to change in area for a given total load.

(iii) At liquid-helium temperatures (1°–4° K), the conductance varies as T^2 but becomes less temperature-sensitive at higher temperatures, the change in conductance between 64° and 77° K being only about 10 per cent.

(iv) Some typical figures given by Berman for the thermal conductance of the contact between (*a*) two copper rods, (*b*) two 0·001 in steel disks are:

(*a*) $1{\cdot}02 \times 10^{-2}$ W deg^{-1} at 4·2°K; $32{\cdot}5 \times 10^{-2}$ W deg^{-1} at 77°K;

(*b*) $0{\cdot}54 \times 10^{-2}$ W deg^{-1} at 4·2°K; 26×10^{-2} W deg^{-1} at 77°K;

in each case the contact was made at room temperature and the load applied was 100 lb.

Later measurements on metals by Berman and Mate (1958) were in substantial agreement but showed the conductivity to vary a little more slowly with temperature, as $T^{1\cdot 3}$ for copper–copper and gold–gold. For sapphire–sapphire contacts under 100-lb load they found $\dot{Q} \approx 10^{-5} T^3$ W degK^{-1} at liquid-helium temperatures. The most remarkable feature from a practical viewpoint was that the magnitude of the heat conductance for gold–gold contacts was about twenty times better than copper–copper whereas sapphire–sapphire was at least ten times worse (below 10° K). Therefore for good heat transfer in a mechanical switch, the contacts should be gold-plated. Conversely to achieve good insulation in a support system, hard ceramic contacts should be used; for example, a sapphire ball resting on a sapphire plate.

Mate (1965) has done further measurements on the conductance of dielectric crystals and powders pressed in contact at low temperature.

The experimental work on heat conduction through insulating supports by Mikesell and Scott (1956) is of less fundamental interest than Berman's experiments but of great practical significance in cryogenic design. Their work was concerned with the internal support of large dewar vessels by using stacks of thin disks. Although the mechanical strength of the stack is great, the thermal contact resistance between disks is such that the total thermal resistance of a stack may be a hundred times greater than that of a solid rod of the same material and dimensions. For example, measurements of the thermal conductance of a stack of about 315 stainless-steel plates of thickness 0·0008-in under a pressure of 1000 lb/in^2 showed that: with end temperatures of 296° and 76° K, the conductance is 0·93 W cm^{-2} and with end temperatures of 76° and 20° K it is 0·062 W cm^{-2}; assuming the length of the stack is

$$315 \times 0{\cdot}0008\,\text{in} = 0{\cdot}25\,\text{in},$$

these figures correspond to effective mean conductivities of 2·7 mW/cm degK and 0·71 mW/cm degK respectively. These may be compared with values of $\bar{\lambda} = 123$ mW/cm degK (300°–77° K) and $\bar{\lambda} = 55$ mW/cm degK (77°–20° K) from Table XIII (§ 7.3) for stainless steel in solid form. Mikesell and Scott found that the conductance of the stacks of disks could be decreased substantially by using a thin layer of manganese dioxide dust between each pair of plates.

Such stacked disks provide an excellent means of transmitting high pressures into a cryostat, as a stack can have very high compressional strength, and yet retain a low heat conductivity (see, for example, their use by Berman 1956).

Experiments on contact resistance also confirm the reasons for the frequent difficulties that are encountered in thermally anchoring electrical leads effectively at low temperatures. When wires are brought down through a pumping tube into a cryostat, they are often wrapped around a copper post or tube which is in close thermal contact with the refrigerant; despite this, they often seem to be the source of heat inflow into a specimen (calorimeter, etc.) to which they are connected and which is otherwise isolated. The only method of avoiding this trouble appears to be to wrap as great a length of the wire as possible in close contact with the anchoring pillar and then to cement it firmly in place; glyptal, bakelite varnish, Formel varnish, or nail polish are all useful cements for this purpose.

Although the heat transfer between liquid helium and a solid is not properly a problem of pressed contacts, it seems appropriate to mention here some experimental values obtained for the flow of heat $\dot{Q}$ across a liquid–solid interface. As part of an investigation of the heat conductivity of liquid helium below 1° K, Fairbank and Wilks (1955) observed that between copper and the liquid helium, $\dot{Q}$ was proportional to the temperature gradient and varied nearly as the square of the temperature; in fact

$$\dot{Q} = 2{\cdot}20 \times 10^{-2} T^2 \text{ W cm}^{-2} \text{deg}^{-1}.$$

Further measurements on heat transfer between solids and liquid He II have been made (e.g. Challis, Dransfeld, and Wilks 1961) as the origin and extent of the boundary resistance are both of fundamental interest and of practical concern.

The studies of boundary resistance at liquid-helium temperatures have been made for lead–copper and tin–copper interfaces prepared by casting the lead or tin *in vacuo* onto cleaned copper (Barnes and Dillinger 1966). When the metals are each in the normal state, the thermal resistance is extremely small, less than 0·1 deg cm^2/W, but when one is in the superconducting state there is appreciable resistance. The authors showed that a Pb–Cu–Pb–Cu sandwich could therefore act as a useful thermal switch.

8. Heat switches

General

The simplest and most commonly used heat 'switch' is still helium exchange gas. By changing the pressure of gas in the space between two walls from 10^{-1} mm Hg to 10^{-5} mm Hg, the heat transfer between them is altered by many orders of magnitude. The chief drawbacks to this 'switch' are that (i) it takes a long time to turn it 'off' completely, particularly at liquid-helium temperatures, (ii) some gas always remains adsorbed on a surface and may be desorbed later and upset the energy balance, and (iii) at very low temperatures there is no gas left; even ^{3}He has negligible vapour pressure below 0·1° K. Therefore various alternative switches have been evolved, viz.:

(i) Mechanical switches which can be turned 'on' and 'off' easily but which generate some heat in the switching and are therefore most suitable above *ca.* 0·5° K.

(ii) Superconducting 'switches' for which the ratio of heat conductance in the 'on' position to the conductance in the 'off' position only becomes sufficiently high at temperatures below $\sim$ 1° K.

(iii) Graphite 'switches' or links which represent a particular case of materials whose thermal conductivity varies rapidly with temperature; it is low enough at $\sim$ 1° K to act as a thermal insulator but high enough at 10° K to conduct away thermal energy at an appreciable rate.

(iv) Various other devices which make use of helium gas or liquid helium in tubes and can be turned 'on' or 'off' without too much delay.

Mechanical switches

These have become the accepted method of cooling calorimeters in the temperature region from 1° to 10° K; below this the heat evolved on breaking the contact becomes relatively large. The first type of mechanical switch described by Keesom and Kok (1933) and Ruehrwein and Huffman (1943) used the mating of conical surfaces for heat transfer. However, on breaking, these produce rather a lot of frictional heating and have not been used much at liquid-helium temperature. Two designs which are used down to 1° K are:

(i) the jaw-type based on that of Webb and Wilks (1955) and refined by Manchester (1959) and Phillips (1959);

(ii) the plate-type first used by Ramanathan and Srinivasan (1955) and Rayne (1956).

Of the examples of these which are described in the literature, the jaw-type have generally given less heating on breaking the contact but not so high a conductance when closed as the plate-type. Phillips (1959) achieved release energies of ~ 100 ergs by careful design of the jaws and by using rigid support for his specimen; the thermal conductance was of ~ 10^{-3} W degK^{-1} between 1° and 4° K; this might have been improved by gold-plating the jaws and the wire connected to the specimen.

Hill and Pickett (1966) have also described a switch rather similar to that of Manchester in which the jaws are gold-plated rather than covered with indium on the contacting surfaces. Their heat conductance was higher and varied less rapidly with T than did that of Manchester, particularly below 2° K. They found conductance varied as $T^{1\cdot 7}$ and was *ca.* 10^{-4} W degK^{-1} at 2°K which is still three or four times less than Phillips reported.

J. A. Birch (1966, private communication) has used the plate-type down to 0·3° K and achieved a release energy of ~ 100 ergs by eliminating any roughness and twisting action in the breaking of the contact. Birch has simply modified Rayne's switch (see Fig. 83) by ensuring that the push rod P is in close contact with the main helium bath at some point and using a hydraulic system with small hand-pump to move the push rod. Contact with liquid helium is achieved by having the bellows N down in the liquid at about the same level as the valve E. The contacting plates above and below the calorimeter are lapped flat and gold-plated to improve thermal conductance.

Aslanian and Weil (1960) have used a hydraulic control system in which the bottom plate-contact sits on the top of flexible bellows and is pushed upwards onto the sample by inflating with helium; they also report energies on release of *ca.* 100 ergs.

The plate-type might also be improved by using a thin tension wire acting against a spring rather than a push rod. Of course Seidel and Keesom (see Fig. 77) have used a wire to lift their specimen and have relied simply on its weight to give the contact pressure. Their contact plates were not gold-plated so that conductance was small and cooling-down time relatively long.

Superconducting switches

The use of a superconducting link as a switch was first suggested by Heer and Daunt (1949), Gorter (1948), and Mendelssohn and Olsen (1950) as a result of experiments on the heat conductivity of superconducting elements. This switch depends on the fact that the thermal conductivity λ_n of a superconducting element in the normal state (induced by a magnetic field exceeding the critical field), is larger than its conductivity λ_s in the superconducting state. In the normal state, conduction electrons are chiefly responsible for the conductivity and at temperatures near or below 1° K, λ_n is limited by physical or chemical impurities so that $\lambda_n \propto T$. For example, consider a metal of electrical resistivity $\rho_{295} \simeq 10^{-5}\,\Omega\,\mathrm{cm}$ and of sufficiently high purity that $\rho_r/\rho_{295} \simeq 10^{-3}$; from the Wiedemann–Franz–Lorenz law: $\rho\lambda/T = L \simeq 2{\cdot}45\times10^{-8}\,\mathrm{W\,cm^{-1}\,deg^{-1}}$, it follows that $\lambda_n \simeq 2{\cdot}45T\,\mathrm{W\,cm^{-1}\,deg^{-1}}$ at low temperatures.

In the superconducting state, at temperatures well below the transition temperature, only the lattice waves are responsible for thermal conduction and λ_s is limited by presence of grain boundaries: then Casimir's formula gives

$$\lambda_s \simeq 1{\cdot}6\times10^3 A^{\frac{2}{3}} T^3 D,$$

where D is the average dimension of the crystal grains.

Since
$$C_v = AT^3\ \mathrm{J/cm^3\,deg},$$
$$= 1940\frac{T^3}{\theta^3}\frac{\rho}{\text{at. wt.}},$$

therefore
$$\lambda_s = 2\times10^5\left(\frac{\rho}{\text{at. wt.}}\right)^{\frac{2}{3}}\frac{T^3}{\theta^2}D\ \mathrm{W/cm\,deg}.$$

If, for example, $\theta \simeq 100°\,\mathrm{K}$,

$$\rho \simeq 10\ \mathrm{g\,cm^{-3}},$$

and the atomic weight $\simeq 100$, then

$$\lambda_s = 4DT^3\ \mathrm{W/cm\,deg}.$$

Hence in a case where the grain dimensions are about 0·1 cm, it is clear that $\lambda_n/\lambda_s \simeq 6T^{-2}$; for $T = 0{\cdot}1^\circ$ K this gives $\lambda_n/\lambda_s \simeq 600$.

Experimental work on heat conduction below 1° K confirms that $\lambda_s \propto T^3$ at very low temperatures although the constant of proportionality found experimentally may be smaller by a factor of 2 or 3 than that suggested by the grain size (see, e.g., review by Mendelssohn 1955, Graham 1958). This merely enhances the actual value of the ratio, λ_n/λ_s, compared with that derived from Casimir's equation and the Wiedemann–Franz–Lorenz law. Most recently March and Symko (1966) have reported on the use of indium and lead for thermal switches below 1° K. At temperatures in the vicinity of 0·1° K their heat conductivities can be changed by a factor of 1000 or more by merely applying a field of a few hundred oersteds. They constitute a simple and effective heat switch and have been used in many cryostats; these include linking a specific heat sample to a cooling stage (Phillips 1959; Martin, 1961; Martin, Zych, and Heer 1964) and linking two cooling stages (see § 9.6).

Graphite switches

The word 'switch' is perhaps a misnomer for the use of a material which merely has a very rapid variation of heat conductivity with temperature. This can be used as mechanical support to help the initial cool-down of a system whose final temperature is to be so low that the conductivity of the link material then becomes negligible. Many types of polycrystalline graphite are useful for this. Their thermal conductivity varies approximately as T^3 and may have a magnitude of 1 W/cm degK at 100° K, $\sim 10^{-3}$ at 10° K, and 10^{-8} W/cm degK at 0·1° K (Berman 1952, 1958). Shore *et al.* (1960) first suggested the use of pitch-bonded graphite as a low-temperature support and a thermal shunt in parallel with a superconducting thermal switch: at liquid-nitrogen temperatures it is an excellent conductor for cooling their paramagnetic salt but becomes an effective insulator at the lowest temperatures (see Fig. 81 in § 5.3). Later Pandorf, Chen, and Daunt (1962) showed that suitable carbon radio-resistors were quite effective as shunts in parallel with superconducting thermal switches. Their tests on Allen–Bradley and Ohmite resistors gave values of *ca.* 10^{-3} W/cm degK at 10° K, and 5×10^{-6} at 1° K; they were not as high as pure graphite at temperatures above 10° K.

This behaviour is not confined to graphite. For any dielectric material at low-enough temperatures the conductivity $\lambda \approx aT^3$, the absolute magnitude being dependent on grain size and phonon velocity (or Debye temperature θ_D). Sufficiently fine-grained samples of the ceramics alumina, magnesia, beryllia, etc., may have $\lambda \approx 10^{-2}$ W/cm degK at 10° K and therefore λ will be $\sim 10^{-8}$ W/cm deg at 0·1° K.

Zimmerman, Arrott, and Skalyo (1958) have found that mica condensers also will serve this purpose.

Other switches

The term 'switch' may also be applied to various tube devices in which gaseous or liquid helium acts as a medium of heat transfer. For example, the thermosiphon principle may be used with the particular gas condensing in the top end of the tube in contact with a cold heat sink and then dropping to the lower end where it evaporates in contact with the source or object to be cooled (e.g. Bewilogua, Knöner, and Kappler 1966). When cooled the gas and liquid can be pumped away.

Another device uses liquid He II in a tube. The liquid is a good conductor at temperatures near 1° K and can be easily displaced by a cylinder of Teflon which is a poor conductor (Shore 1960). Vilches and Wheatley (1966) have described the use of capillary tubes filled with liquid helium, for transferring heat to and from magnetic cooling salts (see § 8.5).

9. References

Ambler, E. and Kurti, N. (1952). *Phil. Mag.* **43**, 1307.
Aslanian, T. and Weil, L. (1960). *Cryogenics* **1**, 117.
Barnes, L. J. and Dillinger, J. R. (1966). *Phys Rev.* **141**, 615.
Berman, R. (1952). *Proc. phys. Soc.* **165**, 1029.
—— (1956). *J. appl. Phys.* **27**, 318.
—— (1958). *Industrial carbon and graphite*, Papers Conf., London 1957, p. 42.
—— and Mate, C. F. (1958). *Nature, Lond.* **182**, 1661.

BEWILOGUA, L., KNÖNER, R., and KAPPLER, G. (1966). *Cryogenics* **6**, 34.

BIRMINGHAM, B. W., BROWN, E. H., CLASS, C. R., and SCHMIDT, A. F. (1955). *Proc. 1954 cryogenic engng conf., N.B.S. Report No. 3517*, p. 27.

BLACKMAN, M. B., EGERTON, A., and TRUTER, E. V. (1948). *Proc. R. Soc.* A**194**, 147.

BOWERS, R. and MENDELSSOHN, K. (1950). *Proc. phys. Soc.* A**63**, 1318.

BREMNER, J. G. M. (1950). *Proc. R. Soc.* A**201**. 305, 321.

CHALLIS, L. J., DRANSFELD, K., and WILKS, J. (1961). Ibid. **260**, 31.

COOKE, A. H. and HULL, R. A. (1942). Ibid. **181**, 83.

CORRUCINI, R. J. (1959). *Vaccuum* **7–8**, 19.

DAHL, A. I. (1963). *Adv. cryogen. Engng* **8**, 544.

DARBY, J., HATTON, J., ROLLIN, B. V., SEYMOUR, E. F. W., and SILSBEE, H. B. (1952). *Proc. phys. Soc.* A**64**, 861.

DEINESS, S. (1965). *Cryogenics* **5**, 269.

DINGLE, R. B. (1952). *Physica* **18**, 985.

—— (1953*a*). Ibid. **19**, 311.

—— (1953*b*). Ibid. 348.

—— (1953*c*). Ibid. 729.

—— (1953*d*). Ibid. 1187.

DIXON, M., GREIG, D., and HOARE, F. E. (1964). *Cryogenics* **4**, 374.

EGGLETON, A. E. J., TOMPKINS, F. C., and WANFORD, D. W. B. (1952). *Proc. R. Soc.* A**213**, 266.

FAIRBANK, H. A. and WILKS, J. (1955). Ibid. A**231**, 545.

FULK, M. M., REYNOLDS, M. M., and PARK, O. E. (1955). *Proc.* 1954 *cryogenic engng conf., N.B.S. Report No.* 3517, p. 151.

GORTER, C. J. (1948). *Physica* **14**, 504.

GRAHAM, G. M. (1958). *Proc. R. Soc.* A**248**, 522.

HEER, C. V. and DAUNT, J. G. (1949). *Phys. Rev.* **76**, 854.

HILL, R. W. and PICKETT, G. R. (1966). *Soumal. Tiedeakat. Toim.* **AVI**, No. 210, p. 40.

JEANS, Sir JAMES (1948). *An introduction to the kinetic theory of gases.* Cambridge University Press.

KEESOM, W. H. (1942). *Helium.* Elsevier, Amsterdam.

—— and KOK, J. A. (1933). *Physica* **1**, 770.

KEMP, W. R. G., KLEMENS, P. G., TAINSH, R. J., and WHITE, G. K. (1957). *Acta metall.* **5**, 303.

KENNARD, E. H. (1938). *Kinetic theory of gases.* McGraw-Hill, New York.

MCADAMS, W. H. (1954). *Heat transmission*, 3rd edn. McGraw-Hill, New York.

MCFEE, R. (1959). *Rev. scient. Instrum.* **30**, 98.

MALAKER, S. F. (1951). *Phys. Rev.* **84**, 133.

MALLON, R. G. (1962). *Rev. scient. Instrum.* **33**, 565.

MANCHESTER, F. D. (1959). *Can. J. Phys.* **37**, 989.

MARCH, R. H. and SYMKO, O. G. (1965) in 'Heat flow below 100° K', p. 57, Annexe 1965–2 Suppl. *Bull. int. Inst. Refrig.*

MARTIN, B. D., ZYCH, D. A., and HEER, C. V. (1964). *Phys. Rev.* **135**, A671.

MARTIN, D. L. (1961). *Proc. R. Soc.* A**263**, 378.

MATE, C. F. (1965) in 'Heat flow below 100° K', p. 101, Annexe 1965–2 Suppl. *Bull. int. Inst. Refrig.*

MENDELSSOHN, K. (1955). *Prog. low Temp. Phys.* **1**, 184.

—— and OLSEN, J. L. (1950). *Proc. phys. Soc.* A**63**, 2.

MERCOUROFF, W. (1963). *Cryogenics* **3**, 171.

MIKESELL, R. P. and SCOTT, R. B. (1956). *J. Res. natn. Bur. Stand.* **57**, 371.

PANDORF, R. C., CHEN, C. Y., and DAUNT, J. G. (1962). *Cryogenics* **2**, 239.

PHILLIPS, N. E. (1959). *Phys. Rev.* **114**, 676.

POWELL, R. L. and BLANPIED, W. A. (1954). *Natn. Bur. Stand. Circular* 556. U.S. Govt. Printing Office, Washington, D.C.

—— ROGERS, W. M., and RODER, H. M. (1957). *Proc. 1956 cryogen. Engng Conf., N.B.S., Boulder, Colorado*, p. 166.

RAMANATHAN, K. G. (1952). *Proc. phys. Soc.* A**65**, 532.

—— and SRINIVASAN, T. M. (1955). *Phil. Mag.* **46**, 338.

RAYNE, J. A. (1956). *Aust. J. Phys.* **9**, 189.

REUTER, G. E. H. and SONDHEIMER, E. H. (1948). *Proc. R. Soc.* A**195**, 336.

ROBERTS, J. K. (1930). Ibid. A**129**, 146.

—— (1940). *Heat and thermodynamics*, 3rd edn, Blackie & Sons, London.

RUEHRWEIN, R. A. and HUFFMAN, H. M. (1943). *J. Am. chem. Soc.* **65**, 1620.

SCHÄFER, K. (1952). *Z. Elektrochem.* **56**, 398.

SHORE, F. J. (1960). *Rev. scient. Instrum.* **31**, 966.

—— SAILOR, V. L., MARSHAK, H., and REYNOLDS, C. A. (1960). Ibid. **31**, 970.

SONDHEIMER, E. H. (1952). *Phil. Mag.* Suppl. i. 1.

THOMAS, L. B. and SCHOFIELD, E. B. (1955). *J. chem. Phys.* **23**, 861.

VILCHES, O. E. and WHEATLEY, J. C. (1966). *Rev. scient. Instrum.* **37**, 819.

WEBB, F. J. and WILKS, J. (1955). *Proc. R. Soc.* A**230**, 549.

WESTRUM, E. F., HATCHER, J. B., and OSBORNE, D. W. (1953). *J. chem. Phys.* **21**, 419.

WHEATLEY, J. C., GRIFFING, D. F., and ESTLE, T. L. (1956). *Rev. scient. Instrum.* **27**, 1070.

WILLIAMS, J. E. C. (1963). *Cryogenics* **3**, 234.

WORTHING, A. G. (1941). *Temperature—its measurement and control in science and industry*, Vol. i, p. 1164. Reinhold, New York.

ZIEGLER, W. T. and CHEUNG, H. (1957). *Proc.* 1956 *cryogen. Engng Conf. N.B.S., Boulder, Colorado*, p. 100.

ZIMMERMAN, J. E., ARROTT, A., and SKALYO, J. (1958). *Rev. scient. Instrum.* **29**, 1148.

CHAPTER VIII

TEMPERATURE CONTROL

1. Introduction

CONTROL of temperature may be achieved by a variety of methods and with various degrees of efficiency, a desirable efficiency depending on the particular investigation for which a cryostat is intended. A measure of the efficiency is the width of the temperature region over which control can be maintained and the smallness of the temperature fluctuation achieved at any temperature in this region. In general terms temperature control demands an effective balance within the 'experimental space' between the supply and loss of thermal energy.

The various methods may be divided into two broad categories: first, those in which the experimental space (temperature T) is thermally linked (preferably by a 'poor' thermal link) to surroundings at a higher temperature, T_0, and heat is continuously extracted from the experimental space; secondly, those in which the surroundings or a thermal reservoir to which the 'space' is loosely thermally connected, are at a lower temperature (T_0) than the desired control temperature, T, in which case heat is supplied to the 'space' to maintain control. Examples of the first are the use of controlled gas desorption, controlled adiabatic expansion of a gas, and Joule–Thompson throttling; evaporation of a liquid under a controlled reduced pressure might also be considered to belong to this category, as might the 'Swenson' method of control. In the second class belong the common methods of electrical heating by which a chamber or specimen is raised to a temperature above that of the surroundings by supply of electrical current to a resistance element, the current being controlled either manually or automatically in response to a temperature signal.

The two means most commonly used—vapour-pressure control and electrical heating—are discussed in §§ 2 and 3 below, followed in §§ 4 and 5 by descriptions of the 'controlled refrigeration' systems and the 'Swenson' method.

2. Control of vapour pressure

The relation between the temperature and the vapour pressure of a liquefied gas was discussed in Chapter IV (Temperature Measurement) and the importance of the vapour pressure of liquid helium, hydrogen, oxygen, etc., as secondary thermometers was stressed. Here we shall consider 'manostats' or pressure-control devices by which the vapour pressure of a boiling liquid may be controlled to within certain limits.

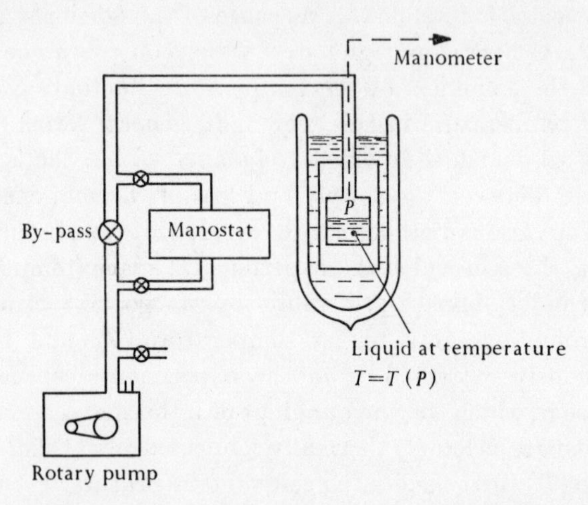

FIG. 105. Control of the pressure above a boiling liquid.

The usual arrangement (Fig. 105) has a manostat in the pumping tube from the liquid whose boiling pressure is to be controlled. The 'by-pass' valve enables the vapour to be pumped directly, e.g. when initially reducing the pressure over the liquid from atmospheric to near the control pressure, or when pumping to pressures below the lower limit of the manostat.

In the writer's experience, the most convenient pressure controller for many cryogenic applications is the Cartesian manostat. The history and theory of the application of the Cartesian diver to automatic pressure control has been discussed in some detail by Gilmont (1946, 1951); such manostats, suitable for use both above and below atmospheric pressure, have been commercially available for some years (e.g. from the Emil

Greiner Company of New York). However, a glass Cartesian manostat similar to that shown in Fig. 106 can be very easily made. A number of these glass devices—made by a competent glass-blower in two or three hours—have been used by the writer for controlling the vapour pressure of liquid helium, liquid nitrogen, and liquid oxygen from atmospheric pressure down to 2 or 3 mm Hg. The glass diver of about 1-in diam. is about $\frac{1}{4}$ in

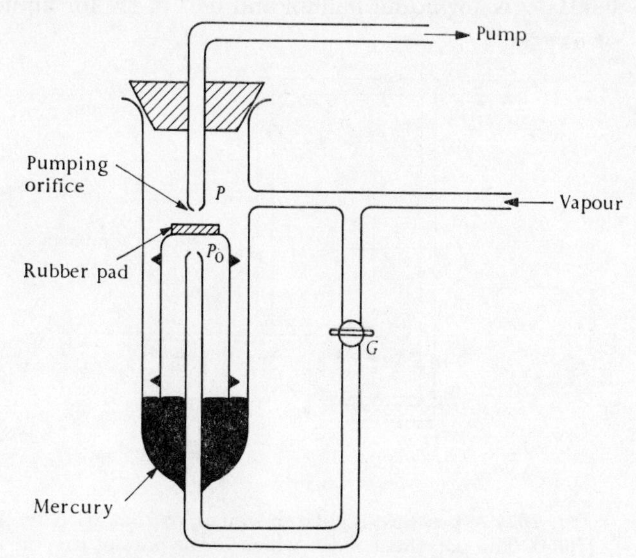

FIG. 106. A Cartesian manostat of glass.

smaller in o.d. than the i.d. of the main tube, and has six glass pips attached for centring; these pips allow the diver to move freely up and down. A valve is formed by the glass orifice seating and unseating against a rubber pad (preferably gum rubber of $\frac{1}{16}$–$\frac{1}{8}$ in thick sheet) or a rubber bung. In operation, the pressure P above the liquid is reduced by pumping through the by-pass valve (Fig. 105) and keeping the glass tap G open until P is near the desired control pressure; then the by-pass valve and G are both closed, and P will fluctuate with a small amplitude ΔP about the control pressure P_0. As shown by Gilmont (1951) and found experimentally, the fluctuation $\Delta P/P_0$ depends on the size of the orifice and varies slightly with P_0. With an orifice of about 1-mm diam. the expected pressure variation is about 0·5 per cent. In practice, the variation or fluctuation is sufficiently rapid that it is difficult to observe any variation in the vapour pressure above the liquid surface, as recorded by a mercury manometer. Over the greater part of the range from 760 mm to 2 mm Hg pressure, such manostats are quite capable of reducing the temperature fluctuations to the order of 0·001 degK for liquid helium and 0·01 degK for liquid nitrogen or oxygen.

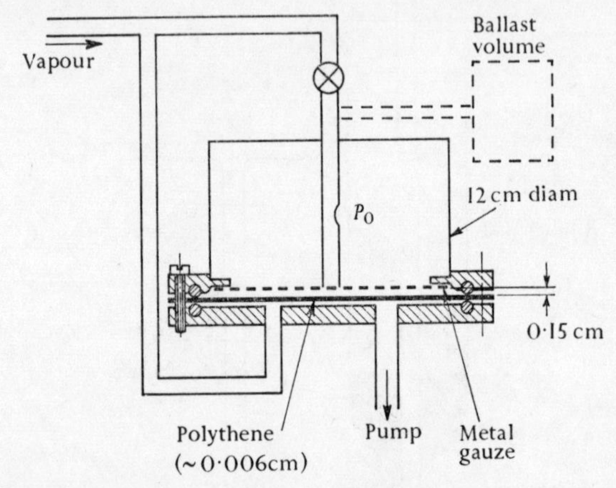

FIG. 107. A pressure regulator similar to that used by Kreitman (1964). The polythene sheet which forms part of the 'on–off' valve is 2–4 mil thick.

A similar type of manostat can be made with a flexible bellows instead of the mercury and float; this is simply an aneroid manostat but very thin-walled metal bellows are needed. Cataland, Edlow, and Plumb (1961) have described an elaborate version of this which has a bellows up to 6 in diam. and a ball-bearing valve; it can keep the vapour pressure of helium baths constant to within *ca.* 0·02 mm Hg, the control pressure being set by adding weights to a balance pan which loads the bellows.

Another manostat which is commonly used has a rubber or plastic diaphragm which moves up or down and thereby opens or closes a valve when the vapour pressure becomes greater or less than the control pressure P_0. Fig. 107 illustrates one used by the writer and some of his colleagues. It is very similar to those described by Kreitman (1964) and by Sommers (1954). The chief difficulty with these controllers is that changes in room temperature cause the control pressure P_0 to drift; this effect can be reduced by adding a large well-insulated ballast volume.

The controller in Fig. 107 can be made with a larger orifice and therefore a higher pumping speed than the Cartesian manostat. Therefore it is more suitable for pumping on liquid in dewars (where the evaporation rate is much higher) than on a small chamber which is protected by a surrounding bath. Another simple manostat with a high pumping speed can be made from thin-walled rubber tubes or 'sheaths' (Walker 1959); their only disadvantage is that the rubber perishes if exposed to ultra-violet light and to most oils and greases.

Manostats such as those of Cataland, Edlow, and Plumb (1961) may control to better than 1 millidegree down to 1·8° K, but their performances deteriorate below this and so manual control by a needle valve becomes necessary unless electrical heating is used.

For very precise control of temperature of liquid-helium baths, it may be more simple to adjust and fix an appropriate pumping speed and then control by automatic electrical heating. Boyle and Brown (1954), Sommers (1954), Blake and Chase (1963) have all described electronic controllers (see also § 3 below) designed for this purpose; they are activated by a carbon resistor in the helium bath, this resistor forming one arm of an a.c. Wheatstone bridge.

It should be noted that in pumping liquid-oxygen vapour, some care should be taken in choosing the type of mechanical pump and pump oil to use. Pumps of the Kinney type should not be used for oxygen as the explosion risk is great. The usual type of rotary vane pump has been used by the writer for pumping oxygen for many years, without any trouble, although it is considered advisable by some competent authorities (e.g. Professor S. C. Collins, private communication) to replace the normal pump oil by tricresyl phosphate.

3. Control by electrical heating

As we have stressed before, regulation of temperature outside the ranges covered by the vapour pressures of available liquefied gases presents difficulties; the magnitude of these difficulties depends on the required degree of precision of the temperature regulation. The usual method is by controlling the supply of electrical energy to preserve a temperature T in the space which is above that of the surroundings (at T_0). As the loss of heat, $-\dot{Q}$, to the surroundings by radiation, conduction, etc., is a monotonically increasing function of $T-T_0$, so the electrical input, $\dot{Q}$ must increase similarly with $T-T_0$. The situation is represented schematically in Fig. 108, where a temperature

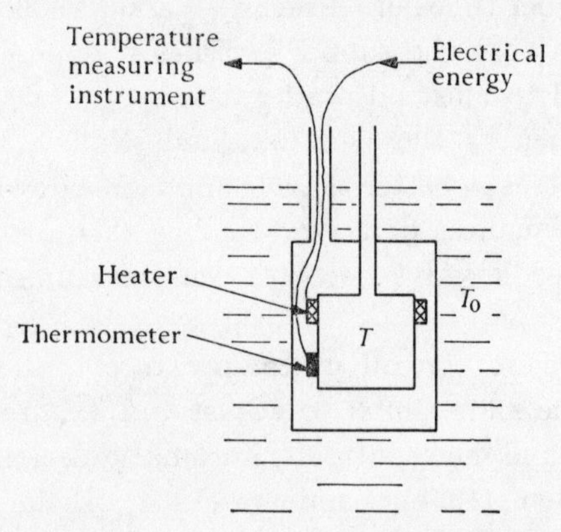

FIG. 108. Temperature control by electrical heating.

detector, e.g. gas thermometer, thermocouple, or resistance thermometer, provides information by which manual or automatic regulation of the electrical current to a heater is determined.

Manual control of the heating current often meets the experimental demands, when physical measurements may be done quickly and when they are not very sensitive to temperature drift; particularly at low temperatures, where heat capacities are small and the time for establishment of thermal equilibrium is short, a correct adjustment of heating current may be found quickly. Unfortunately, any change in the heat loss to the surroundings—due to change in pressure of residual exchange gas, for example—must affect the temperature and await manual adjustment of the heating current.

Most automatically controlled heating systems used at low temperatures are based on continuous control, as opposed to the simple on-off type of controller operated by a bimetallic spiral or expanding mercury column which is often used as a thermostat at higher temperatures. Such continuous controllers are usually actuated by the signal from a thermocouple or resistance thermometer; they amplify this signal and feed it back into the electrical heater. The degree of amplification required depends on the sensitivity of the detector, i.e. the magnitude of the signal, and on the precision required in the control of temperature.

A comparison of a resistance thermometer and thermocouple thermometer as temperature detectors shows each to have certain advantages and disadvantages. As was pointed out in Chapter IV, no available resistance thermometer has a high sensitivity over the entire range from 4·2°–300° K; also by its very nature the resistance thermometer is itself a source of thermal energy which can be a serious disadvantage when it is desired to control T very close to T_0. But resistance thermometers may be operated on a.c. and therefore supply an a.c. signal (e.g. out-of-balance e.m.f. when the detector is one arm of an a.c. bridge) which can be amplified more easily than the d.c. signal from a thermocouple. For temperatures in the range 4°–20° K, a carbon resistance provides a very sensitive thermometric element; for the range from 20°–300° K, a copper resistance winding is suitable and for the range from 150° down to 4° K, manganin wire can provide a means of high-precision control even though its sensitivity is rather less than that of a copper thermometer. Turning to the thermocouple we observe that there is negligible heat dissipation due to the couple itself and a suitable thermocouple, e.g. Au + 2 per cent Co : Cu, will provide a useful signal at temperatures down to 4° K. Although the thermocouple is often less reliable as a thermometer than a resistance element due to the presence of strains and inhomogeneities producing spurious signals, it is nevertheless quite stable for long periods of time when fixed in position in a cryostat and not subject to violent changes in temperature. The major disadvantage is that a high-gain d.c. amplifier is required for precision control. Fortunately, reliable chopper-amplifiers for this purpose have been described and are now commercially available.

A good example of precision control by the thermocouple method is that of Dauphinee and Woods (1955); the regulator shown schematically in Fig. 109 below is based on a chopper-

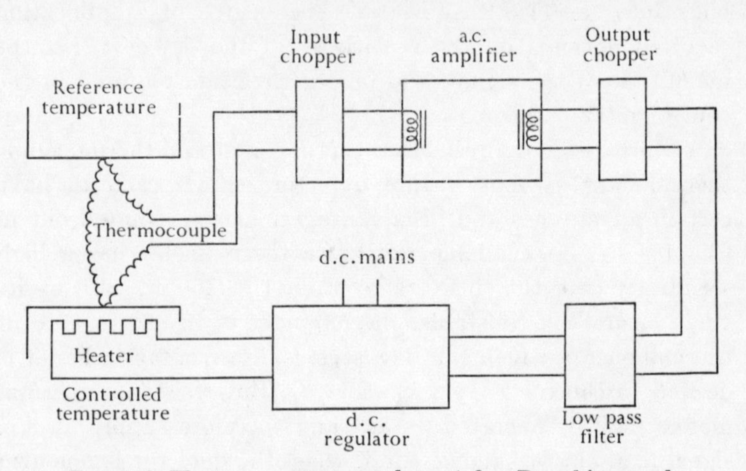

FIG. 109. The temperature regulator (after Dauphinee and Woods 1955).

type d.c. amplifier suitable for detecting thermocouple signals as small as 10^{-8} V. Choppers modulate the signal from a differential thermocouple; it is then amplified about 10^7 times and the output demodulated by further choppers, the input and output choppers being synchronous. The amplifier described is a fairly conventional low audio-frequency amplifier, preceded and followed by high-quality transformers. With a Ag + Au : Au + Co thermocouple, this chopper-amplifier has been used (e.g. Dauphinee, MacDonald, and Preston-Thomas 1954; White and Woods 1955) to control temperatures over the range from 4·2°–300° K to a precision of a thousandth of a degree. As described in § 6.3 (see Fig. 89) the Ag + Au : Au + Co junction is soldered to an inner copper experimental chamber and the Ag + Au : Cu and Au + Co : Cu junctions are thermally anchored to a surrounding vessel which is at 4·2° K, 77° K, or 90° K. The signal resulting when the temperature of the inner chamber is $(T - T_0)$ degK above the outer chamber is suppressed by passing current through a fixed thermal-free resistor in series with the thermocouple; the out-of-balance signal is chopped, amplified, unchopped, and used to control the current through a bank of tubes connected in series with a 1500-Ω constantan heater, wound onto the inner chamber.

An alternative method of amplification is the use of the galvanometer-amplifier principle; a d.c. signal may be fed directly to a galvanometer and the light beam from the galvanometer mechanically chopped by a sector disk before falling on a split photo-cell. The modulated photo-cell output may then be amplified, passed through a phase-sensitive device and fed to the heater. A thermostat for controlling temperatures to 0·01 degC between −40° C and −150° C, using a thermocouple–galvanometer–photomultiplier–amplifier sequence, has been described by Gilchrist (1955). An automatic temperature controller for a calorimeter shield which is also based on a difference thermocouple–galvanometer–photomultiplier sequence has been described by Zabetakis, Craig, and Sterrett (1957).

As we mentioned above, the resistance thermometer operated on an a.c. voltage lends itself to a rather more simple control circuit than the thermocouple. The writer (White 1953) has used a cryostat over the range from 4°–150° K in which a 1000-Ω manganin resistance non-inductively wound on the experimental chamber forms one arm of an a.c. Wheatstone bridge. The bridge is fed with a few volts at standard frequency, and has a small variable capacitance in one arm to balance the capacitance of the manganin thermometer. Any temperature fluctuation in the thermometer produces an out-of balance signal which is amplified in a high-gain audio amplifier, then passed through a phase detector, power-amplifier, and into the heater. This control unit was originally designed and built by Wylie (1948) (similar to the electronic controllers of Sturtevant 1938, and

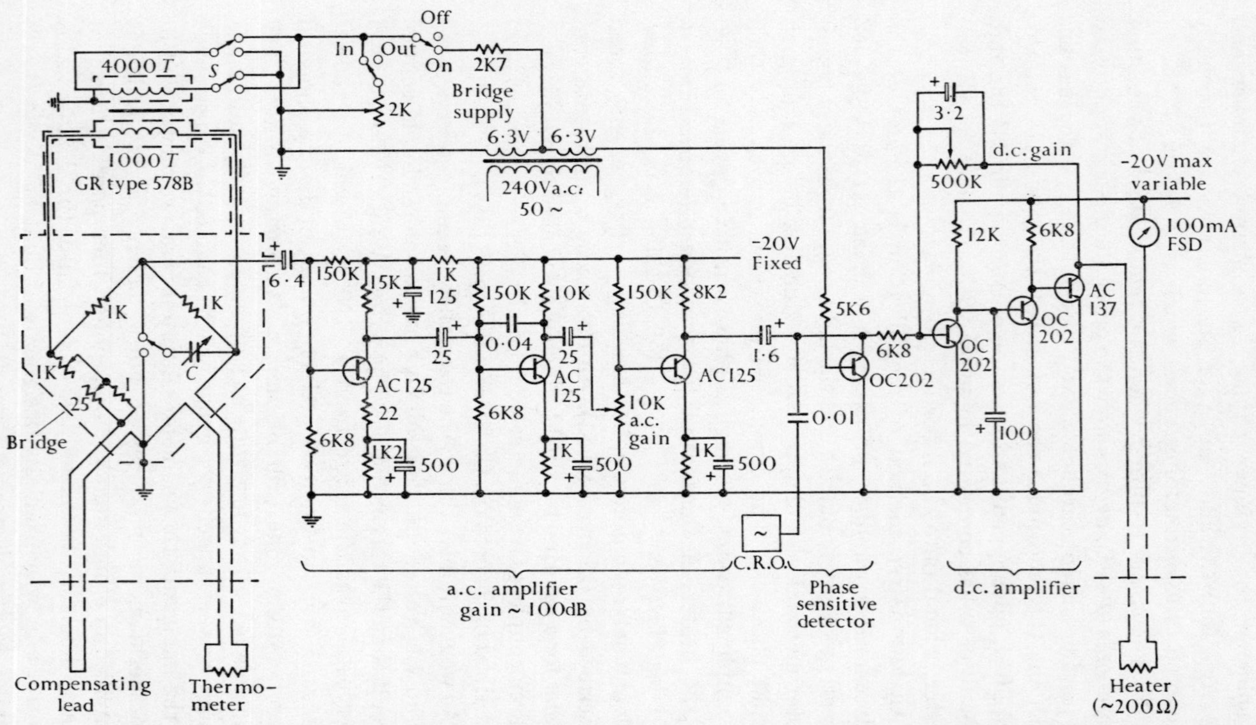

FIG. 110. An electronic temperature controller using an a.c. Wheatstone bridge and phase-selective amplifier.

Penther and Pompeo 1941) for use with a copper thermometer element, to control liquid baths to $\pm 0{\cdot}001$ degC at or near room temperature. More recently the writer and his colleagues at the National Standards Laboratory have used a transistor version (Fig. 110) developed by Messrs. P. Nowakowski, H. Kinnersly, and W. R. G. Kemp. The amplifier responds to a signal from either a carbon thermometer (for which dR/dT is negative) or manganin and copper resistors (dR/dT positive) depending on the position of the switch S. The output goes into a 200-Ω heater in the cryostat. This and most other controllers of this type suffer from the same problems: serious drifts occur in the control point unless the resistance bridge containing the reference resistors is carefully made and kept at constant temperature; also 'hunting' may occur unless the sensing thermometer is in close thermal contact with the heat load and the heater. With care experimental chambers in a cryostat may be controlled to a millidegree above 10° K and a helium bath controlled to 10 or 20 microdegrees but this is not easy. Probably the use of four-terminal sensing elements and ratio transformers in the bridge circuit could make such a degree of control more accessible†.

There are presently no commercial controllers available which satisfy these requirements although it should be possible to adopt good phase-selective amplifiers, which are available, to this purpose.

4. Controlled refrigeration

By 'controlled refrigeration' we mean those methods by which an experimental chamber is maintained at a temperature below its general surroundings, using controlled expansion or desorption of a gas. Such methods have been most widely used to bridge the temperature interval between 4·2° K and that which can be attained with liquid hydrogen.

In the Clarendon Laboratory at Oxford, Simon expansion helium liquefiers have been an integral part of a number of research cryostats; physical investigation of properties in the

† Four terminal sensors and two capacitance arms are used in the bridge of the latest and most successful controller at the N.S.L. (C. P. Pickup and W. R. G. Kemp, 1968, *Cryogenics*, to be submitted).

region between 4° and 10° K is done by carefully controlling the adiabatic expansion of the compressed helium gas (initially at 10° K and about 120 atm) so as to maintain temperatures of say 8°, 7°, or 6° K before the expansion and liquefaction is complete. The high thermal capacity of the compressed gas makes it relatively easy to cool to 7° K and maintain this temperature quite closely for a period of many minutes. Examples of the use of this method of control are afforded by the work of Berman (1951) and Rosenberg (1955) on thermal conductivity, and of MacDonald and Mendelssohn (1950) on electrical resistance.

Simon's desorption liquefier has also been used as an effective temperature controller in the range between 4° and 10° K (see, for example, Simon 1937). The process is very similar to that of controlled expansion, except that the helium-gas pressures involved are much smaller. Instead of controlling the expansion of the compressed helium gas, the rate at which the helium is pumped from its charcoal adsorbent is carefully controlled; the relatively large heat capacity of the adsorbed gas also acts as a thermal reservoir, the temperature of which is therefore insensitive to small changes in heat transfer. Measurements by de Haas, de Boer, and Van den Berg (1933–4) on electrical resistance, and by Bijl (1950) on paramagnetic resonance, were made with the help of this method of maintaining steady temperatures.

Rose-Innes and Broom (1956) have adapted this desorption principle to allow temperatures to be maintained from 4·2° K up to liquid-oxygen temperatures without the use of hydrogen. Fig. 111 illustrates a calorimeter which is cooled to 4·2° K by transferring liquid helium into the inner dewar, the bottom of which contains charcoal (in can D). After completing measurements at 4·2° K and below, this liquid is evaporated by the heater R so that only adsorbed helium remains. The experimental chamber can be warmed up to a temperature of say 20° or 30° K and then maintained there by pumping away some of the adsorbed gas; the heat capacity of the charcoal and gas is sufficient that the bottom of the dewar can be kept, for example, at 21° K for more than 30 min with 100 mW being dissipated in the calorimeter vessel.

It is uncommon to use Joule–Thomson or Linde liquefiers for the control of temperatures above the liquefaction temperature of the gas being circulated. In principle it is clearly possible by regulating the rate of flow of gas through the expansion valve, or perhaps more effectively by regulating the input pressure of the gas stream (regulation at the compressor), to prevent liquid

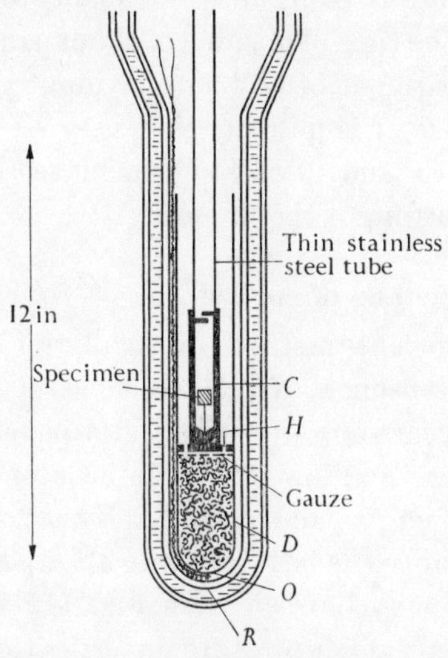

FIG. 111. Temperature control by the desorption method of Rose-Innes and Broom (1956).

being formed at the valve; reducing the flow rate sufficiently allows the inflow of heat from the surroundings to balance the Joule–Thomson cooling at a desired control temperature; alternatively, reducing the pressure of the incoming gas stream reduces the cooling efficiency (of the throttling process), so that, again, a balance may be obtained.

An obvious drawback to this type of control is that fine adjustment of flow rate or input pressure is difficult; in particular the flow rate may change quite abruptly due to a small particle of some solid impurity (e.g. solid air in a cold helium-gas stream) collecting at or being dislodged from the throttling valve. Since the available heat capacity of the 'region' near the throttling valve is usually small (in contrast to the heat capacity of the compressed gas in a Simon expansion liquefier or of the adsorbed gas in a desorption apparatus), small fluctuations in the rate of flow or input pressure will result in a fairly rapid change in temperature. It has proved possible in the Collins type of helium liquefier to 'halt' the temperature in the space below the expansion valve, before the liquefaction temperature is reached; in this machine adjustment may be made to the cam-settings which control the time of opening of the expansion engine valves, as well as adjustment of flow rate or input pressure. However, even in this case, it is unlikely that very accurate temperature control at say 10° and 40° K can be achieved in an experimental space in the bottom of the 'liquefier'.

5. Swenson's method of control

A rather efficient method of regulating temperature was evolved by Swenson at the Massachusetts Institute of Technology a few years ago (described in a paper by Swenson and Stahl 1954), a method in which both the latent heat of vaporization of a liquefied gas and part of the heat capacity of the gas are used to balance the heat inflow to an experimental chamber. The principles may be seen from Fig. 112 which illustrates a cryostat used for high-pressure investigations. Temperatures between 4° and 80° K were maintained in the experimental enclosure by drawing liquid helium from the reservoir through a $\frac{1}{8}$-in thin-walled cupro-nickel tube to the heat exchanger. The exchanger consists of three concentric layers of $\frac{3}{16}$-in o.d. copper tubing, wound around and solder bonded to the copper-walled enclosure which is to be kept at constant temperature. The rate at which the liquid helium is drawn through this tubing is controlled by the setting of a fine needle valve in the pumping tube between the heat exchanger and a mechanical pump. When controlling the enclosure at some temperature T ($> 4{\cdot}2°$ K), the liquid helium evaporates in the copper coil extracting about 20 cal/mole from the system; it is warmed up to a temperature close to the control temperature T as it circulates through the coil, extracting further heat equal to about $5 \times (T - 4{\cdot}2)$ cal/mole. It is found that to maintain temperatures of 20° or 30° K, a very careful setting of the needle valve is necessary to adjust the small pressure drop (< 1 cm Hg) between the inlet and outlet of the coil so that the flow rate of coolant is not too great. The addition of an electrical heater, e.g. a carbon resistor in a copper sleeve, cemented to the outside of the heat exchanger, simplifies the problem of control; the needle valve can then be used for comparatively coarse temperature adjustment and the heater current as a fine adjustment.

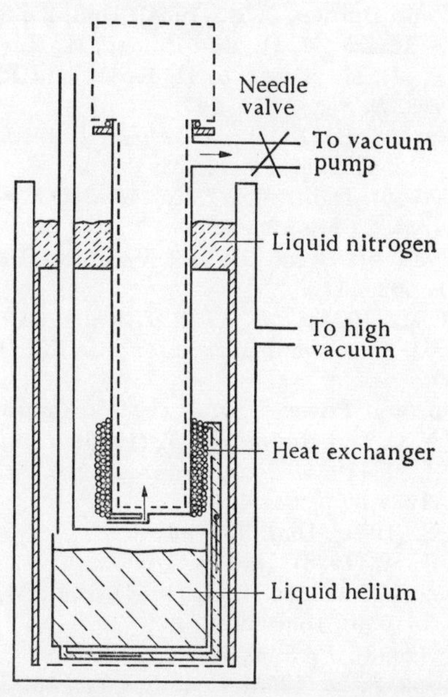

FIG. 112. Schematic diagram of a cryostat in which the temperature is controlled by the 'Swenson' method (after Swenson and Stahl 1954).

Temperatures between 80° K and room temperature can be regulated in the same manner, replacing liquid helium in the reservoir by liquid nitrogen and dispensing with the outer liquid-nitrogen shield.

Such a cryostat can hardly achieve the same precise degree of temperature regulation as an electronically controlled chamber of the type described in § 8.3 or as can be achieved by careful control of liquid-vapour pressure (§ 8.2). However, it does make efficient use of the cooling capacity of the refrigerant, and it would be quite possible to use automatically controlled electrical heating to counterbalance the somewhat coarse nature of the needle-valve flow-control mechanism.

6. References

BERMAN, R. (1951). *Proc. R. Soc.* A**208**, 90.
BIJL, D. (1950). *Proc. phys. Soc.* A**63**, 405.
BLAKE, C. and CHASE, C. E. (1963). *Rev. scient. Instrum.* **34**, 984.
BOYLE, W. S. and BROWN, J. B. (1954). Ibid. **25**, 359.
CATALAND, G., EDLOW, M. H., and PLUMB, H. H. (1961). Ibid. **32**, 980.
DAUPHINEE, T. M., MACDONALD, D. K. C., and PRESTON-THOMAS, H. (1954). *Proc. R. Soc.* A**221**. 267.
—— and WOODS, S. B. (1955). *Rev. scient. Instrum.* **26**, 693.
GILCHRIST, A. (1955). Ibid. **26**, 773.
GILMONT, R. (1946). *Ind. Engng Chem. analyt. Edn* **18**, 633.
—— (1951). *Analyt. Chem.* **23**, 157.
DE HAAS, W. J., DE BOER, J., and VAN DEN BERG, G. J. (1933–4). *Physica* **1**, 609, 1115.
KREITMAN, M. M. (1964). *Rev. scient. Instrum.* **35**, 749.
MACDONALD, D. K. C. and MENDELSSOHN, K. (1950). *Proc. R. Soc.* A**202**, 103.
PENTHER, C. J. and POMPEO, D. J. (1941). *Electronics* **14**, April 20.
ROSE-INNES, A. C. and BROOM, R. F. (1956). *J. scient. Instrum.* **33**, **31**.
ROSENBERG, H. M. (1955). *Phil. Trans. R. Soc.* **247**, 441.
SIMON, F. E. (1937). *Physica* **4**, 879.
SOMMERS, H. S. (1954). Ibid. **25**, 793.
STURTEVANT, J. M. (1938). Ibid. **9**, 276.
SWENSON, C. A. and STAHL, R. H. (1954). Ibid. **25**, 608.
WALKER, E. J. (1959). Ibid. **30**, 834.
WHITE, G. K. (1953). *Proc. phys. Soc.* A**66**, 559.
—— and WOODS, S. B. (1955). *Can. J. Phys.* **33**, 58.
WYLIE, R. G. (1948). *Divn. of Physics Report PA*-2. C.S.I.R.O. Nat. Stand. Lab., Sydney.
ZABETAKIS, M. G., CRAIG, R. S., and STERRETT, K. F. (1957). *Rev. scient. Instrum.* **28**, 497.

CHAPTER IX

ADIABATIC DEMAGNETIZATION

1. Introduction

At the beginning of this book it was pointed out that methods of cooling and entropy reduction are inextricably linked. Because the entropy or disorder of a gas is a function of volume as well as of temperature, the volume forms a useful physical variable by which the entropy of the gas can be reduced; if the initial alteration in volume V_1 to V_2 be made isothermally in a direction such as to reduce the entropy S, a subsequent adiabatic or isentropic change of V_2 to V_1 results in a drop in temperature.

To reach temperatures below that available by liquefying and pumping on helium, it is necessary to find some system which remains partially disordered and which can be ordered by a change in some available physical parameter. It was suggested by Giauque (1927) and independently by Debye (1926) that a disordered assembly of magnetic dipoles which may still occur in a paramagnetic substance at liquid-helium temperatures, should constitute such a system. Using the magnetic field H, as the variable quantity, successive processes of isothermal magnetization and adiabatic demagnetization should produce cooling. As is now well known, Giauque and MacDougall (1933) at Berkeley and de Haas, Wiersma, and Kramers (1933) at Leiden proved this experimentally; soon afterwards Kurti and Simon (1934) at Oxford also carried out successful demagnetizations.

Soon after this Gorter (1934) and Kurti and Simon (1935) suggested that temperatures of a microdegree might be attained by demagnetization of nuclear spins; however, the required starting condition that $H/T_0 \sim 10^7$ Oe/deg made this technically impractical for some years to come. Eventually in 1956 Kurti *et al.* (1956) successfully demagnetized copper from initial fields of 20–28 kOe and an initial temperature of 0·012° K to reach about 20×10^{-6}° K.

Analogous cooling processes have been demonstrated for an assembly of electric dipoles: application of an electric field to a material containing a disordered array of electric dipoles decreases the entropy and when the field is switched off isentropically, a cooling or electrocaloric effect occurs (e.g. Hegenbarth 1965; Känzig, Hart, and Roberts 1964). In the case of a superconductor, adiabatic magnetization produces a marked cooling (e.g. Yaqub 1960). Neither of these two processes is likely to displace the demagnetization of electronic or nuclear spin systems as means of cooling below 0·1° K.

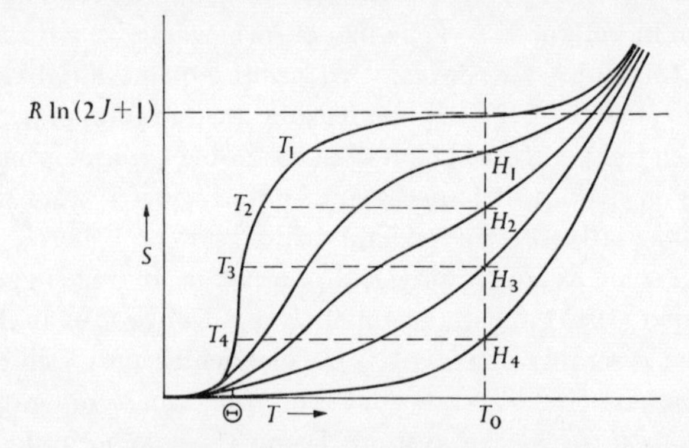

FIG. 113. Entropy of a paramagnetic salt used in cooling experiments (after de Klerk 1956).

The history of magnetic cooling and some of the later developments have been given in monographs of Casimir (1940) and Garrett (1954). There are also detailed reviews by Ambler and Hudson (1955), de Klerk and Steenland (1955), de Klerk (1956), Mendoza (1961), and Little (1964). For the uninitiated, the published lectures delivered by Simon (1937) at the British Museum and by Kurti (1952) at the Royal Institution are very instructive introductions to the principles of magnetic cooling. The cooling process is illustrated in terms of entropy reduction by Fig. 113. This figure shows what is happening thermodynamically during the isothermal magnetization (change from $H = 0$ to $H = H_1$ at T_0) and subsequent isentropic demagnetization (change from $H = H_1$ at T_0 to $H = 0$ at T_1). We may also ask what happens to the alignment of the individual magnetic dipoles during these changes. First, let us consider the dipoles initially at T_0 as a non-interacting assembly, or more properly as an assembly with such a small interaction that the dipoles have no preferred orientation. If the paramagnetic ion has a total angular momentum J, then its moment may take up any one of $2J+1$ directions with respect to the vanishingly small interaction field, the energy difference between adjacent orientations (or energy levels) being U, where U is assumed to be much less

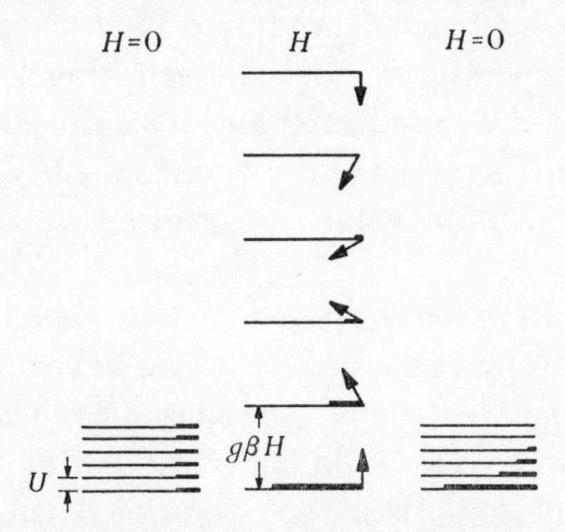

FIG. 114. Energy diagram of a paramagnetic ion (after Kurti 1952).

than kT_0. The probability of the magnetic moment of an ion having a certain orientation is governed by the Boltzmann distribution, i.e. the factor $\exp(U/kT)$; since $U = g\beta H_{\text{int}} \ll kT_0$, any of the $2J+1$ orientations are equally probable. However, on applying a field H_1 such that $g\beta H_1 \geqslant kT_0$, the ions will go into the lower energy levels which correspond to dipoles pointing in the direction of the applied field (see Fig. 114). Here g is the Landé splitting factor

$$1+\frac{J(J+1)+S(S+1)-L(L+1)}{2J(J+1)}$$

and β is the Bohr magneton. In Fig. 114 the degree of occupation of the various energy levels (here $J = 5/2$ so $2J+1 = 6$) is indicated by the shading on the lines. On removing the applied field H_1 isentropically, the degree of order must remain constant but the difference U returns to its former small value. For this to occur the distribution functions involving $\exp(g\beta H_1/kT_0)$ $(H = H_1$ at $T = T_0)$ and $\exp(g\beta H_f/kT_1)$ $(H = H_f \rightarrow 0$ at $T = T_1)$ must be equal, hence the final temperature

$$T_1 = \frac{H_f T_0}{H_1}.$$

Lest it be supposed that on reducing the external field H to zero, T falls to zero and thereby violates the third law, it should be remembered that the dipoles do interact, just as atoms in a gas interact and eventually condense into a liquid or solid state at sufficient low temperatures. So the small interactions (which may be represented as producing an internal field, H_{int} or H_f) cause the fine splitting of levels by an amount U which is sufficient to give T_1 a finite value and prevent an 'attainability catastrophe'.

A further condition for effective magnetic cooling is that the non-magnetic entropy of the system should be small in comparison with the magnetic entropy S_m, or more strictly in comparison with the possible reduction in magnetic entropy. This non-magnetic entropy comprises the vibrational entropy of the lattice, e g.

$$S_0 = \int_0^{T_0} \frac{465(T/\theta_D)^3}{T}\, dT = 155\left(\frac{T}{\theta_D}\right)^3 \text{cal/mole deg},$$

which is usually negligible at or near 1° K in comparison with ΔS_m.

Summarizing for the paramagnetic ion:

(i) $S_m = R\ln(2J+1)$ when $H = 0$,

(ii) ΔS_m should be appreciable in a field H, and

(iii) we require $S_{\text{lattice}} \ll \Delta S_m$.

Experimentally the magnetic cooling process may be represented schematically by Fig. 115. The time spent at stage (b) must be sufficiently long for the heat of magnetization $Q = T_0 \Delta S_m$ to be removed through helium exchange gas to the surrounding helium bath at temperature $T_0 \sim 1°$ K, i.e. for the magnetization to be effectively isothermal. (c) represents the period during which the exchange gas is removed. Whether or not this exchange gas is removed completely, the process of demagnetization to a temperature $T_1 \ll T_0$ is sufficiently rapid that residual gas is adsorbed onto the cold surface of the salt pill.

Provided that liquid helium and suitable pumps, dewar, magnet, etc., are available, cooling to $\frac{1}{10}°$ or $\frac{1}{100}°$ K can no longer be considered a particularly difficult or exclusive art, thanks

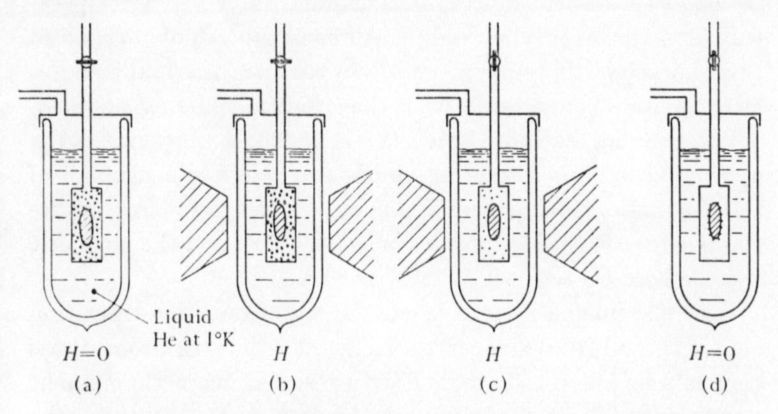

FIG. 115. Schematic diagrams of the experimental procedure used in magnetic cooling.

to the pioneering efforts at Berkeley, Leiden, Oxford, and Cambridge. However, the problems of knowing what temperatures are attained by demagnetization, of maintaining those temperatures for appreciable periods, and of ensuring that the temperature at one point in a salt pill is also the temperature at another point or in some attached specimen are formidable.

2. Paramagnetic salts

Salts suitable for magnetic cooling must contain paramagnetic ions which have a non-zero resultant angular momentum or magnetic moment. The interaction between the magnetic ions (dipole–dipole or exchange interaction) should be sufficiently small that the energy level splitting U is much less than kT at 1° K; but in a field of a few thousand oersteds the splitting $g\beta H$ should be greater than kT. In addition, it is desirable that any higher energy levels should be already so high in zero external field that their influence on the distribution function is negligible, i.e. they should be 'quenched'.

Then we may assume that the entropy and magnetic moment M are functions of H/T only, e.g.

$$M = \text{constant}\, H/T \quad \text{(Curie's law)}$$

or magnetic susceptibility $\chi = \text{constant}/T$ down to a temperature $T \sim \theta = U/k$. When temperatures in the neighbourhood of this characteristic temperature θ are reached, thermal energies have become comparable with the residual splitting or interaction energies, so that Curie's law is no longer obeyed; θ is the approximate limit of cooling and is also near the maximum of the magnetic specific heat anomaly. At temperatures appreciably above θ, the ion interaction is negligible and the magnetic specific heat $C_{\mathrm{m}} = \text{constant}/T^2$.

Suitable magnetic ions come largely from the transition elements and the rare earth elements; that is, from those elements in which the atoms have a residual magnetic moment due to an unfilled but partially shielded inner shell. Of these, the most commonly used have been gadolinium and cerium among the rare earths and those transition elements with an unfilled 3*d* shell, viz. Ti, Cr, Mn, Fe, Co, Ni; certain cupric salts have also been found useful.

These elements, present in the form of magnetic ions, are normally used in mixed salts in which the interaction is reduced by the presence of non-magnetic ions and water of crystallization. For transition elements the energy levels of interest are generally the spin levels as the orbital magnetism—if present—is quenched by the crystalline electric fields, i.e. Stark splitting of the orbital levels such that the energy gap between them is much greater than kT at 1° K. The magnetic moment due to electron spin is not directly affected by these electric fields but only via a coupling of the orbital and spin momenta.

The properties of a number of the most commonly used salts for which magnetic cooling experiments, susceptibility measurements, specific heat measurements, and paramagnetic resonance studies have provided data, are given in reviews by Cooke

(1955), Ambler and Hudson (1955), de Klerk (1956), Mendoza (1961), and Little (1964). For example, in the case of ferric ammonium alum, the chemical formula is $FeNH_4(SO_4)_2.12H_2O$. The ground state of the ferric ion is ${}^6S_{5/2}$ from which $2S+1 = 6$ and therefore the total spin entropy $S_m = R\ln 6$. The susceptibility closely follows Curie's law with a Curie constant of 4·37 per mole and resonance studies give $g = 2{\cdot}00$. Measurements of specific heat show

$$C/R \simeq 0{\cdot}013_5/T^2.$$

Some of these salts are listed in Table XVI below, with values for the temperature T_1° K reached by adiabatic demagnetization from $T_0 \simeq 1{\cdot}1$° K, $H_1 \simeq 5000$ Oe (de Klerk 1956). Also shown

TABLE XVI

Magnetic cooling with $H_1 = 5000$ Oe, $T_0 = 1{\cdot}1$° K

Salt	T_1 °K	θ_m °K
$Cr(NO_3)_3.9H_2O$	0·21	—
$FeNH_4(SO_4)_2.12H_2O$	0·090	0·061
$Mn(NH_4)_2(SO_4)_2.6H_2O$	0·165	0·11
$CuK_2(SO_4)_2.6H_2O$	0·099	—
$Gd_2(SO_4)_3.8H_2O$	0·28	0·21
$CrK(SO_4)2.12H_2O$	0·20	0·095
$Cr(NH_3CH_3)(SO_4)_2.12H_2O$†	0·21	—
$Ce_2Mg_3(NO_3)_{12}.24H_2O$	~0·01	—

† Chromium methylamine alum.

are some values of $\theta_m = U/k$ from Kurti and Simon (1935); they have assumed that the basic level of each ion is split into $2J+1$ levels, separated by equal energy differences due to interaction of $U = k\theta_m$. This characteristic temperature θ_m then governs the final temperature that may be obtained by demagnetization:

$$\frac{T_1}{T_0} = \frac{14{\cdot}9}{g}\frac{\theta_m}{H_1},$$

where H_1 is in kOe. For many magnetic ions of interest, $g \simeq 2$, so that

$$\frac{T_1}{T_0} \simeq 7\cdot5\frac{\theta_{\mathrm{m}}}{H_1}.$$

A property of obvious interest when using a salt as a cooling medium is its specific heat. Not only is this a fairly direct criterion of the splitting of the energy levels, and hence of the temperatures which may be obtained by demagnetization, but it also governs the warming-rate of the salt arising from a given 'heat leak'. If the electrical resistance of a material is to be measured near 0·1° K, it is pointless to use a cooling salt such as cerium magnesium nitrate which has a single maximum in the specific heat below 0·01° K and hence a very small heat capacity between 0·1° and 1·0° K.

The specific heat curves shown in Fig. 116 have been plotted from data given by de Klerk (1956) in his review. Not shown is that for cerium magnesium nitrate, for which $C/R \simeq 6\times10^{-6}T^{-2}$ (Daniels and Robinson 1953); this may be compared with $C/R = 193\times10^{-4}T^{-2}$ for chromium methylamine alum.

Other paramagnetic salts have been used for cooling, but the salts mentioned in Table XVI and Fig. 116 include those which have received the greatest attention in the past thirty years and therefore about which we have the greatest knowledge of entropy, specific heat, energy levels, etc.

As suggested above, these properties may be altered by diluting a paramagnetic salt so as to increase the distance between magnetic ions, thereby decreasing the magnetic interaction and hence the splitting of the lower energy levels. As de Haas and Wiersma (1935) found, dilution of chromium potassium alum with aluminium potassium alum allows much lower values of T_1 to be obtained provided that H_1 is increased. But dilution also reduces the magnetic entropy S_{m} so that the total heat capacity per unit volume is reduced and likewise the capacity of the salt for cooling other objects is reduced.

As we shall discuss further in §§ 4 and 5 below, the cooling salt is usually mounted in a cryostat in one of four forms: (i) single crystal, (ii) powder compressed into a semi-solid rod, (iii) loose powder packed into a glass, plastic, or metal container to which can be added either glycerol, oil, or helium to improve the heat transfer, (iv) a solid cylinder formed by mixing salt powder with epoxy resin. In any of these forms, the salt must be protected against loss of its water of crystallization. It is usually advisable to store 'salt pills' in a refrigerator or in liquid nitrogen and to avoid pumping a vacuum in the space around them until

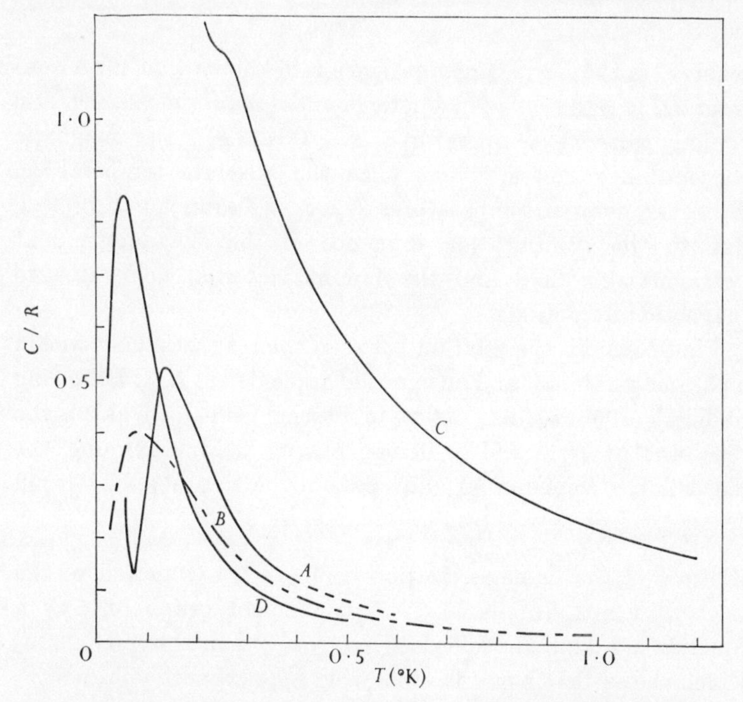

FIG. 116. Heat capacity of some salts used in magnetic cooling. *A*, chromium methylamine alum; *B*, chromium potassium alum; *C*, gadolinium sulphate; *D*, ferric ammonium sulphate.

they have been cooled well below room temperature. Some paramagnetics which are not hydrated have been used for cooling but they have not won general acceptance for use below 0·1° K: they include the double fluorides such as $(NH_4)_3CrF_6$ (Ambler and Hudson 1956), acetyl acetonates (Daunt and Pillinger 1955), ruby (Heer and Rauch 1955), and some rare-earth alloys (Parks 1963).

3. Temperature measurement

Relationship between temperature and magnetic susceptibility

For paramagnetic salts there is an observable magnetic property, the susceptibility, which is temperature sensitive and provides a thermometric parameter. We may define a temperature scale T^* such that the observed volume susceptibility is related to it by

$$\chi = C/T^*,$$

where C is the Curie constant, since it is known that for a non-interacting assembly of magnetic dipoles (and for most magnetic cooling salts at temperatures $T \sim 1°\,\mathrm{K} \gg \theta_{\mathrm{m}}$) the magnetic susceptibility varies inversely as the absolute temperature. However, even at temperatures $T \gg \theta_{\mathrm{m}}$ we cannot put $T^* = T$ due to the complicating influence of the shape-dependent demagnetizing field and the Lorentz internal field—due to magnetic interactions.

First consider the relation between the magnetic field within a paramagnetic salt and an external applied field H_{ext}. Following de Klerk (1956) we may define an internal field H_{int} which is the resultant of H_{ext} and a demagnetizing field. Confining the discussion to a spheroidal (ellipsoid of revolution) shaped salt pill

$$H_{\mathrm{int}} = H_{\mathrm{ext}} - \varepsilon M/V,$$

where ε is the demagnetization coefficient, a function of the ratio of length to diameter. A convenient graph of ε as a function of l/d for ellipsoids is given by Kurti and Simon (1938 a) which shows that ε has the following approximate values:

l/d	1	2	3	4	6	8	10
ε	4	2·2	1·4	1	0·5	0·3	0·2

In the particular case of a sphere ($l/d = 1$), ε has the value $4\pi/3$.

We may also define a 'local' field H_{loc} in the salt which is the sum of H_{int} and the field due to magnetic interaction between the ions, sometimes called the Weiss field:

$$\begin{aligned} H_{\mathrm{loc}} &= H_{\mathrm{int}} + H_{\mathrm{weiss}} \\ &= H_{\mathrm{ext}} + H_{\mathrm{weiss}} - \varepsilon M/V. \end{aligned}$$

In the simplest case Lorentz has shown that the dipole interaction produces a field $4\pi M/3V$ (see, however, de Klerk 1956, Ambler and Hudson 1955, for more complete discussion of this and the Onsager and Van Vleck treatments of internal fields). To a first approximation in a cubic lattice the Lorentz treatment gives a correct answer, provided that the temperature is not so low as to be comparable with the characteristic temperature θ_{m}.

Thus
$$H_{\mathrm{loc}} = H_{\mathrm{ext}} + \left(\frac{4\pi}{3} - \varepsilon\right) M/V,$$

from which
$$\chi_{\mathrm{loc}} = \frac{\chi_{\mathrm{ext}}}{1+\left(\frac{4\pi}{3} - \varepsilon\right)\chi_{\mathrm{ext}}/V},$$

which for a spherical sample reduces to

$$\chi_{\mathrm{loc}} = \chi_{\mathrm{ext}}.$$

T^* was defined with respect to the external field

$$T^* = C/\chi_{\mathrm{ext}},$$

which means that T^* values will be shape-dependent; Kurti and Simon introduced a quantity $T^{(*)}$ defined as

$$T^{(*)} = C/\chi_{\mathrm{loc}},$$

so that for a spherical pill, $T^{(*)} = T^*$.

However, more generally using

$$\chi_{\mathrm{loc}} = \chi_{\mathrm{ext}}\left\{1+\left(\frac{4\pi}{3} - \varepsilon\right)\chi_{\mathrm{ext}}/V\right\}^{-1}$$

we have

$$\begin{aligned} T^{(*)} &= \frac{C}{\chi_{\mathrm{ext}}}\left\{1+\left(\frac{4\pi}{3} - \varepsilon\right)\chi_{\mathrm{ext}}/V\right\} \\ &= T^* + \left(\frac{4\pi}{3} - \varepsilon\right) C/V \\ &= T^* + \Delta. \end{aligned}$$

For most paramagnetic salts used in cooling experiments it may be assumed that $T^{(*)} = T$ provided $T \gg \theta_{\mathrm{m}}$.

Therefore $$\chi_{\text{ext}} = \frac{C}{T-\Delta} = \frac{C}{T-\left(\frac{4\pi}{3}-\varepsilon\right)fC/V};$$

here f is a filling factor which may be required, in the case of loosely packed salt pills, to take account of their density being appreciably less than the density of a single crystal. For example, let us consider potassium chrome alum. From Kurti and Simon (1938*a*), the Curie constant per cubic centimetre is 0·0066 so that if a pill has an approximately spheroidal shape, ratio of length to diameter of four, and density about 97 per cent of the single-crystal density, then

$$\Delta = \left(\frac{4\pi}{3}-1\right) \times 0{\cdot}0066 \times 0{\cdot}97$$
$$= 0{\cdot}019_5,$$

so that the measured susceptibility $\chi \propto (T-0{\cdot}02)^{-1}$. Some values of the Curie constant per cubic centimetre and Δ for an ellipsoid of $l/d = 3$ are given below (from Kurti and Simon 1938*a*).

TABLE XVII

Some values of the Curie constant and the quantity Δ (*assuming* $f = 1$)

Substance	C/V	Δ °K
Gadolinium sulphate	$5{\cdot}85 \times 10^{-2}$	0·166
Manganous ammonium sulphate	$2{\cdot}00 \times 10^{-2}$	0·057
Ferric ammonium alum	$1{\cdot}55 \times 10^{-2}$	0·044
Chromic potassium alum	$0{\cdot}66 \times 10^{-2}$	0·019
Titanium caesium alum	$0{\cdot}13 \times 10^{-2}$	0·0037

Before considering the discrepancies between the thermodynamic temperature and the 'magnetic' temperature which may occur at temperatures below 1° K let us consider the usual methods of measuring the susceptibility.

Susceptibility measurement

Although the balance method (see Chapter VI) of measuring susceptibility was used in Leiden in their early experiments on demagnetization (see also Mendoza and Thomas 1951) it has generally been superseded by the induction bridge. Two obvious reasons are that measurement by a microbalance of the force exerted on the salt pill by an inhomogeneous field requires (i) a comparatively strong magnetic field so that the salt cannot be completely demagnetized, and (ii) suspension of the pill from the balance which makes the problems of thermal isolation rather difficult. As a result most laboratories use a mutual inductance method—a ballistic bridge or a.c. Hartshorn bridge —or a self-inductance method such as the a.c. Anderson bridge used at Berkeley.

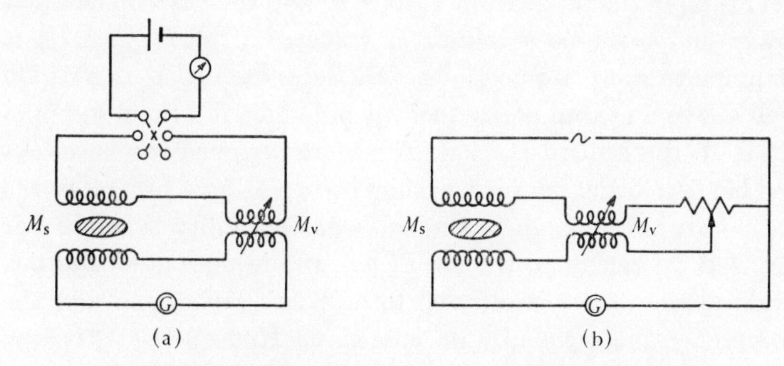

FIG. 117. (a) Ballistic bridge; (b) a.c. Hartshorn bridge.

In the ballistic method, a measured current supplied by an accumulator (Fig. 117 (a)) is reversed through the primary coil of a mutual inductance M_s around the salt pill; this induces a current pulse through the secondary which is measured by the deflexion of a ballistic galvanometer G—usually of period 10 to 20 s. The primary of M_s consists of a few hundred turns of copper wire wound onto the tail of the helium dewar and therefore immersed in liquid nitrogen, or it may be wound on the outer jacket of the experimental chamber and be in liquid helium; the secondary coil of some thousand or more turns of fine copper wire lies inside the primary. Part of the inductance may be cancelled by the variable inductometer M_v which is external to the cryostat. As discussed by de Klerk in his review (1956) practices differ in various laboratories: M_v may be used to create a null method whereby changes in the setting of M_v record the relative change in susceptibility of the salt pill; or M_v may be set so that the ballistic deflexion δ is approximately zero at say 4° K and as the temperature is lowered δ is recorded as a function of temperature.

In the latter case

$$\delta = \text{constant}\,\chi + b = \frac{a}{T-\Delta} + b.$$

The constants a, b which depend on the coil geometry and on the Curie constant are determined by plotting δ against the reciprocal of $(T-\Delta)$ in the range from 4° to 1° K, and then extrapolated at lower temperatures to obtain a value of $T^* = T^{(*)} - \Delta$. It is common to wind two portions of the secondary of M_s in a region well above and well below the salt pill and wind them in opposition to the central portion. These are designed to cancel out the ballistic deflexion (or a.c. signal when using a bridge) in the temperature range where the salt's susceptibility is small. For detailed references to the use of ballistic bridges (and also a.c. bridges) the reader is referred to de Klerk (1956) and also the comprehensive article by de Klerk and Hudson (1954) on the equipment for adiabatic demagnetization at the National Bureau of Standards.

The a.c. methods have gradually become more popular than the ballistic, partly because sensitivity can be increased more easily and components have become commercially available. The Hartshorn bridge (Fig. 117 (b)) has been used frequently; it requires an audio-frequency supply of between 20 and a few hundred cycles per second and a good detector-amplifier. Details of such bridges have been given by de Klerk and Hudson (1954), Erickson, Roberts, and Dabbs (1954), Pillinger, Jastram, and Daunt (1958), and others. Commercial models operating at either 17 or 155 c/s have been made by Cryotronics Incorporated of New Jersey from the design of Pillinger *et al.*; for example Abel, Anderson, and Wheatley (1964) have described the use of this bridge to determine temperatures using cerium magnesium nitrate as the paramagnetic substance. Two other sensitive a.c.

methods make use of frequency counting in a tuned circuit (Betts *et al.* 1964) and ratio-transformers to provide tapping points on a mutual inductance bridge (Maxwell 1965). A possible disadvantage of the a.c. method is that stray heating may arise in metal parts due to eddy currents and in the salt due to relaxation effects. This latter can be made negligible by working at low frequencies since the heat adsorption amounts to $\frac{1}{2}H_0^2\omega\chi''$, where H_0 is the amplitude of the alternating field, ω is its angular frequency, and χ'' is the complex part of the susceptibility

$$\chi = \chi' - i\chi''.$$

For details of the use of the Anderson inductance bridge the reader is referred to Giauque and MacDougall (1935).

T–$T^{(*)}$ relations at very low temperatures

As yet we have not discussed the relation of the measured T^* or $T^* + \Delta = T^{(*)}$ to the thermodynamic temperature T. In the liquid-helium region above 1° K, the temperature T of the salt is assumed to be that of the surrounding helium bath with which it is in thermal contact, and a very good estimate of T may be obtained from the helium vapour-pressure tables. Thus the constants in the equation $\chi_{\mathrm{ext}} = C/(T-\Delta)$ (or in practice, perhaps $\delta = a/(T-\Delta)+b$) may be determined assuming that the measured susceptibility or inductance is inversely proportional to $(T-\Delta)$. In this case we may assume that at temperatures just below 1° K, the magnetic temperature T^* determined from the measured susceptibility or inductance will, after addition of the quantity Δ, be a good estimate of T, i.e.

$$T^{(*)} = T.$$

However, as the temperature is decreased further, and in particular as $T \to \theta_{\mathrm{m}}$, this state of affairs may not be expected to continue.

From the second law of thermodynamics we define

$$T = \frac{dQ_{\mathrm{rev}}}{dS} = \frac{dQ_{\mathrm{rev}}/dT^{(*)}}{dS/dT^{(*)}};$$

it follows that by a series of demagnetizations from a known field H and known temperature T, and an allied series of specific heat determinations (warming the salt to determine $T^{(*)}$ as a function of Q), the numerator and denominator in this equation may be determined. Hence the relation between $T^{(*)}$ and T may be found for a particular salt.

Both the principles and the practice of the determination of the absolute scale below 1° K have been discussed at length by Kurti and Simon (1938*b*,) Simon (1939), Cooke (1949), Hull (1947), and in the reviews by Ambler and Hudson (1955), and de Klerk (1956). Fig. 118 illustrates the principles of the determi-

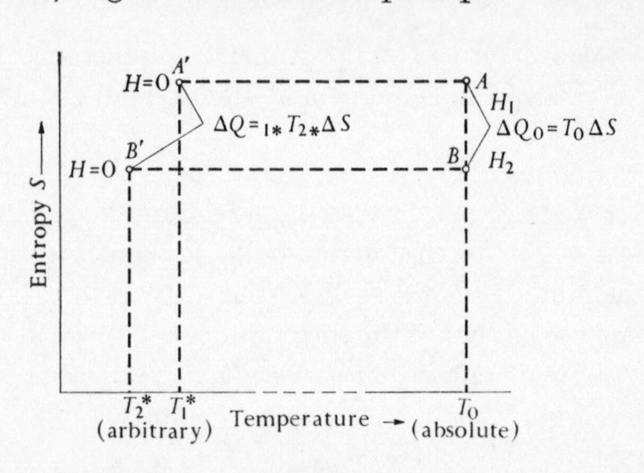

FIG. 118. Establishment of a thermodynamic temperature scale below 1° K (after Kurti 1952).

nation. At temperature T_0 (in the liquid-helium region), the difference in entropy ΔS of the salt in fields H_1 and H_2 may be determined theoretically from the properties of the salt or directly by measuring ΔQ_0; heat ΔQ_0 is evolved in changing from H_1 to H_2 and may be determined from the quantity of helium evaporated. If two isentropic demagnetizations from H_1 and from H_2 to $H = 0$ take the salt to temperature T_1^* and T_2^* respectively, the difference in entropy is still ΔS between the positions A' and B'. Alternatively, by supplying heat ΔQ to the salt, it may be warmed from the temperature T_2^* to T_1^* and hence the difference ${}_{1*}T_{2*} = \Delta Q/\Delta S$ may be found in degrees absolute. This implies a knowledge of ΔQ which must be obtained by calibration of the heating source (γ-ray heating is used in

Oxford and a.c. electrical heating at Leiden) at a temperature in the vicinity of T_0; here the salt is known to obey Curie's law and hence the thermodynamic temperatures are known. By a series of demagnetizations, a fairly accurate picture of the T–T^* relation may be obtained.

In Fig. 119 is shown the relation determined experimentally for potassium chrome alum (after Cooke 1949) together with theoretical curves based on treatments of the internal interaction field by Lorentz, Onsager, and Van Vleck. In Table XVIII are some values at various thermodynamic temperatures

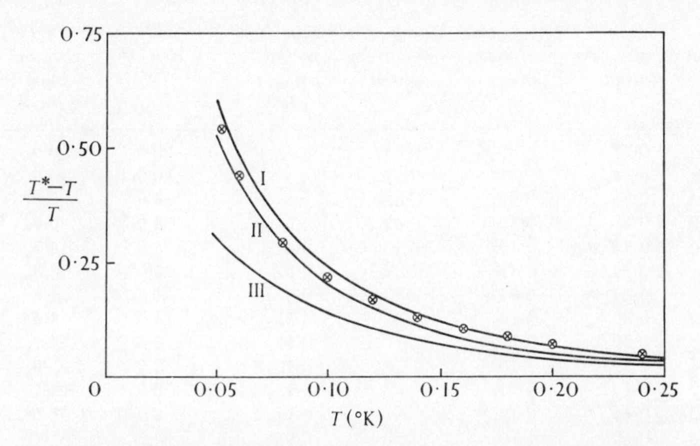

Fig. 119. Relation between magnetic temperature and absolute temperature for chromium potassium alum. Curve I, Onsager's theory; curve II, Van Vleck's theory; curve III, Lorentz's theory. Crosses are experimental results (after Cooke 1949).

of the magnetic temperature $T^{(*)} = T^* + \Delta$, i.e. in terms of a spherical salt pill. The values listed for the first four salts, that is excepting CMN (cerium magnesium nitrate) are taken from Ambler and Hudson (1955, 1957), Cooke, Meyer, and Wolf (1956). Iron ammonium alum must be regarded with some suspicion and is little used nowadays as a thermometer.

CMN has become a favoured thermometer as $T = T^{(*)}$ down to temperatures near 0·006° K. Below this the T–T^* relations have been determined by Daniels and Robinson (1953), Frankel, Shirley, and Stone (1965), and Hudson and Kaeser (1967). The average of the two more recent determinations are listed in Table XVIII. Note that below 0·005° K these authors assess their random error as $\pm 0{\cdot}3$ mdeg and indeed the two sets of data differ by 0·5 to 0·6 mdeg between 0·002 and 0·004° K. For example, according to Hudson and Kaeser, $T = 0{\cdot}003$° K corresponds to $T^{(*)} = 0{\cdot}0042$ while according to Frankel *et al.* it corresponds to $T^{(*)} = 0{\cdot}0037$° K.

TABLE XVIII

T (°K)	$T^{(*)}$ *Iron ammonium alum*	$T^{(*)}$ *Chromium potassium alum*	$T^{(*)}$ *Manganese ammonium alum*	$T^{(*)}$ *Chromium methyl-amine alum*	T (°K × 10^3)	$T^{(*)}$ (°K × 10^3) *Cerium magnesium nitrate*
0·8	$0{\cdot}81_7$	$0{\cdot}80_0$	$0{\cdot}81_7$		20·0	20·0
0·7	$0{\cdot}72_5$		$0{\cdot}72_3$		10·0	10·0
0·6	$0{\cdot}63_2$	$0{\cdot}60_4$	$0{\cdot}62_9$		8·0	8·0
0·5	$0{\cdot}53_5$	$0{\cdot}50_5$	$0{\cdot}53_3$		6·0	$6{\cdot}0_5$
0·4	$0{\cdot}43_7$	$0{\cdot}40_6$	$0{\cdot}43_7$	$0{\cdot}41_8$	5·5	5·7
0·3	$0{\cdot}34_2$	$0{\cdot}31_0$	$0{\cdot}34_0$	$0{\cdot}31_5$	5·0	5·3
0·2	$0{\cdot}24_7$	$0{\cdot}21_5$	$0{\cdot}23_5$	$0{\cdot}21_7$	4·5	4·9
0·18	$0{\cdot}22_8$	$0{\cdot}19_5$	—	$0{\cdot}19_7$	4·0	4·6
0·16	$0{\cdot}20_9$	$0{\cdot}17_4$	—	$0{\cdot}17_7$	3·5	$4{\cdot}2_5$
0·14	$0{\cdot}19_0$	$0{\cdot}15_6$	—	$0{\cdot}15_7$	3·0	$3{\cdot}9_5$
0·12	$0{\cdot}17_2$	$0{\cdot}13_8$	—	$0{\cdot}13_7$	2·5	3·7
0·10	$0{\cdot}15_2$	$0{\cdot}12_1$	—	$0{\cdot}11_7$	2·0	3·5
0·08	$0{\cdot}13_3$	$0{\cdot}10_3$	—	0·098	—	—
0·06	$0{\cdot}11_3$	$0{\cdot}08_6$	—	$0{\cdot}077_5$		
0·05	$0{\cdot}10_5$	$0{\cdot}07_9$	—	0·067		
0·04	$0{\cdot}10_2$	$0{\cdot}07_5$	—	0·060		

It is clear from Table XVIII that the paramagnetic susceptibility of CMN can serve as a useful thermometer down to $T \approx 0{\cdot}003$° K. At lower temperatures, the nuclear-spin system provides a magnetic thermometer. Waltsedt *et al.* (1965) have shown how magnetization of nuclear spins may be measured by means of pulsed nuclear magnetic resonance. Another group at the Clarendon Laboratory have also shown that the anisotropy of γ-ray emission (G. V. H. Wilson, private communication) provides a useful measure of the nuclear alignment and hence of the temperature; an obvious difficulty at temperatures of

$\sim 10^{-3}$ °K or less is that the nuclear spin system whose temperature is measured by these techniques is not necessarily the same as that of the electrons or lattice due to relaxation times becoming too long.

Before concluding this brief discussion of temperature measurement below 1° K, we should note that a number of secondary thermometers can be useful in this temperature range. These are in the electrical resistance category and have been discussed already (§§ 8.7, 8.8) in the section dealing with semiconducting resistance thermometers and metallic alloy resistance thermometers. With any secondary thermometer, i.e. other than the susceptibility of the salt itself, the problems of thermal contact and thermal equilibrium arise, and are considered below in § 5.

4. Cryostats

Introduction

For effective cooling by demagnetization certain provisions must be met in the cryostat:

(i) It must allow the salt to be cooled to ~ 1° K by thermal contact with a pumped liquid-helium bath.

(ii) At the salt there must be an adequate magnetic field to reduce the entropy appreciably, and thermal contact must be maintained to allow the heat of magnetization ($Q = T_0 \Delta S_{\mathrm{m}}$) to be transferred to the helium bath.

(iii) Before demagnetization this thermal contact must be broken, and after demagnetization thermal contact must be sufficiently bad that the leakage of heat to the salt pill is small compared with its heat capacity.

(iv) A mutual inductance in the cryostat should register the change in susceptibility of the salt and not be unduly influenced by other materials present.

In any laboratory the particular design of adiabatic demagnetization cryostat must depend somewhat on the means of attaining $T_0 \simeq 1$° K, that is whether a small helium liquefier is incorporated in the cryostat or whether liquid helium may be transferred into it from a storage dewar. However, this is relatively unimportant in determining the general design of the experimental space around the salt pill and the method of suspending the pill.

Leiden and Oxford cryostats

In view of the long experience acquired at Leiden and Oxford the two cryostats illustrated in Figs. 120 (a) and (b) are of particular interest. In the first figure (Fig. 120 (a) of a Leiden

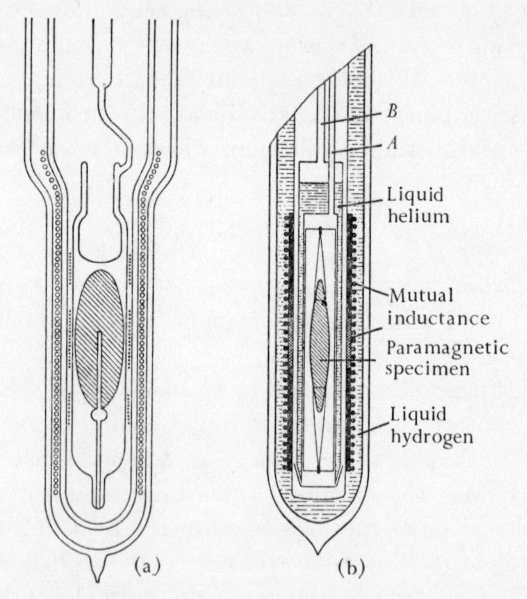

FIG. 120. Demagnetization cryostats. (a) Leiden (after Casimir, de Haas, and de Klerk 1939); (b) Oxford (after Kurti 1952).

cryostat) the ellipsoidal salt is mounted on a glass pillar within a small vacuum jacket, also of glass. The secondary coil of the mutual inductance is wound onto the vacuum jacket and is in the liquid helium contained by the inner dewar. The primary coil, wound onto the tail of the helium dewar, is immersed in the liquid hydrogen or liquid nitrogen of the outer dewar. Helium exchange gas is used in the vacuum jacket for cooling the salt pill and afterwards pumped away. The N.B.S. installation (de Klerk and Hudson 1954) uses a very similar demagnetization cryostat, and in both Leiden and the N.B.S. the cryostat is suspended from a long rotatable arm so that the whole cryostat may be moved out of the poles of the magnet after demagnetization.

In the Oxford pattern (Fig. 120 (b)) the salt pill is tautly suspended by silk or nylon threads inside a cylindrical metal cage; this cage fits into the German silver vacuum chamber which is sealed at the bottom by a greased cone and socket joint.† Formerly all these Oxford cryostats had a small Simon expansion liquefier in the upper part of the apparatus (see, for example, Hull 1947, for description of whole cryostat and details of demagnetization procedure). The upper part of tube A (Fig. 120 (b)) being soldered to the wall of the expansion vessel, allows helium gas to be cooled and condensed into the double-walled German silver chamber. The mutual inductance is wound onto a former of cloth-bonded bakelite or similar insulator, which fits over the outside wall of the vacuum case and is therefore surrounded by liquid hydrogen. The liquid helium is pumped through tube B and exchange gas is admitted and removed through tube A. It should be noted that if metal chambers surround the salt these should be of a metal such as German silver which does not become ferromagnetic at low temperatures. When an a.c. induction method is used for determining susceptibility it is preferable to use glass throughout the 'tail' assembly.

In Berkeley, Giauque and his collaborators have used spheroidal glass containers filled with powdered paramagnetic salts, the container being suspended by a glass tube inside a glass vacuum jacket (Fritz and Giauque 1949). They have also used successfully plastic containers (Giauque *et al.* 1952, Geballe and Giauque 1952) in which the salt pills are mounted.

† As sealing agent for such a ground joint high-vacuum greases such as Apiezon L or Apiezon K are frequently used, but it has been found that fluids which form glasses may be more satisfactory. Hudson and McLane (1954) describe the use of an alcohol-glycerine (2 parts glycerine and 1 part of *n*-propyl alcohol) mixture which gives leakproof joints in the helium region, provided that it is cooled slowly to liquid-nitrogen temperatures.

A metal cryostat

In Fig. 121 (White 1955) is a variation of the Oxford pattern in which the inner helium chamber C can be filled with liquid

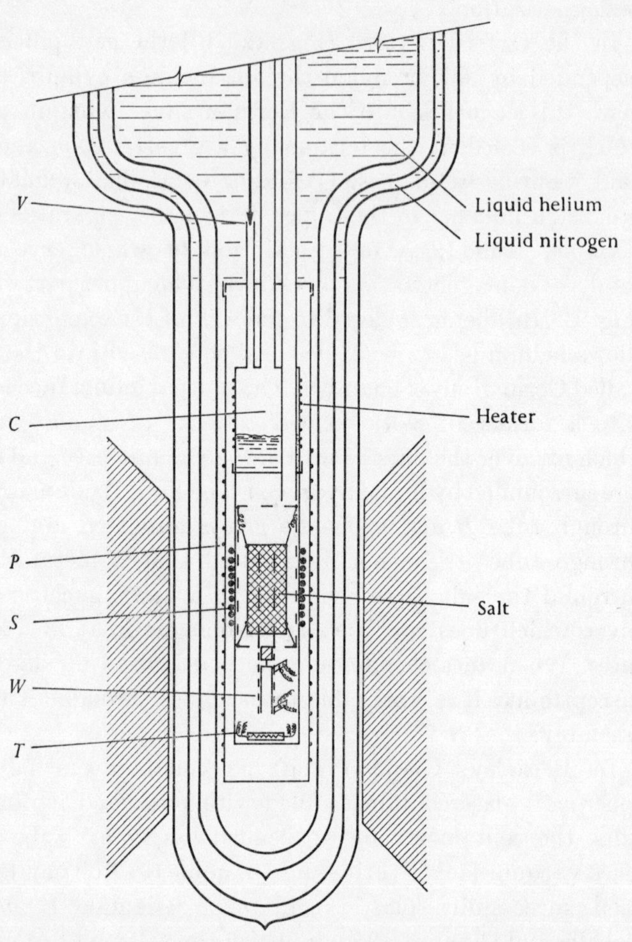

FIG. 121. Diagram of a demagnetization cryostat used for electrical resistance measurements (after White 1955).

from the surrounding dewar through valve V, and may be reduced in temperature to about 1·1° K by pumping. The inner vacuum jacket is of German silver but has a copper flange at the top to facilitate soldering it with Wood's metal to the chamber

C; vertical copper ribs are silver-soldered to the German silver to ensure temperature equilibrium, and yet avoid eddy currents being induced during the mutual inductance measurement. The salt pill of potassium chrome alum was pressed into a hard cylinder, painted with glyptal or nail varnish and suspended by nylon threads from a small German silver frame, made to slide inside the inner vacuum case. The primary P of the mutual inductance is wound onto the outer brass vacuum jacket and the secondary S wound onto the inner German silver jacket. A modification (Dugdale and MacDonald 1957) has been to place a platinum–glass seal in the bottom of the chamber C and take electrical leads out through this seal; the leads go through the pumped liquid helium and up the helium pumping tube rather than up the vacuum pumping tube; this provides more effective thermal anchoring and reduces the 'warming-up' rate considerably.

The experimental procedure with such a cryostat consists of (i) filling the inner chamber C with liquid helium and the space around the salt with a small pressure ($\sim 10^{-2}$ mm Hg) of helium exchange gas, (ii) cooling in steps from 4·2° to 3°, 2·5°, 2°, 1·5°, 1·1° K and measuring the ballistic deflexion or mutual inductance at each temperature, (iii) placing the dewar tail between the poles of a magnet, switching on the magnetic field and allowing a few minutes for heat of magnetization to be removed before pumping away the exchange gas, (iv) after pumping the exchange gas for a period of 10–30 min, the field is reduced to zero and the cryostat removed, (v) magnetic temperature is then measured by the inductance bridge.

The cryostat shown in Fig. 121 is mounted on a relay rack which, with the aid of wheels and a track, can be wheeled into the magnet gap very easily.

Salt pills

The pressed variety may be easily made by partially filling a cylindrical hole in a steel block with the powdered salt, the bottom end of the hole being closed by a short length of steel rod which is free to slide in the hole. With a sliding steel piston inserted above the salt, a hand-operated or hydraulic press is then used to form a compact pill. The pill can then be pushed out fairly easily, although fracturing can occur during the removal which has led to the occasional use of a split-steel block, in which the two halves are bolted together during the pressing operation. A little vacuum grease mixed with the powdered salt in a pestle and mortar before pressing often results in a stronger pill. With reasonable care, such salt pills may be ground or filed to a required shape and then surface-sealed with nail varnish or other cement to retard hydration. Strong well-bonded pills can be made by mixing the powdered salt with a cold-setting epoxy resin or other plastic; thermal contact with the salt is facilitated by including copper wires or foil in the mixture. Geballe and Giauque (1952) have described a technique of shaping salt pills into ellipsoids of revolution.

Alternatively, large single crystals of many paramagnetics may be grown from solution and can be formed onto a metal mesh or gauze in the process, e.g. onto silver, gold, or platinum gauze (Dugdale and MacDonald 1957, Wheatley, Griffing, and Estle 1956).

As may be seen from the figures in this section, narrow-tail glass dewars have been the usual containers for liquids in adiabatic demagnetization cryostats. The narrow tail allows the use of a smaller pole-gap in the magnet and hence higher fields for the same magnet current; the top section of the dewar is usually enlarged to give greater liquid-storage capacity. Metal dewars are now used frequently as they can either be made or bought with narrow tails (see also § 2.1). Of course superconducting magnets, which are immersed in the liquid helium, make the narrow tail unnecessary and inconvenient.

5. Heat transfer and thermal equilibrium below 1° K

Introduction

Two major problems in carrying out measurements of physical properties at temperatures below 1° K are those of maintaining the temperature of a specimen sensibly constant and uniform, and knowing what the temperature is. Obviously this is particularly the case when the specimen under investigation is neither the cooling salt itself nor a paramagnetic salt whose temperature can be inferred from a susceptibility measurement. In most investigations on the properties of paramagnetic salts, the primary problem is to ensure a small heat leak and hence slow warming-up rate so that they are in thermal equilibrium during readings; they provide their own thermometer. But when investigating physical properties of materials which do not act as cooling agents themselves, the additional problem arises of thermal contact with the cooling agent and with the primary thermometer—the salt.

As we discussed in Chapter VII, the heat leakage is due largely to (i) radiation, (ii) residual gas conduction, (iii) conduction through solid supporting rods, wires, etc., (iv) gas desorption or adsorption, and (v) mechanical vibration. These factors have also been discussed in the review articles on adiabatic demagnetization.

Very useful experimental studies of heat leaks and transfer below 1° K have been done by Wheatley and his colleagues at the University of Illinois (e.g. Wheatley, Griffing, and Estle 1956, Anderson, Reese, and Wheatley 1963, Connolly, Roach, and Sarwinski 1965, Abel *et al.* 1965, Vilches and Wheatley 1966, Wheatley 1966).

Radiation

In so far as radiation is concerned, the heat leak is negligible if the salt is surrounded by a metal wall near 1° K, provided that no radiation from surfaces at higher temperatures can enter the 'salt' chamber via pumping tubes. In glass demagnetization cryostats, radiation protection is necessary in the form of a coating on the glass chamber which is at or near 1° K. Wheatley *et al.* found that an adsorbent film of 'Aquadag' and layer of aluminium foil was the most efficient; with this protection, rather than chemical silvering, a flashlight beam did not noticeably change the normal heat leak of 8 ergs/min from other sources.

Gas adsorption

It seems dubious whether residual gas conduction and adsorption can be separated as heat-transfer factors, when a salt pill or specimen is at 0·01°–0·1° K and is surrounded by a 1° K enclosure. However, the dependence of heat-leak rate on the exchange-gas pressure† and pumping time before demagnetization is quite marked. Exchange-gas pressures of $\sim 10^{-4}$ mm appear to be too small for the heat of magnetization to be removed efficiently, and exchange-gas pressures of 10^{-2} mm are rather large as they result in large heat leaks (after demagnetization) even when pumping times of more than 1 h are used. This is partly due to adsorption of helium on the surfaces at 1° K, which cannot be completely removed by pumping but which transfers slowly to the salt pill after demagnetization; such effects are much less marked with glass than with metal cryostats. In the 'baking-out' or 'outgassing' technique for metal cryostats, often used at Oxford (Hull 1947), the exchange gas is pumped away with the helium bath kept at 1·5° K rather than 1° K and the temperature is finally reduced to 1° K just before demagnetization. The effect of adsorbed exchange gas is most noticeable when salt pills or specimens have to be warmed up again to 0·5° or 0·6° K for the purpose of physical measurements, as the vapour pressure of liquid helium is relatively high ($\sim 10^{-4}$ mm Hg) at such temperatures, and the vacuum deteriorates rapidly due to desorption from the walls of the pill. In the experiments of Wheatley *et al.* with a glass cryostat, they confirmed that a suitable exchange-gas pressure is $\sim 10^{-3}$ mm and to ensure minimum heat leak later, a pumping time of about 2 h is needed. With a pumping time of 1 h the heat leak 10 min after demagnetization was 18·5 ergs/min compared with 2·3 ergs/min when pumped for $2\frac{1}{4}$ h.‡ For many experiments, such small heat leaks are an unnecessary luxury and a pumping of a few minutes followed by demagnetization (and the resulting

† For methods of removing heat of demagnetization, i.e. heat transfer, without the use of exchange gas, see discussion on heat switches in § 7.8.

‡ To achieve such small heat leaks a second 'guard-ring' salt pill was attached to the supporting glass tube.

adsorption of exchange gas on the salt pill) may be all that is necessary.

Heat conduction through supports

Conduction in solids is discussed in more detail in § 12.4 and data for the materials which are used in cryogenic apparatus are given in Table I in the Appendix. Here we are concerned with the behaviour below 1° K where much of our knowledge comes from the experiments of Wheatley and his colleagues. Anderson *et al.* (1963) found that for a number of 'amorphous' insulating solids the conductivity at temperatures between 0·1° and 1° K could be expressed in the form

$$\lambda = aT^n,$$

where $1{\cdot}8 < n < 3$. For example,

$$\lambda \approx 0{\cdot}38 \times 10^3 T^{2{\cdot}4} \text{ ergs/s cm deg for Teflon},$$

$$\lambda \approx 1{\cdot}0 \times 10^3 T^{2{\cdot}6} \text{ ergs/s cm deg for Epibond 104},$$

$$\lambda \approx 0{\cdot}26 \times 10^3 T^{1{\cdot}75} \text{ ergs/s cm deg for nylon},$$

and $$\lambda \approx 1{\cdot}53 \times 10^3 T^{1{\cdot}75} \text{ ergs/s cm deg for Pyrex glass}.$$

For comparison they quote the following for a sample of superconducting lead:

$$\lambda \approx 45 \times 10^3 T^3,$$

and for manganin:

$$\lambda \approx 5{\cdot}5 \times 10^3 T \text{ ergs/s cm deg}.$$

Manganin is an example of a high-resistivity alloy for which approximate values of heat conductivity can easily be calculated from its electrical resistivity, ρ using the Wiedemann–Franz–Lorenz relation

$$\lambda = LT/\rho \quad (L = 2{\cdot}45 \times 10^{-8} \text{ V}^2 \text{deg}^{-1}).$$

The conductivity of pure metallic elements at low temperatures is often 10^4 to 10^5 times larger than that of manganin. The heat conduction in the normal (non-superconducting) state is predominantly electronic and therefore proportional to T, and its magnitude is determined by impurity scattering. For high-purity copper having a resistance ratio of $R_{273}/R_{4{\cdot}2} \approx 10^3$,

$$\rho_{4{\cdot}2} \approx 1{\cdot}7 \times 10^{-9}\ \mu\Omega \text{ cm},$$

therefore
$$\lambda \approx 2{\cdot}45 \times 10^{-8}\, T/1{\cdot}7 \times 10^{-9}$$
$$= 14 \times 10^{7} T \text{ ergs/s cm deg.}$$

The conductivity of insulating crystals varies much more rapidly at low temperatures where the size of the conductor or crystal grains limits the heat flow. Casimir (1938) has shown that
$$\lambda_B \sim 1{\cdot}6 \times 10^{3} D A^{2/3} T^{3} \text{ W/cm deg,}$$
where $C_v = AT^3$ J/cm^3 deg is the lattice specific heat and D is the average grain size. For example, in a diatomic crystal for which $\theta_D = 200°$ K, density $\rho = 2$ g cm^{-3} and molecular weight is 100.
$$\lambda_B = 0{\cdot}16 D T^{3} \text{ W/cm deg}$$
$$= 1{\cdot}6 \times 10^{6} D T^{3} \text{ ergs/s cm deg.}$$
Obviously at the temperatures prevailing in a cooling salt, crystalline materials whether metallic or not have a much higher conductivity than the glasses and plastics mentioned above.

Two materials commonly used as supports for salt pills are glass rods and nylon filaments. We can calculate the approximate heat-flow along a rod of either of these from the data of Wheatley *et al.* or of Berman, Foster, and Rosenberg (1955). Assume that the rod is 10 cm long, 0·01 cm^2 in cross-section and has its ends at temperatures of 1° K and *ca.* 0° K respectively: a rod of Pyrex would conduct *ca.* 1 erg/s while nylon would conduct only *ca.* 0·1 erg/s.

This heat-leakage down supports may be reduced by using a ^{3}He cooling stage at *ca.* 0·3° K or by using a double-pill technique, mounting the second 'guard-ring' pill of a paramagnetic salt at a point on the supporting tube or thread. In a further refinement, Nicol (1955) and Dugdale and MacDonald (1957) thermally connected a light metal shield to the 'guard' salt so that a copper-plated German silver shield (at $T \simeq 0{\cdot}5°$ K) surrounded the other 'salt' pill or specimen being investigated. In this case not only is conduction down the supporting tube or thread reduced, but heat leaks due to gas conduction or adsorption are reduced also; by such means heat leaks have been reduced to $\sim$ 1 erg/min.

Vibrations

The results of investigations of mechanical vibrations as a source of heat inflow have been discussed in Chapter VII. Suffice it to say that, with glass supports, the natural period of vibration is sufficiently high that normal laboratory or pump vibrations do not set the pill into vibrations of large amplitude, although heat leaks of a few ergs per minute might be traced to such a cause. With salt pills suspended by threads, the resonant period is often much closer to or identical with the period of some local vibrations and may result in heat inputs of hundreds to thousands of ergs per minute; to counteract this, taut suspension threads, efficient damping of the pump vibrations and isolation of the cryostat from pumps and building vibrations may be used.

Thermal contact and equilibrium in the salt

As most heat leaks to a salt are to the surface or to points on the surface, and any heat leaks arising from an attached specimen are also to isolated regions within the salt pill, their effect on the temperature equilibrium of the salt pill is important.

For a single crystal of an insulator we may assume fairly rapid establishment of thermal equilibrium; the thermal conductivity at these low temperatures is due to lattice waves of fairly long wavelength and is limited by crystal or grain boundaries. The Casimir expression for λ_B as a function of grain size, θ_D, T, etc., is fairly well obeyed for many crystals at liquid-helium temperatures (e.g. reviews by Berman 1953, Klemens 1958). The first measurements below 1° K by Kurti, Rollin, and Simon (1936) for a ferric ammonium alum crystal of 7-mm diam. showed

$$\lambda \approx 250 \text{ ergs/s cm deg at } 0{\cdot}07^\circ \text{K}$$

and for a potassium chrome alum crystal of the same size

$$\lambda \approx 1000 \text{ ergs/s cm deg at } 0{\cdot}18^\circ \text{K}.$$

These values are a little smaller than Casimir's expression predicts but confirm that these salts are sufficiently good conductors that temperature gradients within the crystal will be small unless, either the heat leaks are intolerably large, i.e. $\dot{Q} \gg 1$ erg/s *or* the temperature is extremely low, i.e. $T \leqslant 10^{-3}$° K.

In salt pills made of compressed powder, the conductivity will be correspondingly smaller as it is limited by grain size. In fine-grained material there may be a serious lack of thermal equilibrium at the lowest temperatures unless metal foil, wire, or gauze are included in the pill. If this pill is to be used to cool another material, then such metal inserts are useful for making thermal contact between the pill and the object to be cooled. The contact or boundary resistance between a metal and salt crystal becomes of extreme importance. Experiments by Mendoza (1948) indicated that for metal (copper) fins in a pressed pill of powdered salt, the heat flow

$$\dot{Q} \simeq 100A(T_1^3 - T_2^3) \text{ ergs/s},$$

where A is the area of contact and T_1 and T_2 are the temperatures of salt and metal respectively; later Goodman (1953) found a rather higher value for the multiplicative constant. With crystals grown from solution onto silver wire mesh (Dugdale, MacDonald, and Croxon 1957; see also Dugdale and MacDonald 1957), experimental results suggest a heat flow across the metal-crystal interface which is about ten times larger than is deduced from the Mendoza equation. In the case of large single crystals cemented to quartz or copper, Wheatley *et al.* found the heat transfer to be about a hundred times better than that predicted by the Mendoza equation.

More recent experimental values for the conductance $\dot{Q}/A \,.\, \Delta T$ between a salt crystal and other materials, in the range from *ca.* 0·04° to 0·3° K, have been summarized by Vilches and Wheatley (1966):

(i) Potassium chrome alum and copper sheet with Apiezon 'N' grease in between: $2{\cdot}5 \times 10^5 T^3$ ergs/s cm² deg;

(ii) Potassium chrome alum and copper sheet with silicone vacuum grease between: $4 \times 10^4 T^3$;

(iii) Two crystals of CMN with GE7031 cement (+ toluene) between: $1{\cdot}4 \times 10^4 T^{2{\cdot}4}$;

(iv) ferric ammonium alum and quartz with Apiezon 'J' oil between: $1 \times 10^5 T^{2{\cdot}7}$;

(v) Iron ammonium alum or potassium chrome alum and liquid ^{4}He: $2\times10^5T^3$;

(vi) Copper and liquid ^{3}He: $5\times10^4T^3$;

(vii) Copper and liquid ^{4}He: $2\times10^5T^{3\cdot4}$ ergs/s cm^2 deg.

These authors have also given a very useful account of thermal contact problems at very low temperatures and the use of liquid helium as a means of contact. Liquid helium is not a new method of improving the thermal equilibrium within a salt or the heat transfer from the salt to its surrounding; Kurti *et al.* (1936; also Hull 1947) describe a thick-walled metal tube containing a salt powder which was filled to 100 atm pressure of helium gas at room temperature and sealed off. Hudson, Hunt, and Kurti (1949) used an open-capsule technique: the salt container is connected by a capillary tube (0·2–0·3 mm i.d.) to a helium supply from which helium is condensed into the container at 4° K; surface creep of the helium film at temperatures below the λ-point (2·17° K) and recondensation of the evaporating film causes an additional heat influx to the container which is approximately proportional to the diameter of the capillary tube. Although the resultant heat leak may be more than 100 ergs/min, this open capsule technique has been useful for some investigations.

What is apparent from these various experiments on heat contact between the salt and embedded fins? Above about 0·1° K, as Mendoza suggested in 1948, the temperature of a specimen should be within a few hundredths of a degree of the temperature of the salt, provided that the specimen is connected via a good thermal link (copper wires) to extended fins which are firmly embedded in a compressed salt pill or are bonded to a single crystal. This presupposes that heat leaks to the specimen are maintained at a level of the order of 10 ergs/min or less. An interesting means for maintaining low temperatures despite an appreciable heat input is that of 'incomplete demagnetization'; this was originally suggested and used by Kurti, as far as the writer is aware. In this procedure the salt is first demagnetized to a field of, say, 1000 Oe. During the course of the experiment, the field is slowly reduced from 1000 Oe at a rate sufficient to keep the salt temperature sensibly constant; this takes advantage of the relatively large specific heat of the salt in a magnetic field (cf. Fig. 113).

6. Two-stage cooling, cyclic magnetic refrigeration, and nuclear demagnetization

Thermal switches and two-stage processes

The demagnetization processes described in this section consist of two stages in 'series' which must be linked by a heat switch. Various types of switch were discussed earlier in § 7.8. We may illustrate schematically the use of a thermal switch in a 'two-stage' magnetic cooling process (Fig. 122) and examine its

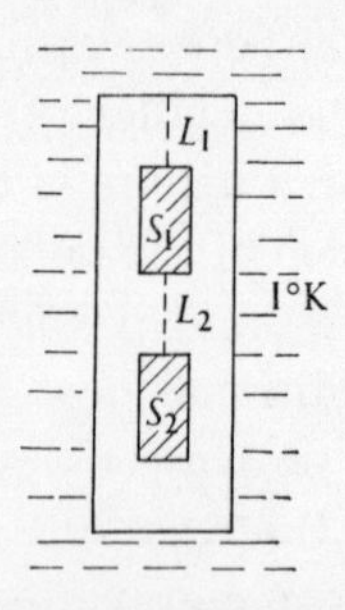

FIG. 122. The use of thermal switches in two-stage cooling processes.

application to three particular problems: (i) two-stage cooling involving electron paramagnetics, (ii) cyclic magnetic refrigeration, (iii) nuclear demagnetization. As illustrated, L_1 and L_2 are thermal links, which when 'closed' allow heat flow; S_1 and S_2 are cooling salts.

In case (i) a two-stage demagnetization may be used to obtain temperatures of $\sim 0{\cdot}001°$ K with comparatively small magnetic fields, e.g. 5000 Oe. A salt S_1 with a comparatively high characteristic temperature θ_m is magnetized with L_1 closed; L_1 is opened and S_1 is demagnetized to obtain a temperature of about $0{\cdot}1°$ K. Salt S_2 of much lower θ_m (e.g. diluted salt) is in a magnetic field during this procedure and is therefore magnetized at $T_0 \sim 0{\cdot}1°$ K. L_2 is now opened and S_2 is demagnetized to give a final temperature $T_1 \sim 0{\cdot}001°$ K.

In (ii) S_2 is a salt which acts as a thermal reservoir and is kept cool by cyclic removal of heat to S_1 and from S_1 to the helium bath: with L_1 closed, S_1 is magnetized; then after opening L_1, S_1 is demagnetized and L_2 is closed to partially cool S_2. Subsequently L_2 is opened and L_1 is closed and the process is repeated. Once the system has reached equilibrium, and provided that the heat leak to S_2 is small in comparison with its heat capacity, its temperature will oscillate with small amplitude about a mean temperature T with the period of the magnetic cycle.

In principle the nuclear cooling (iii) is similar to (i) but S_2 is now a nuclear paramagnetic while S_1 remains an electronic paramagnetic.

Two-stage cooling of electronic paramagnetics

Considering these processes in a little more detail, the first was ably demonstrated by the experiments of Darby *et al.* (1951) in which a final thermodynamic temperature of about 0·003° K was obtained with a field $H \simeq 4200$ Oe; with $H \simeq 9000$ Oe, they reached $T \simeq 10^{-3}$° K and maintained the temperature below 0·01° K for about 40 min.

In Fig. 123 is shown their experimental arrangement and warm-up rate after demagnetization from 4200 Oe. In these experiments L_1 was helium exchange gas and L_2 a superconducting thermal link of lead wire 3 cm long and 0·3 mm diam. The consecutive processes of demagnetizing S_1 (iron ammonium alum), opening link L_2, and then demagnetizing S_2 (diluted mixed crystals of potassium chrome alum and potassium aluminium alum in the ratio of about 20 : 1) were achieved by gradual lowering of the magnet.

A further demonstration of this technique of reaching the millidegree region with small magnets has been given by Nicol (1955); in this case S_1 was manganous ammonium sulphate, and S_2 was chromium potassium alum. A temperature of a few millidegrees was reached, and by surrounding the second salt with a small radiation shield—cooled by the first salt—heat leaks of about 4 ergs/min were maintained to S_2.

Cyclic magnetic refrigeration

An interesting development in cryogenics was the construction of the first cyclic magnetic refrigerator at Ohio State University (Heer, Barnes, and Daunt 1954, Daunt 1957). This could maintain temperatures in the range 0·2–1·0° K by continuous extraction of heat. A few copies were made commercially

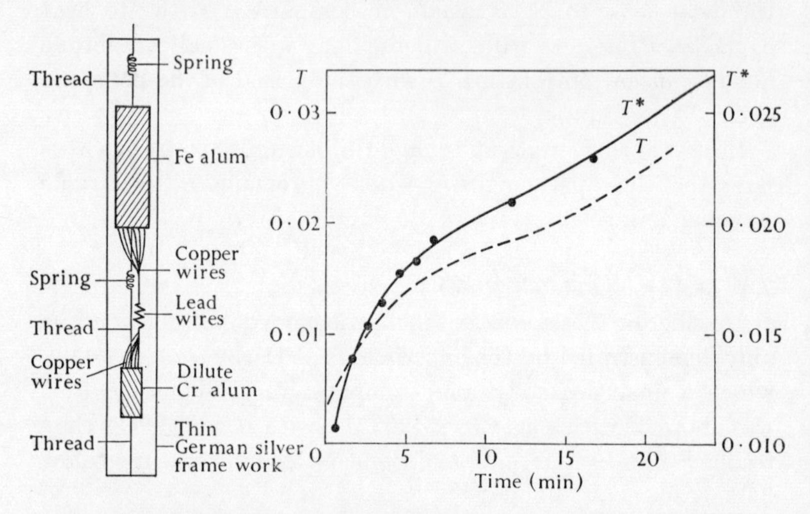

Fig. 123. Diagram of the arrangement used by Darby *et al.* (1951) in a two-stage cooling experiment; the temperature of the dilute chromium salt after demagnetization is shown in the graph (T^*= magnetic temperature, T = absolute temperature).

by the Arthur D. Little Corporation but were soon superseded when ^{3}He became available in sufficient quantities to use as a refrigerant for this temperature range.

However this refrigerator is still of interest as a prototype for later designs which use superconducting solenoids (see Zimmerman, McNutt, and Bohm 1962, for details of a refrigerator with superconducting magnets) and different cooling salts to maintain temperatures below those reached by liquid ^{3}He.

In Fig. 124 (a) the salt R (potassium chrome alum), equivalent to S_2 in our earlier discussion, is the thermal reservoir which is cooled via the lead link (V_R) to salt A. The salt A consisting of iron ammonium alum is magnetized by magnet 2 with the link

V_B to the 1° K helium bath closed and the link V_R open; V_B is then opened and A is demagnetized; V_R is now closed to cool R and then opened before the cycle of 2 min duration is repeated. Magnets 1 and 3 control the superconducting heat links. With the form of lead links used by Heer *et al.* the net rate of heat extraction obtained is about 70 ergs/s at 0·26° K, 170 ergs/s at 0·35° K, and 290 ergs/s at 0·45° K.

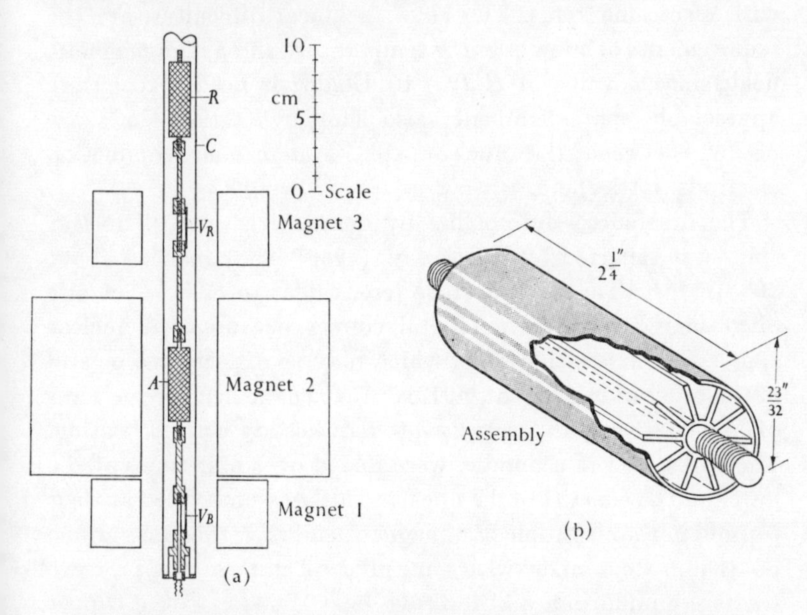

FIG. 124. (a) The basic components of the magnetic refrigerator. (b) The paramagnetic salt container of which the barrel is brass and the central rod and fins are copper (after Heer, Barnes, and Daunt 1954).

Such a machine is particularly useful for investigating physical properties which necessitate continuous heat inputs of 1000 ergs/min or more. A salt container with internal copper fins is illustrated by Fig. 124 (b); thermal equilibrium is assisted by mixing the salt with $\frac{1}{8}$-in lengths of fine copper wire (0·002–0·003 in diam.) and silicone vacuum grease; mixed in the ratio of 1 g of wires and 1 g of grease to 15 g of salt, this is pressed under 3000 lb/in^2 pressure into the container.

Nuclear demagnetization

The limiting temperature which may be reached by demagnetizing a system of magnetic spins is determined by the interaction between the spins. Nuclear spins have magnetic moments which are at least a thousand times smaller than those of electronic spins, so that this interaction is reduced and temperatures can be reached which are a thousand times smaller than with electronic spins. The chief technical difficulties are the requirements of a low starting temperature and a high magnetic field, since a value of $H/T_0 \sim 10^7$ Oe/deg is needed to ensure appreciable spin alignment; also short relaxation times are needed between the nuclear spin system and conduction electrons, lattice, etc.

The first successful cooling by demagnetization of nuclear spins was reported by Kurti *et al.* (1956). They reached about 20×10^{-6}° K. The nuclear stage (equivalent to S_2 in schematic diagram of Fig. 122) is of a metal, copper, because of its nuclear spin–lattice relaxation time, which may be expected to be still of the order of minutes at 0·01° K; 1540 enamelled copper wires of only 0·005-in diam. were used to reduce eddy-current heating. The wires, tied in a bundle, were folded over four times at the lower end to constitute the nuclear 'salt' and embedded at their top end in the electronic paramagnetic salt (S_1); this was chrome potassium alum mixed with some glycerol and water to promote thermal equilibrium and improve heat transfer to the copper wires, and was held in a Perspex container. The heat link L_2 thus consists of the copper wires themselves in order to avoid the complexity of a magnetically operated heat switch for these preliminary experiments. This whole assembly of $S_2+L_2+S_1$ was mounted inside a double-walled brass tube, the top part of which contained 25 g of manganous ammonium sulphate cooled by the initial demagnetization (when S_1 is demagnetized). The lower part of the double-walled shield was kept in good thermal contact with the manganous salt by liquid helium and thin copper wires between the brass walls.

In later experiments liquid ^{3}He was used to cool the heat shield and temperatures of a few microdegrees were achieved from starting conditions of $H/T_0 \approx 2 \times 10^6$ Oe/deg (e.g. Kurti 1963).

These temperatures have probably been reached only by the spin system and not by the lattice. This seems likely from the thermometry experiments of Waldstedt *et al.* (1965) which have shown that the spin–lattice relaxation time τ is given by $1{\cdot}1T^{-1}$ s for copper; for platinum this is appreciably shorter, $\tau \approx 0{\cdot}03T^{-1}$ s, so that at temperatures of *ca.* 10^{-4}° K demagnetization of Pt spins might be used to cool other systems such as possible superconductors.

7. Magnets

Magnetic fields play an essential role in the cooling processes which were discussed in this chapter. Magnets and cryogenics are also interdependent in other areas: magnetic fields are necessary to the solid-state physicist who studies either the motions of free electrons in metals at low temperatures where thermal scattering is reduced, or the spin–energy levels of bound electrons and nucleons. In return, low temperatures produce the superconducting state in some metals and its convenient corollary, the superconducting magnet.

The various types of magnets which are commonly used in cryogenics may be classified as follows:

(i) the iron magnet which has a ferromagnetic core of high permeability, has relatively small power requirement, and gives a field which is limited by the saturation magnetization of the iron;

(ii) the normal coil magnet whose field is only limited by the available electric power, removal of Joule heat, and the mechanical strength of its components;

(iii) the pulsed coil in which the limiting field is set more by mechanical strength than by power or heating;

(iv) the superconducting coil which avoids both power and heating problems completely but is limited by the critical field of the superconducting component.

All these magnets have been discussed in some detail in the monographs by de Klerk (1965) and Parkinson and Mulhall

(1967), a review by Montgomery (1963) and the Proceedings of the International Conference on High Magnetic Fields held at M.I.T. in 1961 (*High magnetic fields.* M.I.T. Press and Wiley, 1962).

Iron-core magnets

The magnetization of iron becomes saturated when the domains are fully aligned which requires a field of *ca.* 20 kOe. Fields which are a little greater than this can be obtained with iron magnets by careful shaping of the pole-faces in order to concentrate the field locally. Fields of between 40 and 50 kOe have been reached with pole gaps of 2 cm in the large electromagnets at Leiden (14 tons in weight), Bellevue near Paris (nearly 100 tons), and Uppsala (37 tons). Further refinements in the design of the yoke and pole-faces have allowed fields of nearly 40 kOe to be reached in a 2 cm gap in magnets of only 2 or 3 tons weight, for example in that made by the Arthur D. Little Corporation and designed by Bitter and Reed (1951) and the Swiss Oerlikon magnet.

With the advent of iron-free solenoids, both normal and superconducting, these massive electromagnets have gone out of favour. But there will continue to be a need for more modest iron-core magnets which produce fields up to say 15 kOe; for example, relatively inexpensive magnets such as the Weiss-type made by Newport Instruments Ltd., Buckinghamshire, U.K. (see also Hudson 1949), and also for the closely regulated homogeneous fields that are needed in N.M.R. and E.S.R. measurements and available in magnets produced by Varian Associates, Pacific Electric Motor Company, Magnion Incorporated, Japan Electron Optics Laboratory Co. Ltd., Newport Instruments Ltd. (see also Appendix).

Normal coil magnets

Fig. 125 (from Bitter 1962) shows the difference between iron-core magnets and normal solenoids insofar as their field-power characteristics are concerned. Above the cross-over region of 20–30 kOe, iron magnets are less effective.

For the solenoid the field H and power dissipation W are related by the Fabry formula

$$H = G(W\eta/a_1\rho)^{\frac{1}{2}},$$

where η is the filling factor for the winding, ρ is the electrical resistivity, a_1 is the inner radius; G is a dimensionless factor

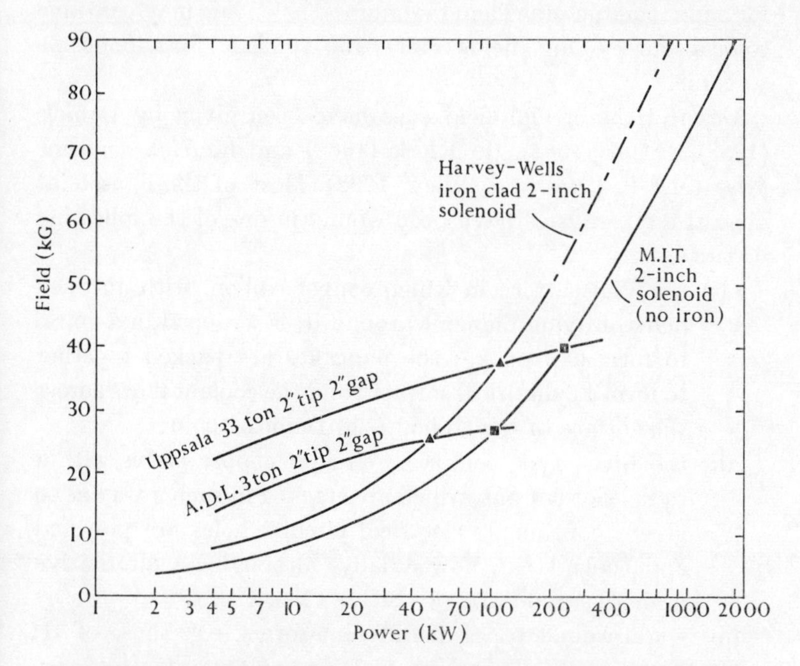

FIG. 125. Comparison of the field–power relationships for iron-core magnets (Uppsala and A.D.L.) and solenoids (after Bitter 1962).

depending on the shape of winding and current distribution, and has values generally between 0·15 and 0·25. If we assume that $G = 0{\cdot}2$, $\eta = 0{\cdot}7$, and $\rho = 2\times10^{-6}\,\Omega\,\text{cm}$,

$$H_{\text{kOe}} = 120(W_{\text{MW}}/a_1)^{\frac{1}{2}},$$

so that for $a_1 = 5$ cm, 5 MW will give a field of 120 kOe and 1 MW is needed to give 50 kOe. Such power loads are not inconsiderable, whether they come from a motor generator controlled by excitation of the field winding or from a rectifier. In the past a limited number of redundant d.c. generators have been obtained cheaply from such sources as defunct electric tramways, submarines, etc. Nowadays, the obvious way to produce fields in the region from 20 to 120 kOe is with a superconducting magnet. The range above 120 kOe is still not so accessible and indeed 200 kOe may prove to be the approximate upper limit for super conductors. Therefore normal solenoids may continue to be a source of high fields (above 150–200 kOe) for a long time to come.

Descriptions of high-field coils have been given by Daniels (1953), Bitter (1962), de Klerk (1965) and in *High magnetic fields* (M.I.T. Press and Wiley, 1962). Most of those used at normal temperatures have been wound in one of the following ways:

(i) the Tsai-pattern in which copper ribbon, with an open helix of nylon filament around it, is wrapped in a spiral to form a pancake; the pancakes are packed together to form a cylindrical solenoid and the coolant flows across the surface of the ribbon (e.g. Daniels 1950);

(ii) the Bitter-type comprises circular copper plates with a radial slot cut out, which are stacked in such a way as to give a continuous electrical circuit: holes are punched and coolant can flow axially, although an alternative design uses radial flow (Bitter 1939, 1962);

(iii) spiral-wound tapes of different forms, e.g. those of H. Kolm at M.I.T. and M. F. Wood at Oxford, which are both described in *High magnetic fields* (1962, pp. 91 and 387 respectively).

(iv) a Leiden design (de Klerk 1965) and a Berkeley design (Giauque and Lyon 1960) use a coil wound from square-section copper wire: spacers allow axial flow of the coolant.

Of the coils mentioned above, the type designed by M. F. Wood has been available commercially from the Oxford Instrument Company Ltd.

For cooling the fluid used most commonly is demineralized water. It gives a more satisfactory heat transfer than the other fluids such as kerosene and orthodichlorobenzene which are sometimes used. Most of the above coils operate at high current and moderately low voltage ($\approx$ 100 V) so that demineralized water is suitable. Whichever coolant is used, a large capacity cooling tank or heat exchanger is necessary to transfer heat to the local water supply, river, harbour, etc. The circulation rate for water coolant to a 5-mW coil will be *ca.* 100 m^3/h.

Cooling with liquid nitrogen, hydrogen, or helium reduces the resistivity of the winding substantially and therefore reduces the *immediate* power requirements. However, the total power, which includes that used to produce the refrigerant, will not be reduced significantly, if at all. In some instances it may be an advantage to cool a coil of copper, aluminium, or sodium with liquid hydrogen if this liquid happens to be available in large quantities as at the Los Alamos Scientific Laboratory (e.g. H. Laquer in *High magnetic fields*, 1962, p. 156). But cooling with liquid helium has no extra merit for a continuously operated magnet coil made of a normal metal; remember that a part of the advantage of low resistivity which is gained by cooling a very pure metal to low temperatures is lost in high-field operation due to magneto-resistance.

Pulsed coil

In the B.S. days, that is 'before supermagnets', pulsed operation of magnet coils provided a very useful source of high fields without the necessity for large generators. Presumably they will continue to be useful for attaining fields above the limit of supermagnets, and applying them to experiments which can be performed during the brief life of the pulse.

Kapitza pioneered the use of high-pulsed fields by using batteries (chemically stored energy) and generators with large flywheels (mechanically stored energy) to provide a large pulse of electric power. More recently storage capacitors have been used also for this purpose (de Klerk 1965). Olsen (1953) has described how a small copper coil immersed in liquid helium achieved fields of nearly 200 kOe by discharge of a 3000-μF bank of condensers, charged to 300 V.

Superconducting magnets

As a result of the efforts of G. B. Yntema, S. H. Autler, J. E. Kunzler, and others, it has been shown that some 'hard' superconductors can carry appreciable currents in the presence of magnetic fields which are far in excess of the thermodynamic critical fields. This does not seem to be a general feature of type-II superconductors which are homogeneous and strain-free, but only of samples which are lacking in homogeneity or contain many physical defects which act as pinning points for the flux lines which penetrate the solid.

Since these high-field materials have been produced in wire form, they have become a conventional winding material for solenoids where relatively high magnetic fields are required. The supermagnets have limitations: firstly they require liquid helium as a coolant; operating at low fields ($\leqslant$ 10 kOe), some show irregularities in their magnetization, that is, the 'flux-jump' instability; also there are small remanence effects. But they are a very convenient alternative to either iron-core magnets or normal solenoids when fields up to 100 or perhaps even 150 kOe are needed.

It is possible either to buy a standard magnet plus power supply, sweep unit, dewar, and the accessories or to have a magnet made commercially to particular specifications or to buy suitable wire (or tape) and wind it in the laboratory. Suppliers are listed in the Appendix.

Superconducting magnets have been the subject of many review articles (e.g. Montgomery 1963, Kim 1964, Berlincourt 1963, Laverick 1965, 1966, de Klerk 1965) papers and conference reports (e.g. *High magnetic fields*, M.I.T. Press and Wiley, 1962; *Adv. cryogen. Engng* Vol. 8, 9, and 10. Plenum Press).

Materials. The metals which remain superconducting at high fields are either ductile alloys such as NbZr, NbTi, MoRe, or compounds of the β-wolfram structure. The latter have the formula Y_3X where Y is a transition element, Nb, V, Ta, and X is a group IIIA or IVA element: Al, Ga, In, Si, Ge.

The alloys which are currently available in wire form are Nb with 25, 33, or 50 atomic per cent Zr ($T_C = 10°–11°$ K) and

Nb + *ca.* 55 atomic per cent Ti ($T_C \simeq 10°$ K). The most commonly used is the Nb + 25 per cent Zr in the form of 0·010-in diam. wire, for which Fig. 126 (from Berlincourt 1963) shows critical current–magnetic field characteristics. Wires of higher zirconium content retain superconductivity in slightly higher fields but

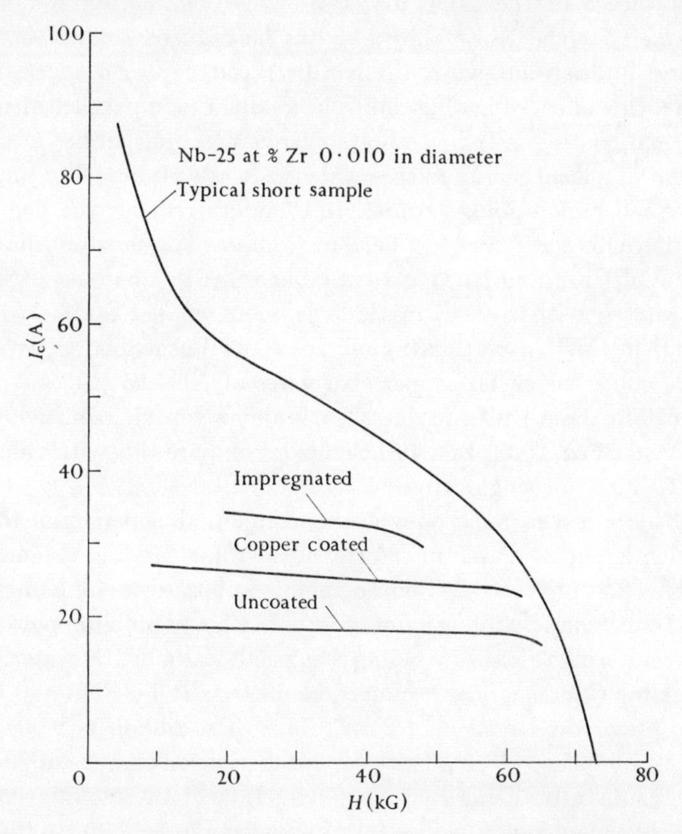

FIG. 126. Critical current against transverse magnetic field at 4° K for samples of Nb-25 atomic per cent Zr (after Berlincourt 1963).

values for the critical I_C are rather smaller for intermediate values of the field. Coils of Nb + 25 per cent Zr commonly reach *ca.* 60 kOe but NbZr coils of mixed composition have reached 70 kOe (Riemersma *et al* 1962). The NbTi alloy remains superconducting in still higher fields but can show greater flux-jump instability at low fields. Coffey *et al.* (1965) have made a 100 kOe magnet using Nb + 56 per cent Ti in the inner section of the coil and Nb + 25 per cent Zr in the outer part.

Fig. 126 also illustrates current degradation, which is the occurrence of much smaller critical currents in coils than in short test pieces of the same materials This degradation is quite marked for the NbZr alloy but can be reduced somewhat by using multistrand wires (in parallel) and copper-coating the wire. One effect of having multiple conductors in parallel with a normal coating is that it allows currents to redistribute in the event of a local energy release which may arise from a flux-jump or small region going normal. It therefore reduces the degradation effect and the low-field instability. The 0·010-in diam. wires of NbZr and NbTi are available either in a bare condition or coated with a 0·001 in thick layer of copper and/or with 0·001 in thick insulation (nylon, epoxy, or Formvar). An available cable has seven copper-clad wires of Nb + 25 per cent Zr (0·010-in diam.) in a Mylar sheath and a reported quenching current of *ca.* 200 A in a 40-kOe field; compare this with values of 20–30 A for single wires.

Nb_3Sn is the only one of the compound superconductors which is yet available in a form suitable for winding magnets. It was first produced at Bell Telephone Laboratories by Kunzler and colleagues by filling a fine tube with tin and niobium powder and sintering this after winding; the result was a brittle material. Flexible ribbon is now produced commercially by RCA and by the Supercon Division of NRC. The RCA ribbon is 0·004 in thick × 0·09 in wide, made by vapour deposition onto a stainless-steel substrate, using hydrogen reduction of the niobium and tin chloride vapours. Supercon ribbon is made *ca.* 0·002 in thick and in widths from 0·125 in by coating niobium foil with molten tin and reacting them to form a thin layer of Nb_3Sn. The ribbons are available with a coating of either copper or silver but without insulation; thin Mylar sheets are suggested as an insulation between each successive layer.

Coil construction. The major points to note in making a solenoid (Fig. 127) are as follows.

(i) The former can be any rigid material such as brass, copper, aluminium, stainless steel, Micarta. For larger coils there is some advantage in choosing a material whose expansion coefficient is similar to the wire, e.g. RCA suggest stainless steel for large coils. On the other hand, a high-conductivity material for the cylinder can cushion the effect of a coil going normal as the inductive coupling lengthens the decay time.

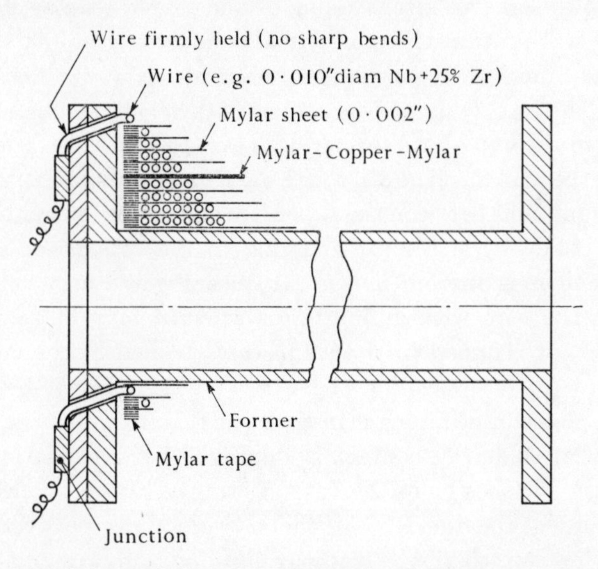

FIG. 127. Schematic diagram of a NbZr wire solenoid. Successive layers are separated by Mylar sheet and every fourth or fifth layer, this separation includes a thin copper (or aluminium) sheet.

(ii) The winding, whether wire or tape, should be firm because any movement induced by electromagnetic forces may cause quenching of the current. The firms which manufacture the ribbon recommend a tension of *ca.* 1 kg during winding.

(iii) Adequate insulation is needed between layers and between the coil and former to prevent electrical breakdown if and when the coil suddenly becomes normal. If this happens, the energy stored in coil, $LI^2/2$ or approximately $VH^2/8\pi$, must be dissipated quickly and can produce quite large voltages between different layers. Mylar sheet of *ca.* 0·002-in thickness wrapped around the former and between the layers will provide protection against breakdown. Stability, particularly at low fields, seems to be improved by having a sheet of copper or aluminium foil every few layers; sandwich sheets of Mylar–copper–Mylar can be obtained for this purpose. When using bare wire, the neighbouring turns should be separated to prevent any danger of circulating currents persisting after the magnet current is reduced to zero; an interwinding of fine copper wire or copper-plating will prevent this and will also improve stability at low fields and increase the current-carrying capacity (see Fig. 126).

(iv) The two ends of the coil are taken out through holes drilled in the end of the former with care to avoid sharp bends; anchor the wire mechanically with a potting compound or epoxy resin. Junctions between the superwire and copper current-leads can be made by either (a) soldering to the copper- or silver-plate if there is any on the wire, (b) cleaning and ultrasonically tinning the wire with indium after which it may be soldered, clamped, or crimped to copper pieces attached to the current leads, or (c) just cleaning the wire and clamping it between copper plates or crimping it in a copper tube to which the leads have been soldered; contact is improved by gold-plating the clamping surfaces or in the case of crimped tubes by cleaning and tinning the inside. Satisfactory junctions between two superwires can usually be made by cleaning, and crimping them side by side in a copper tube. Avoid placing the junctions in a high-field region. More details of winding and joining are given by Riemersma *et al.* (1962), Gugan (1963), Laverick (1965), and in the instruction sheets distributed by some wire manufacturers, e.g. RCA and Supercon.

Calculations of optimum dimensions may be made with the help of graphs and calculations given by Boom and Livingston (1962), Wood (1962), or the design charts produced by RCA and Supercon.

The price of wire or tape for magnets varies with composition, cladding, insulation, total unbroken length, manufacturer and with time, but a rough estimate of the present cost of winding a magnet can be made: about 2000 m of 0·010-in diam. wire is required for a typical 10-cm-long solenoid giving 40 kOe. The NbZr wire would cost *ca.* 50 cents (U.S.) per metre, so that the total amount would be *ca.* $\$10^3$. By comparison, 0·1-in wide Nb_3Sn tape costs *ca.* \$4 per metre but has a critical current of $\approx$ 100 A, which is four to five times higher than the wire.

Performance. The problem of current degradation in coils has already been mentioned and is illustrated in Fig. 126. This is alleviated somewhat by copper-cladding of the wire or the use of a number of wires in parallel, separated only by a normal conductor such as copper. Fig. 126 suggests that impregnation of the coil such as potting or painting with epoxy resin also helps. The value of this impregnation seems uncertain as it makes repair or reuse of the wire difficult and in the case illustrated it may have been partly due to better mechanical constraint. Evidence on this point is not clear.

Lowering the temperature of the helium bath around the coil can also have an odd effect. In many instances it has been observed that the critical current I_C is $\sim$ 1 A *smaller* when the temperature is reduced from 4·2° to say 2° K.

Remanent fields of up to 100 Oe may exist in a coil whose current has been cycled up and back to zero (see Gugan 1962, Anderson and Sarwinski 1963).

In some coils training is needed to get optimum performance after each time that they are cooled into the superconducting region. Some coils of NbZr may quench at 8 or 9 A when the current is first increased but will quench at successively higher currents until their maximum or optimum of say 20 A is reached.

If the coil is provided with a superconducting 'shorting' switch then it may be operated in the persistent mode which allows a very steady field to be maintained; also the current from the external source can be cut off, which stops Joule heating in the leads. The 'shorting' switch can consist of a few feet of the superwire which are in thermal contact with a heater (radio resistor) which drives the wire normal to open the switch; 1 ft of NbZr of 0·010-in diam. has an electrical resistance of about 2 Ω in the normal state.

Current supply. Wire coils requiring maximum current of *ca.* 25 A can easily be supplied by storage batteries with a carbon-pile rheostat or from a transistor circuit (e.g. McAvoy 1963). For many experiments it is preferable to have the current increased from zero to I_{max} at a very steady rate and this requires mechanical or electronic sweeping of the current. Some commercial power-supplies incorporate such sweep units whose speed is adjustable.

The leads which carry current to the magnet in the cryostat should be of optimum size to minimize the combined effect of Joule heating and heat conduction. This was discussed in an earlier chapter (§ 7.5) and it was pointed out that selection of optimum lead size also depends on two rather uncertain factors: first, the heat transfer from the leads to the evaporated helium-gas stream, and second the duty cycle of the magnet. A usual compromise for magnets needing a maximum of 25 A is to use a bunch of 10 or 12 copper wires of 34B and S gauge (each 0·016-cm diam.) or a single wire of *ca.* 0·06-cm diam.; the use of many fine wires exposed to the helium gas gives better heat transfer.

Obviously if we use seven-strand cable for our magnet with $I_C \approx 200$ A or Nb_3Sn tape with $I_C \approx 100$ A, the selection of optimum leads becomes more important. It is desirable to anchor the leads thermally with liquid air and to use two different sizes for the section from room temperature to 80° K and from 8° to 4° K.

The problem of supplying large currents to coils made of either multi-strand cable or superconductors of large cross-section has stimulated great interest in superconducting generators which are included in the cryostat. Many versions of the flux pumps, as they are called, have been developed. Some of these are discussed by de Klerk (1965) and Berlincourt (1963).

8. References

Books and review articles

Ambler, E. and Hudson, R. P. (1955). *Rep. Prog. Phys.* **18**, 251.
Berman, R. (1953). *Phil. Mag. Suppl.* **2**, 103.
Casimir, H. B. G. (1940). *Magnetism and very low temperatures* Cambridge University Press.

COOKE, A. H. (1955). *Prog. low Temp. Phys.* **1**, 224.

GARRETT, C. G. B. (1954). *Magnetic cooling.* Harvard University Press.

KIM, Y. B. (1964). *Physics today*, **17**, p. 21.

KLEMENS, P. G. (1958). *Solid St. Phys.* **7**, 1.

DE KLERK, D. (1956). *Handb. Phys.* **15**, 38.

—— (1965). *The construction of high-field electromagnets.* Newport Instruments Ltd., Buckinghamshire.

—— and STEENLAND, M. J. (1955). *Prog. low Temp. Phys.* **1**, 273.

KURTI, N. (1952). *Low temperature physics, four lectures*, p. 30. Pergamon Press, London.

LAVERICK, C. (1966). *Argonne National Laboratory Reviews*, **3**, 11.

LITTLE, W. A. (1964). *Prog. Cryogen.* **4**, 99.

MENDOZA, E. (1961). In *Experimental cryophysics* (ed. HOARE, JACKSON, and KURTI), p. 165. Butterworths, London.

MONTGOMERY, D. B. (1963). *Rep. Prog. Phys.* **26**, 69.

PARKINSON, D. H. and MULHALL, B. E. (1967). *The Generation of High Magnetic Fields.* Plenum Press, New York.

SIMON, F. E. (1937). *Very low temperatures*, p. 58. Science Museum Handbook, H.M.S.O., London.

—— (1939). *Science Prog.* No. 133, p. 31.

Papers

ABEL, W. R., ANDERSON, A. C., and WHEATLEY, J. C. (1964). *Rev. scient. Instrum.* **34**, 444.

—— —— BLACK, W. C., and WHEATLEY, J. C. (1965). *Physics* **1**, 337.

AMBLER, E. and HUDSON, R. P. (1956). *Phys. Rev.* **102**, 916.

—— —— (1957). *J. chem. phys.* **27**, 378.

ANDERSON, A. C., REESE, W., and WHEATLEY, J. C. (1963). *Rev. scient. Instrum.* **34**, 1386.

—— SALINGER, G. L. and WHEATLEY, J. C. (1961). Ibid. **32**, 1110.

—— and SARWINSKI, R. J. (1963). Ibid. **34**, 288.

BERLINCOURT, T. (1963). *Br. J. appl. Phys.* **14**, 9.

BERMAN, R., FOSTER, E. L., and ROSENBERG, H. M. (1955). Ibid. **6**, 181.

BETTS, D. S., EDMONDS, D. T., KEEN, B. E., and MATTHEWS, P. W. (1964). *J. scient. Instrum.* **41**, 515.

BITTER, F. (1939). *Rev. scient. Instrum.* **10**, 373.

—— (1962). Ibid. **33**, 342.

—— and REED, F. E. (1951). Ibid. **22**, 171.

BOOM, R. W. and LIVINGSTON, R. S. (1962). *Proc. Inst. Radio Engrs* **50**, 274.

CASIMIR, H. B. G. (1938). *Physica* **5**, 495.

—— DE HAAS, W. J., and DE KLERK, D. (1939). Ibid. **6**, 241.

COFFEY, H. T., HULM, J. K., REYNOLD, W. T., FOX, D. K., and SPAN, R. E. (1965). *J. appl. Phys.* **36**, 128.

CONNOLLY, J. I., ROACH, W. R., and SARWINSKI, R. J. (1965). *Rev. scient. Instrum.* **36**, 1370.

COOKE, A. H. (1949). *Proc. phys. Soc.* A**62**, 269.

—— MEYER, H., and WOLF, W. P. (1956). *Proc. R. Soc.* A**233**, 536.

Daniels, J. M. (1950). *Proc. phys. Soc.* B**63**, 1028.
—— (1953). Ibid. B**66**, 921.
—— and Robinson, F. N. H. (1953). *Phil. Mag.* **44**, 630.
Darby, J., Hatton, J., Rollin, B. V., Seymour, E. F. W., and Silsbee, H. B. (1951). *Proc. phys. Soc.* A**64**, 861.
Daunt, J. G. (1957). Ibid. B**70**, 641.
—— Heer, C. V., McMahon, H. O., Reitzel, J., and Simon, L. (1955). *Conference de physique des basses températures*, p. 362. *Suppl. Bull. int. Inst. Refrig.*
—— and Pillinger, W. L. (1955). *Suppl. Bull. int. Inst. Refrig.*, Annexe 1955–3, p. 158.
Debye, P. (1926). *Ann. Phys.* **81**, 1154.
Dugdale, J. S. and MacDonald, D. K. C. (1957). *Can. J. Phys.* **35**, 271.
—— —— and Croxon, A. A. M. (1957). Ibid. **35**, 502.
Erickson, R. A., Roberts, L. D., and Dabbs, J. W. T. (1954). *Rev. scient. Instrum.* **25**, 1178.
Frankel, R. B., Shirley, D. A., and Stone, N. J. (1965). *Phys. Rev.* **140**, A1020; ibid. **143**, 334.
Fritz, J. J. and Giauque, W. F. (1949). *J. Am. chem. Soc.* **71**, 2168.
Geballe, T. H. and Giauque, W. F. (1952). Ibid. **74**, 3513.
Giauque, W. F. (1927). Ibid. **49**, 1864.
—— Geballe, T. H., Lyon, D. N., and Fritz, J. J. (1952). *Rev. scient. Instrum.* **23**, 169.
—— and Lyon, D. N. (1960). Ibid. **31**, 374.
—— and MacDougall, D. P. (1933). *Phys. Rev.* **43**, 768; **44**, 235.
—— —— (1935). *J. Am. chem. Soc.* **57**, 1175.
Goodman, B. B. (1953). *Proc. Phys. Soc.* A**66**, 217.
Gorter, C. J. (1934). *Phys. Z.* **35**, 923.
Gugan, D. (1963). *Cryogenics* **3**, 220.
de Haas, W. J. and Wiersma, E. C. (1935). *Physica* **2**, 335.
—— —— and Kramers, H. A. (1933). *Nature, Lond.* **131**, 719.
Heer, C. V., Barnes, C. B., and Daunt, J. G. (1954). *Rev. scient. Instrum.* **25**, 1088.
—— and Rauch, C. J. (1955). *Suppl. Bull. int. Inst. Refrig.* Annexe 1955–3, p. 218.
Hegenbarth, F. (1965). *Physica Status Solidi* **8**, 59.
Hudson, R. P. (1949). *J. scient. Instrum.* **26**, 401.
—— Hunt, B., and Kurti, N. (1949). *Proc. phys. Soc.* A**62**, 392.
—— and Kaeser, R. S. (1967). *Physics*, **3**, 95.
—— and McLane, C. K. (1954). *Rev. scient. Instrum.* **25**, 190.
Hull, R. A. (1947). *Rep. Cambridge Conf. Low Temp.*, p. 72. Physical Society, London.
Känzig, W., Hart, H. R., and Roberts, S. (1964). *Phys. Rev. Lett.* **13**, 543.
de Klerk, D. and Hudson, R. P. (1954). *J. Res. natn. Bur. Stand.* **53**, 173.
Kurti, N., Robinson, F. N. H., Simon, F. E., and Spohr, D. A. (1956). *Nature, Lond.* **178**, 450.

KURTI, N. ROLLIN, B. V., and SIMON, F. E. (1936). *Physica* **3**, 266.
—— and SIMON, F. E. (1934). *Nature, Lond.* **133**, 907.
—— —— (1935). *Proc. R. Soc.* A**149**, 152.
—— —— (1938 a). *Phil. Mag.* **26**, 849.
—— —— (1938 b). Ibid. 840.
—— (1963). *Adv. cryogen. Engng* **8**, 1.
LAVERICK, C. (1965). *Cryogenics* **5**, 152.
MCAVOY, B. R. (1963). *Rev. scient. Instrum.* **34**, 200.
MAXWELL, E. (1965). Ibid. **36**, 553.
MENDOZA, E. (1948). *Les phénomènes cryomagnétiques*, p. 53. Collège de France.
—— and THOMAS, J. G. (1951). *Phil. Mag.* **42**, 291.
NICOL, J. (1955). *Nat. science foundation conf. low temp. phys. chem.* Paper K-4. Baton Rouge, Louisiana.
OLSEN, J. L. (1953). *Helv. phys. Acta* **26**, 798.
PARKS, R. D. (1963). *Proc. 8th Int. Conf. Low Temp. Phys.* p. 427. Butterworths, London.
PILLINGER, W. L., JASTRAM, P. S., and DAUNT, J. G. (1958). *Rev. scient. Instrum.* **29**, 159.
RIEMERSMA, H., HULM, J. K., VENTURINO, A. J., and CHANDRASEKHAR, B. S. (1962). *J. appl. Phys.* **33**, 3499.
VILCHES, O. E. and WHEATLEY, J. C. (1966). *Rev. scient. Instrum.* **37**, 819.
WALSTEDT, R. E., HAHN, E. L., FROIDEVAUX, G., and GEISSLER, E. (1965). *Proc. R. Soc.* A**284**, 499.
WHEATLEY, J. C. (1966). *Soumal. Tiedeakat. Toim.* **AVI**, 210, p. 15.
—— GRIFFING, D. F., and ESTLE, T. L. (1956). *Rev. scient. Instrum.* **27**, 1070.
WHITE, G. K. (1955). *Can. J. Phys.* **33**, 119.
WOOD, M. (1962). *Cryogenics* **2**, 297.
YAQUB, M. (1960). Ibid. **1**, 101.
ZIMMERMAN, J. E., MCNUTT, J. D., and BOHM, H. V. (1962). Ibid. **2**, 153.

CHAPTER X

VACUUM TECHNIQUES AND MATERIALS

1. Introduction

IT is clear from a consideration of heat transfer that one of the most important requirements in low-temperature research is that of maintaining a high vacuum. Whether this vacuum be in a space around the refrigerant liquid or in an enclosure within the liquid, pressures of the order of 10^{-6} mm Hg or less must often be maintained to ensure thermal isolation or thermal equilibrium. The general techniques of producing high vacua have been discussed in such books as Strong (1938), Dushman (1949), Yarwood (1955), Pirani and Yarwood (1961), and Barrington (1963). Here we shall merely point out some features of a suitable pumping system and summarize the equations for the pumping speed of tubes and orifices, so that pumps and pumping-tube dimensions may be chosen which are commensurate with one another and which may meet such practical requirements as the removal of exchange gas from a cryostat.

The books by Strong (1938) on laboratory techniques and by Pirani and Yarwood (1961) discuss the common methods of making seals, e.g. metal–glass seals, sealing cements, solders, etc. In low-temperature work some special problems arise as many cement seals and rubber gasket or rubber O-ring seals are no longer practicable for vacuum-tight assemblies. In § 3 of this chapter a brief account and discussion is given of available low-melting-point solders which are particularly useful in making vacuum-tight joints which can be cooled and can also be 'broken' and remade fairly readily. Some other types of vacuum seals and cements as well as the use of metal–glass seals at low temperatures are described later in the chapter.

2. Vacuum technique

High-vacuum system

In most low-temperature equipment, a vacuum system is required to produce and maintain pressures of the order of 10^{-6} mm

Hg in a relatively small chamber. As there are or should be no permanent gas leaks into the chamber (cf. a nuclear accelerator or ion source into which a small regulated leak of hydrogen, deuterium, or helium is maintained), pumping speeds are not of primary importance in maintaining a low pressure. However, in many cases a small pressure (say 10^{-1} mm Hg) of helium gas may be admitted to the chamber for purposes of obtaining thermal equilibrium. The time required for the subsequent removal of this exchange gas will depend on the effective speeds of the pumps and connecting tubes; unless an unusually short time of the order of seconds is required for this removal, the pumping speed of the system need only be of the order of a litre per second, i.e. much smaller than the speeds of hundreds or thousands of litres per second required in ion accelerators. It is fortunate that pumping speed requirements are rather modest, since the design of an experimental cryostat usually restricts the possible size of pumping tubes which connect the experimental chambers (enclosed as they are in a dewar vessel) to the pumping system. A conventional pumping system is illustrated in the schematic diagram of Fig. 128. Pure helium for use as exchange gas (or for filling gas thermometers) is kept in the flask; this can be replenished by taking helium gas from a cylinder through a liquid-nitrogen-cooled charcoal trap or using evaporated helium gas from a liquid-helium storage dewar.

G_1 and G_2 are vacuum gauges for determining the approximate pressure in the system. G_1 is conveniently a discharge gauge in which a high voltage from a Tessla coil or a Ford coil produces a visible discharge at pressures from $\sim 10^{-2}$ mm Hg to ~ 10 mm Hg or higher depending on the gas present. The character of the discharge is determined by the gas pressure and type of gas. With a little experience, helium gas (giving a pinkish discharge turning pale green at lower pressures), nitrogen (reddish), oxygen (straw yellow), carbon dioxide (whitish blue), and water or organic vapours (bluish colour) can be distinguished. For measurement of lower pressures (the function of G_2 in Fig. 128) a number of different types of gauge can be made or obtained from manufacturers of high-vacuum equipment. Among those commonly used are (i) the thermionic ionization gauge with a lower limit of 10^{-8} to 10^{-9} mm Hg, (ii) the mercury McLeod gauge which has a lower limit of about 10^{-6} mm Hg and does not register the pressure of condensable vapours, (iii) the Knudsen gauge which registers the true pressure of gas and vapour irrespective of their composition down to 10^{-6} mm Hg, (iv) the hot-wire Pirani gauge with a lower limit of 10^{-3} to 10^{-4} mm Hg, (v) the Philips (or

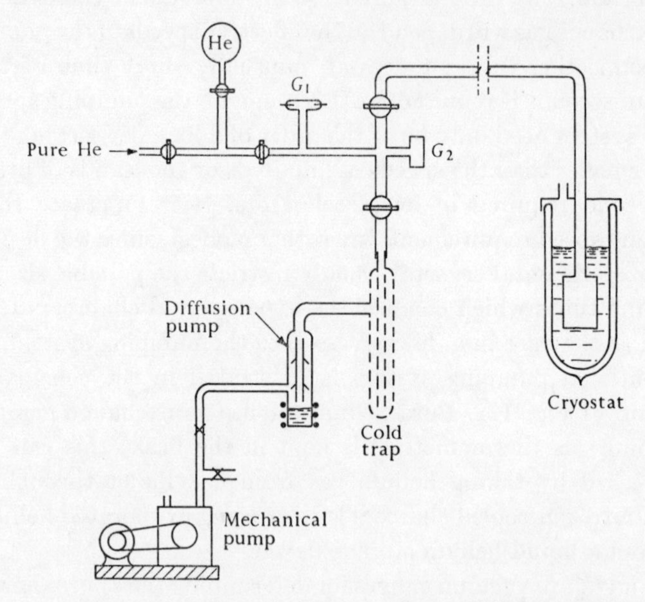

Fig. 128. Schematic diagram of a pumping system.

Penning) cold cathode ionization gauge which is available with a lower limit of 10^{-6} to 10^{-7} mm Hg. Of these the Philips gauge has many advantages in range, simplicity, and reliability of operation but its sensitivity varies somewhat with the composition of the gas present so that it only provides an approximate measure of the pressure; the usual calibration supplied with a commercial model is approximately correct for air. Whichever gauge is used, a record of the pressure in the cryostat chamber is only obtained if the gauge is in a position close to the chamber, i.e. if the resistance of the pumping line (reciprocal of its pumping speed) is much less between the point of attachment of the gauge and the chamber than it is between this point and the throat of the diffusion pump. A further problem arises when the cryostat chamber is surrounded by liquid helium as the vapour pressure of gases other than helium will be negligible at this temperature. When the system is in equilibrium, and if there is no helium leak into the cryostat, the gauge may register a much higher pressure than exists in the cryostat chamber; the gauge is at room temperature, and is exposed to gases and vapours arising from the nearby tube walls, tap grease, back diffusion through the pump, etc., but these gases (other than helium) do not affect the static pressure in the chamber which is at liquid-helium temperatures. As Garfunkel and Wexler (1954) have pointed out, one of the only reliable means of measuring the helium-gas pressure in such a chamber, which may be as low as 10^{-8} or 10^{-9} mm Hg, is to connect a helium mass spectrometer leak detector on the high-pressure side of the diffusion pump and determine the rate at which helium gas is being 'ejected' by this pump.

The choice of metal or glass for the pumping system depends partly on the desired pumping speed and on individual preference as well as the availability of materials.

If large pumping speeds and hence large pumping tubes are required, metal tubes and large-bore metal valves—necessarily rather expensive or difficult to make—are needed. If neither a large pumping speed nor a mechanically robust system is necessary and a competent glass-blower is available, the entire pumping system (excepting the rotary mechanical pump) may be constructed relatively quickly from glass.

In Fig. 128 a cold trap (usually liquid-nitrogen cooled) is shown in dashed lines. With a mercury diffusion pump, this trap is necessary to reduce the ultimate vacuum below the vapour pressure of mercury ($\sim 10^{-3}$ mm Hg) at room temperature. In an oil-diffusion pump the ultimate pressure without a cold trap varies from 10^{-4} mm Hg to about 10^{-7} mm Hg depending on the type of oil used and the design of the pump. Oil-diffusion pumps are, however, rather less robust in operation due to the danger of cracking the hot oil by exposure to air, and they generally require a somewhat lower backing pressure for efficient operation; while many mercury diffusion pumps operate satisfactorily with a maximum backing pressure in the range 1 to 5 mm, most oil diffusion pumps require that the backing pressure be kept well below 0·5 mm Hg.

These questions are discussed in much greater detail in the standard texts on vacuum technique by Yarwood, Dushman, Strong, etc.

Pumping speed

The pumping speed of an orifice is expressed as the volume of gas passing through the orifice per unit time measured at the pressure which exists at the orifice; denoted by S this speed is usually expressed in litres per second. When the mean free path of the gas molecules is appreciably greater than the dimensions of the orifice or width of pumping tubes, so-called Knudsen conditions apply, i.e. the molecules diffuse without being affected by the presence of other gas molecules. Since the mean free path is inversely proportional to pressure and in most gases at room temperature is 100 cm for a pressure of 10^{-4} mm Hg, the condition for Knudsen flow is seen to be usually satisfied at the pressures produced by a diffusion pump, unless the dimensions of the pumping tubes and orifices are very great.

Under such high-vacuum conditions, the pumping speed or volume of gas escaping per unit time through an orifice is given by

$$S = 3{\cdot}64A\sqrt{(T/M)}\ \text{l/s},$$

where A is the area in square centimetres, T is the absolute temperature, and M is the molecular weight of the gas. Hence, for air at room temperature

$$S = 11{\cdot}7\ \text{l/s cm}^2.$$

Thus we may expect the pumping speed at the throat of a diffusion pump to be partly controlled by the available area of the annular aperture through which the gas diffuses before it is entrapped by the streaming vapour of the pump. In fact the efficiency of diffusion pumps is rather less than unity so that the effective speed for air is given by

$$S_P = \text{speed factor} \times A \times 11{\cdot}7 \text{ l/s},$$

where the 'speed factor' has a value of about 0·4 for a good oil diffusion pump and about 0·2–0·3 for a mercury diffusion pump.

In the case of a cylindrical pumping tube for which the length l is much greater than the diameter d, the speed or conductance is given by

$$S_L = 3{\cdot}82\sqrt{\left(\frac{T}{M}\right)}\frac{d^3}{l} \text{ l/s}.$$

For air at room temperature this reduces to

$$S_L \simeq 100\frac{r^3}{l} \quad (\text{radius } r = \tfrac{1}{2}d)$$

$$= \frac{r_{\text{mm}}^3}{l_{\text{mm}}} \text{ l/s}.$$

For helium gas, $M = 4$ and therefore the speed of an orifice, pump, or tube is $\sqrt{(29/4)} \simeq 2{\cdot}7$ times greater than for air. If a section of the pumping tube is at a low temperature T the speed of this section will be reduced proportionately to $T^{\frac{1}{2}}$ due to the lower kinetic velocity of the gas molecules at this temperature; this refers to the speed measured at the pressure existing in the cold tube, not to the mass flow or equivalent room-temperature speed which is increased as $T^{-\frac{1}{2}}$. Otherwise it might appear that for a cryostat pumping system, tubes of larger bore are required for the colder section than for the room temperature section. This is, of course, not the case, as a detailed analysis shows, since the pressure of a given mass of gas being taken through the system will be proportional to the temperature at any point assuming it is in equilibrium. Garfunkel and Wexler (1954) have calculated the effective room-temperature speed S_r for a number of tubes in series in a temperature gradient; if each of these tubes is of length l_i and radius r_i and they are connected at one end to an enclosure (temperature T_1, pressure P_1) and at the other to a 'pump' at room temperature T_r where it is assumed $P = 0$,

$$S_r = \frac{4(2\pi RT_r)^{\frac{1}{2}}}{3\sqrt{M}\sum\frac{l_i}{r_i^3}\left(\frac{T_r}{T_i}\right)^{\frac{1}{2}}}$$

which, if $T_r \simeq 295°$ K, reduces to

$$S_r \simeq \frac{9000}{\sqrt{(MT_1)}\Sigma \frac{l_i}{r_i^3}} \text{ l/s},$$

where l_i and r_i are in centimetres. In this derivation it is assumed that Knudsen conditions apply, that $l_i \gg r_i$, and that the gradient of P/T is small. Thus the effective speed varies inversely as the square root of the temperature T_1 at the cold end.

If we wish to calculate the time required to reduce the pressure in a volume V from P_1 to P_2, where the limiting low pressure that may be reached is P_0, time t is given by

$$t = \frac{V}{S} \ln \frac{P_1 - P_0}{P_2 - P_0};$$

and if we assume that $P_0 \ll P_1, P_2$,

$$t = 2{\cdot}3 \frac{V}{S} \log_{10} \frac{P_1}{P_2}.$$

For example, suppose P is initially 1 mm Hg in a volume of 1 litre and P_2 is to be 10^{-5} mm Hg, then if $S_P = 5$ l/s and $S_L = 0{\cdot}5$ l/s,

$$\frac{1}{S} = \frac{1}{5} + \frac{1}{0{\cdot}5}$$

$$= 2{\cdot}2$$

and

$$t = 2{\cdot}3 \frac{1}{1/2{\cdot}2} \log_{10}(10^5)$$

$$= 25 \text{ s}.$$

In practice the time required would be longer than this as the speed S_P of the diffusion pump would be small until the pressure was reduced to $\sim 10^{-3}$ mm Hg. From the initial pressure of 1 mm Hg down to 10^{-3} mm Hg, S_P would be largely governed by the speed of the backing pump and the tubing connecting the backing pump to the diffusion pump. Also, the speed S_L of the high-vacuum line would be much reduced in this higher pressure range as the conditions for Knudsen flow are not satisfied, i.e. collisions between gas molecules would reduce the diffusion rate.

At sufficiently high pressures that the mean free path is very much less than the lateral dimensions of the pumping tubes, we may assume normal laminar flow governed by Poiseuille's equation. In the intermediate region, usually corresponding to a pressure of 10^{-1} to 10^{-2} mm Hg, the mean free path and the lateral dimensions are comparable and precise calculation of the effective conductance of a tube is difficult.

We stated that the speed of a diffusion pump is controlled by the rate of diffusion of gas molecules through the throat of the pump into the vapour stream. In a rotary mechanical pump an oil-sealed eccentric rotor revolves in a cylindrical cavity 'sweeping' out the gas from the high pressure or entry side to the exhaust side at a speed depending on the number of revolutions per unit time and the volume of the space between the rotor and the fixed cylinder. In a pumping system, e.g. that of Fig. 128, the speed of the mechanical pump need be much less than the speed of the diffusion pump. For example, if the speed of the diffusion pump is $S_{\mathrm{P}} = 50$ l/s at a pressure of 10^{-4} mm Hg, and the pump requires a backing pressure of 10^{-1} mm Hg, then the speed of the mechanical pump need be only

$$S_{\mathrm{M}} = \frac{10^{-4}}{10^{-1}} \times 50 = 0{\cdot}05 \text{ l/s}$$

$$= 3 \text{ l/min};$$

it is advisable in practice to increase this figure a few times to allow for the restricting effect of the tubing between the pumps. Backing pump speeds are often expressed in litres per minute rather than in litres per second.

It should be emphasized that in designing a pumping system, consideration should be given to the matching of the pumping speeds of the various components; therefore an approximate estimate should be made of the speed of the proposed pumping tubes between the diffusion pump and the cryostat. It is pointless to make or buy a large 200-l/s diffusion pump, when it is to be connected by a length of 100 cm of 2-cm bore tubing (speed $S_{\mathrm{L}} = r^3_{\mathrm{mm}}/l_{\mathrm{mm}} = 1$ l/s) to the vacuum chamber of a cryostat; the total speed or conductance is the reciprocal of the sum of the reciprocals of the individual speeds, i.e. the resistances are additive so that $S = (\frac{1}{200}+\frac{1}{1})^{-1} \simeq 1$ l/s in this example. Likewise it is rather wasteful to use a 450-l/min mechanical pump to back an 8-l/s diffusion pump as a 50-l/min mechanical pump is just as effective.

An illustration of the dependence of pumping speed on pressure for some common mechanical and diffusion pumps is given in Fig. 129. With dry air the one- and two-stage mechanical pumps

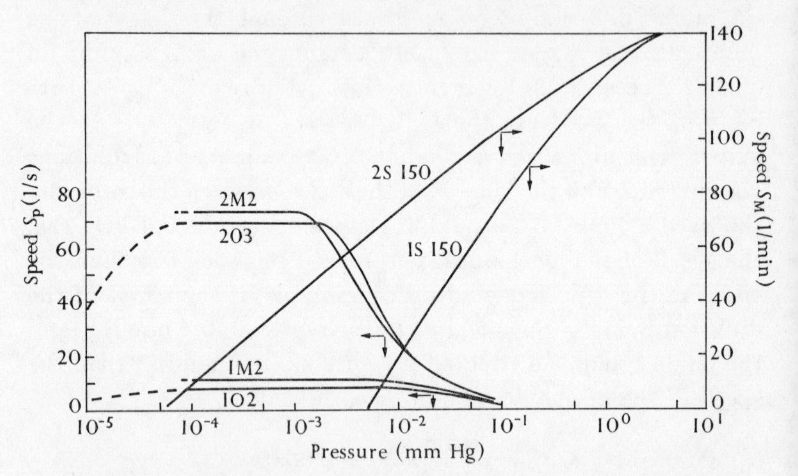

Fig. 129. Speeds of some commercial pumps (Edwards and Company, London) for dry air at room temperature. 1S150, Speedivac rotary pump: 1-stage; 2S150, Speedivac rotary pump: 2-stage; 1M2, mercury-diffusion pump: 2-stage; 2M2, mercury-diffusion pump: 2-stage; 1O2, oil-diffusion pump: 2-stage; 2O3, oil-diffusion pump: 3-stage.

should give ultimate vacua of about 10^{-2} and 10^{-4} mm Hg respectively. This assumes that no condensable vapours are present which may contaminate the oil. Mechanical pumps do not remove condensable vapours efficiently unless they are fitted with a gas ballast system, i.e. an automatic system which injects gas every cycle to aid in 'sweeping' away the condensable vapours. A diffusion pump does pump condensable vapours although the speed will vary with the molecular weight of the vapour. The mercury-diffusion pumps 1M2 and 2M2 will yield an ultimate vacuum of about 10^{-6} mm Hg if used with a cold trap. The oil-diffusion pumps 102 and 203 may be expected to give an ultimate vacuum (without cold trap) of 10^{-5} to 10^{-6} mm Hg, depending on the type of oil and the heating power used.

Low-temperature leaks

Cryogenic apparatus does not usually require a pumping system of high speed. By necessity the size of the connecting tubes into a cryostat limits pumping speed severely. This limitation means that leaks, if present, will be a serious problem. Often they will be serious not only because of low pumping speeds but because the leaks will be exposed to liquid refrigerants at low temperatures, which makes the leak rates much greater. Leak-hunting is usually done at room temperature with a mass spectrometer or by cruder means, and may show no sign of a leak; and yet when the vacuum chamber is covered with liquid helium, a leak may become all too obvious but difficult to locate with any precision. The most annoying leaks are the λ-leaks which only appear when liquid helium II surrounds the offending part. The superfluid helium will creep through fine cracks of 0·1 μm or less in diameter at a rate which makes it impossible to maintain a sufficiently high vacuum for many experiments, e.g. calorimetry. Yet at room temperature the leak rate may be so small that a mass spectrometer detecting 10^{-10} cm^3 s^{-1} will not give any indication of a leak. Such leaks as this are not uncommon in silver–solder joints which have been overheated, or in porous brass plates or in copper which contains oxygen and becomes slightly porous after heating. They can be avoided by careful selection of materials and soldering techniques. Croft (1963) included such leaks in his interesting review of 'Cryogenic difficulties'.

3. Solders

The most commonly used solders may be divided into three main classes: hard solders, soft solders, and low-melting-point solders. The latter have a frequent application in low-temperature equipment as they provide a means of making and breaking a vacuum joint without raising the temperature of the adjacent metals to more than about 100° C. More detailed information on the composition of hard and soft solders than is given below may be obtained from physical tables or a standard handbook such as the *Metals Handbook* (American Society of Metals, Cleveland, Ohio).

Hard solders

These solders generally melt between 600° and 1000° C and are nominally a Cu–Zn brazing alloy with added silver; those having around 50 per cent silver content have the lowest melting points. They are generally used with a borax or boric acid flux, the two components being commonly mixed to a paste with alcohol. Such fluxes are readily available commercially from firms which supply hard solders, e.g. Handy and Harman, Connecticut, or Johnson Matthey, London. Among the alloys are:

'Easy Flo' (Handy and Harman)—a strong solder melting at about 630° C, containing 50% Ag, 15·5% Cu, 16·5% Zn, 18% Cd.

'A.S.T.M. Grade 4'—a strong solder of relatively low melting range (675°–745° C) containing 45% Ag, 30% Cu, 25% Zn.

'A.S.T.M. Grade 2'—flows readily and melts over range 775°–815° C and contains 20% Ag, 45% Cu, 35%Zn.

'A.S.T.M. Grade 7'—high grade malleable and ductile solder containing 70% Ag, 20% Cu, 10% Zn which melts over the range 725°–755° C.

Silver—100% Ag melts at 960° C.

The three solders listed with A.S.T.M. (American Society for Testing Materials) specifications are fairly representative of commonly used silver solders, their compositions being taken from *Metals Handbook* p. 1211 (American Society of Metals, 1939).

Soft solders

Soft solders are generally tin–lead alloys, the eutectic mixture of 63 per cent Sn, 37 per cent Pb having a melting point of 183° C. Mild fluxes such as rosin, rosin in alcohol, or a paste of petroleum jelly, zinc chloride, and ammonium chloride are often used in radio soldering. When a flux with a stronger cleaning action is required as in many metal–metal junctions, a zinc chloride solution is formed by dissolving zinc in hydrochloric acid. For soldering to stainless steel excess hydrochloric acid is desirable or a strong solution of phosphoric acid may be used. Commercially available fluxes include 'Baker's Fluid' (Baker's Soldering Fluxes, Middlesex, England), Kester Soldering Salt (Kester Solder Co., Chicago, Illinois) and a stainless steel soldering flux (Dunton Company, Rhode Island).

The solders include:

37·5% Sn, 60% Pb, 2·5% Sb—melting over a range from 185°–225° C; this is generally used for joining lead pipes and cable sheaths.

40–45% Sn, 60–55% Pb—these are very ductile solders, melting over a range from about 183°–230° C and used generally in radio soldering.

50% Sn, 50% Pb—a good general-purpose solder melting from 180°–225° C.

60% Sn, 40% Pb—a high-quality general-purpose solder melting from 183°–191° C.

100% Pb—pure lead is sometimes useful because of its higher melting point of 327° C.

50% Sn, 32% Pb, 18% Cd—the addition of cadmium reduces the melting point in this alloy ('low-temperature solder') to 145° C.

Low-melting-point solders

These solders, some of which melt well below 100° C, are alloys of bismuth with the elements lead, tin, cadmium (and indium in some instances). Like pure bismuth, those alloys containing a high percentage of bismuth (usually more than about 50 per cent) expand on solidification. The alloys are rather weak and brittle and should be used for joints or seals where they are not subject to stress, particularly bending stress. If used in a situation illustrated by Fig. 130 (a) with copper or brass, then Wood's metal (or other low-melting-point alloy) should give a

21

completely satisfactory vacuum-tight joint, unaffected by repeated cooling to liquid-helium temperatures. It is advisable with these solders, particularly those with melting points at 100° C or less, to 'pre-tin' the metal surfaces which are to be joined, with a soldering iron beforehand. These alloys do not generally 'tin' when at or just above their liquid points. Therefore parts should be pre-tinned before assembly using sufficient heat to activate the flux (zinc chloride solution, Kester's soldering salt, Baker's fluid are satisfactory).

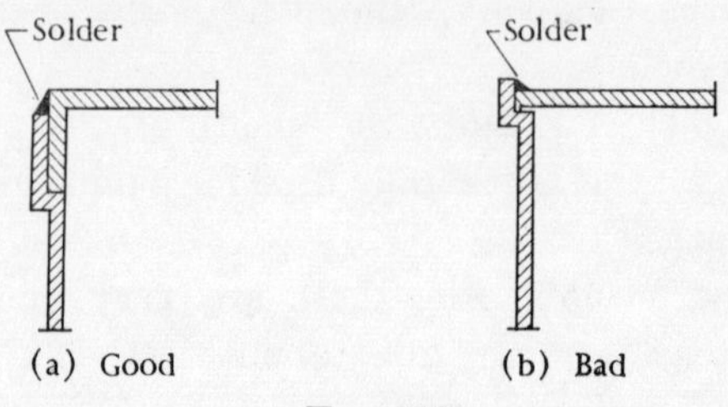

Fig. 130

Apart from Wood's metal, the alloys listed below are Cerro-alloys, commercially available from the Cerro de Pasco Copper Corporation (New York). These latter alloys are chiefly produced for making moulds, castings, fillers to aid tube bending but provide a useful range of low-melting-point solders.

Wood's metal—50% Bi, 25% Pb, 12·5% Sn, and 12·5% Cd melts from 65° to 70° C.

Cerrobend—a eutectic alloy of 50% Bi, 26·7% Pb, 13·3% Sn, and 10% Cd melts at 70° C. It can be conveniently used as a substitute for Wood's metal and expands slightly on casting.

Cerrobase—a eutectic alloy of 55·5% Bi and 44·5% Pb melts at 124° C, contracts slightly on casting.

Cerrotru—a eutectic alloy of 58% Bi and 42% Sn melts at 138° C. This alloy expands slightly on casting but then shows very little further growth or shrinkage, unlike most of these alloys which continue to grow or shrink slightly for some hours after casting.

Cerrolow 117—a eutectic alloy of 44·7% Bi, 22·6% Pb, 8·3% Sn, 5·3% Cd, and 19·1% In melts at 47° C.

Cerrolow 136—a eutectic alloy of 49% Bi, 18% Pb, 12% Sn, and 21% In melts at 58° C and like Cerrolow 117 expands slightly on casting and later shows a slight shrinkage for some hours.

Cerroseal 35—containing approximately equal amounts of Sn and In, this is a special glass-wetting alloy which melts from 116°–126° C.

A large number of other low-melting alloys, e.g. Rose's metal, Lipowitz's metal, are listed with their compositions and melting points in the *Metals Handbook* (American Society of Metals) and many standard physical tables.

Special solders

Listed below are some less common solders of interest in low-temperature research:

Zinc–cadmium solder†—a eutectic alloy of 82·5% Cd and 17·5% Zn melting at 265° C is often used for attaching electrical potential leads to metal specimens as it is not superconducting in the liquid-helium range of temperatures, unlike Wood's metal and most other soft solders which are superconducting below 7° or 8° K. This Zn–Cd solder flows quite freely on many metal surfaces when molten and is used in a similar manner to a tin–lead soft solder.

Bismuth—this is a non-superconducting solder with melting point at 271° C which forms a rather mechanically weak joint as it does not 'wet' a metal surface very thoroughly.

Bismuth–cadmium solder—a eutectic alloy of 60% Bi and 40% Cd melting at 140° C which does not become superconducting above 0·8° K (Cochran *et al.* 1956).

Indium—melting at 155° C; indium is a useful solder as it bonds to many metals more thoroughly than does a tin–lead solder. Due to its high ductility it does not form a very strong joint; it has been used for soldering to many thin metal films on glass (Belser 1954).

† It is advisable to avoid overheating of the cadmium-rich solders as cadmium fumes are toxic.

Thermal-free solder—a solder containing 70·44% Cd, 29·56% Sn, which gives a very low thermo-electric force with respect to copper, is commercially produced by the Leeds and Northrup Company, Philadelphia.

4. Vacuum seals and cements

Metal–glass seals

The technique of making glass–metal vacuum seals has been discussed in many available texts (e.g. Strong 1938, Partridge 1949, Reimann 1952, Pirani and Yarwood 1961). Mention should be made here of their ability to withstand frequent cooling to the temperatures of liquid helium and remain vacuum-tight.

Copper–glass seals of the Housekeeper type appear to be generally reliable and have been used, for example, in the 'tail' of small Linde helium liquefiers for long periods without any cracking. They have been used in sizes at least up to 1-in diameter and remain vacuum-tight in the presence of liquid helium II, the 'super-fluid component' of which has a vanishingly small viscosity and therefore magnifies the effect of any small leak that is present.

Seals made with platinum wire or platinum tubing to soft glass also seem to be quite satisfactory in maintaining complete vacuum tightness at liquid-helium temperatures. If adequately annealed they can withstand repeated cycling between room temperature and the lowest temperatures.

Seals to hard glass can be made with metals having suitable expansion characteristics, the linear expansion coefficient usually being between 4 and 5×10^{-6} degK^{-1} at room temperature. Fernico and Kovar are two such alloys, containing 54% Fe, 28–29% Ni, 17–18% Co ($\alpha\approx4{\cdot}6\times10^{-6}$ degK^{-1} at room temperature), which seal to some of the borosilicate glasses that are made by Chance or Corning.

Many commercial firms now make metal-to-glass and ceramic-to-metal seals but not all are satisfactory for use at low temperatures. For example, some years ago varied experiences were reported with the Kovar seals made by Stupakoff Ceramic and Manufacturing Company, Pennsylvania. Some experimenters (e.g. Lane 1949) reported them satisfactory at liquid-helium temperatures while others (e.g. Corak and Wexler 1953) found small leaks. It appears that home-made seals which are annealed and soldered with care are perfectly satisfactory but some commercial seals do leak when exposed to liquid helium.

Sealing cements and compounds

Apiezon Products Ltd. supply through various distributing firms a large range of products useful in high-vacuum work. These products include: diffusion-pump oils Apiezon A, B, and C, a viscous low-vapour-pressure oil Apiezon J, a very viscous oil Apiezon K, greases L, M, and N for use in conical ground joints and taps, a tap grease T with a low temperature coefficient of viscosity useful for ground joints which may be raised above room temperature (maximum safe temperature of 110° C), and also a sealing compound and waxes. Silicone oils for diffusion pumps and silicone tap grease are also available from many suppliers of high-vacuum equipment.

The Apiezon sealing compound Q is a graphite–grease mixture with the consistency of plasticine which is very useful for temporary vacuum seals. Apiezon W, W100, and W40 are black waxes of low vapour pressure which flow easily onto a solid surface when molten and melt respectively at about 100°, 80°, and 45° C. Other common waxes which can be used for making vacuum-tight joints are shellac, beeswax–rosin mixture (m.p. 57° C), de Khotinsky Cement (shellac and wood tar mixture, m.p. 140° C), and Picein (black wax, melting at 80°–100° C depending on the variety used). Of these waxes, the Apiezon W series are very easy to use but are somewhat brittle and dissolve readily in a number of organic solvents. De Khotinsky wax is available in hard, medium, and soft grades (e.g. from Central Scientific Company) and is a little more difficult to apply than the W waxes but is less brittle and is not affected by many common organic liquids, including butyl phthalate.

These sealing compounds are not of much use in cementing materials which are to be cooled to low temperature, as they become brittle and tend to flake. For securing or thermally anchoring electrical wires to solid surfaces, some liquid cements —particularly those which can be cured by baking—provide a method of attachment which withstands repeated cooling quite well; bakelite varnish, Formel varnish, GE7031, and glyptal lacquer may all be dried and hardened at a temperature of 120°–140° C and used for this purpose. Nail polish is also a useful cement for anchoring wires providing they are not subject to appreciable mechanical stress; it withstands low temperatures quite well.

The most satisfactory materials for making seals between metals (apart from solder) or between metal and non-metals which can be cooled down to the lowest temperatures are the epoxy resins. These are made in various forms by CIBA (Araldite resins), Shell Chemical Company (Epikote resins), and Dow Chemical Company; they are also distributed by many other firms, in some cases with the addition of fillers to reduce the expansion coefficient to match that of common metals. These epoxy resins are usually available (i) in solid or liquid form which cures when heated to 180–200° C for a specified time ($\sim$ 1 h for Araldite Type I), or (ii) as liquid to which a hardener is added before use, after which the resin cures and hardens at room temperature in a few hours; in some cases the hardening is accelerated and the strength is improved by heating to $\sim$ 100° C for an hour or so.

These resins will not provide strong vacuum-tight seals between solids at low temperatures if the resin has a very much larger expansion coefficient than the solid *and* the bond between resin and solid is poor. As mentioned, fillers such as powdered alumina may be added to reduce the expansion coefficient; also the bond can be improved by cleaning and suitably treating the surface.

In the writer's experience many epoxy resins bond strongly to aluminium and stainless steel but bond less strongly to copper and brass. Mylar, nylon, and polythene have all been bonded with epoxy resin but surface treatment is important; rubbing the surface of Mylar with a fine abrasive helps and nylon can be cleaned with a 50 per cent nitric-acid solution. Hickman *et al.*

(1959), Mehr and McLaughlin (1963), Moss, Kellers, and Bearden (1963) and Collins, Mayer, and Travis (1967) have each described bonding of Mylar with epoxy resin.

Balain and Bergeron (1959) have reported the use of epoxy resin for sealing of electrical leads through $\frac{1}{8}$-in copper tubes using a Stycast epoxy which has an expansion coefficient similar to copper at room temperature. This seal was leak-tight at helium temperature.

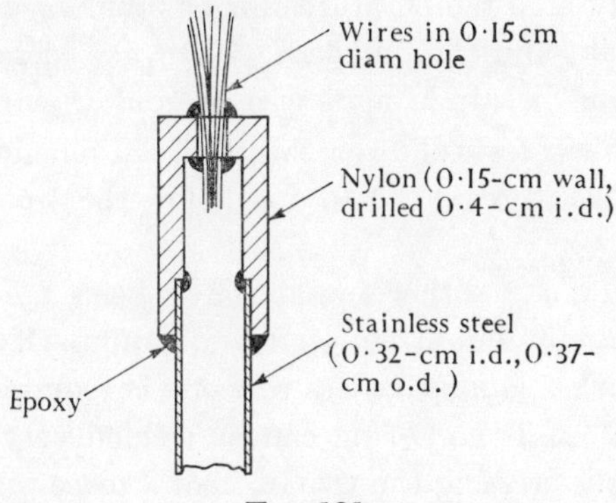

FIG. 131

Wheatley (1964) has described a nylon–epoxy–copper junction in which Epibond 100A provides an intermediate tubular junction-piece between a nylon tube and a copper tube, both the latter tubes being 'feathered' (like a Housekeeper seal) where they join to the epoxy. The excess epoxy is machined away after curing and the join is reported to be leak-tight in liquid helium II.

Epoxy resins have also been used successfully for making seals of the type shown in Fig. 131, in which electrical leads pass from a vacuum chamber into a bath of liquid helium. The writer and his colleagues were introduced to this design by Dr C. A. Swenson and it has proved satisfactory in most cases, provided that the seal is protected from thermal shock during cooling. Care is taken to wet the bunch of wires thoroughly with epoxy resin by warming slightly; the nylon is cleaned for half a minute in 50% nitric-acid solution, and the stainless-steel tube which should be thin-walled is cleaned with oxalic acid.

Demountable vacuum seals

At room temperature, rubber O-rings are the seal most commonly used for vacuum systems, but at low temperatures they are no longer practicable. For this reason solder-joints made with a low-melting-point alloy like Wood's metal have been commonly used by the low-temperature physicist. However, demountable joints sealed with a gasket of ductile metal have proved quite successful at low temperatures just as they have at moderately high temperatures for bakable vacuum systems. Wexler, Corak, and Cunningham (1950) first described a seal made between metal surfaces with a gold O-ring of 0·05-cm diam. wire; their tests showed that the seal remained vacuum-tight at temperatures down to and below the λ-point of liquid helium.

Since then many other metals have been used, including copper, silver, aluminium, tin, lead, and indium. Of these indium has proved most popular as less pressure is required to make it flow and seal, and the O-ring can be welded very easily from wire by simply pressing the two ends of a piece together. Wire of 0·1 or 0·2 cm diam. is frequently used and can be placed as in Fig. 132 (a) (after Seki 1959) or in a shallow groove (Fig. 132 (b)) (after Fraser 1962), or simply between two surfaces which have been ground to fit well together; Fig. 132 (c) illustrates this latter construction used by the writer and his colleagues. In all these cases sufficient tension, compression or spring-loading on the screws is needed to ensure a residual compression on the indium after cooling.

Horwitz and Bohm (1961) and Craig, Steyert, and Taylor (1962) have also described indium seals. Most of the harder metals require more local pressure than exists in Fig. 132 (c) in order to effect a good seal, although we have found tin wire of $\leqslant$0·1-cm diam. will seal quite well in this arrangement. Aluminium gaskets can be sealed with the tooth-like arrangement shown in Fig. 132 (d) (after Atoji 1966).

Seals between smooth flat surfaces, as in Fig. 132 (c), can also be made with a thin layer of Loctite AV, a thread sealant which remains soft (at room temperature) so the seal can be easily broken when necessary. Reeber (1961) and Ashworth and Steeple (1965) have described these. Thorough degreasing or ultrasonic cleaning of the metal surface helps to give a more reliable seal; the seal should be left for an hour or two to harden a little after the initial tightening of the screws and then re-tightened before cooling. Thin polythene gaskets ($\lesssim$ 0·01 cm thick) lightly greased with silicone grease can also be used in

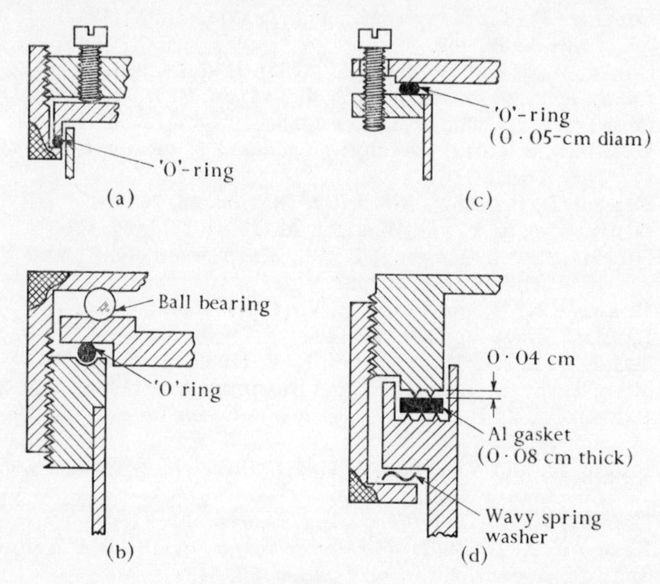

FIG. 132

joints like that in Fig. 132 (c); sometimes they are thoroughly leak-tight in liquid helium but on other occasions they may show a small leak. Similar experiences have been suffered with threaded metal seals in which the screw-thread is covered with silicone grease or a glass-forming mixture of glycerol plus propyl alcohol (see p. 279 in § 9.4); these are often vac-tight for an hour or more after cooling but then begin to leak slowly.

If the experimenter wants to avoid solder joints, he may use indium or another soft metal as a sealant or perhaps try a sealing compound like Loctite AV.

5. References

ASHWORTH, T. and STEEPLE, H. (1965). *Cryogenics* **5**, 267.
ATOJI, M. (1966). *Rev. scient. Instrum.* **37**, 1086.
BALAIN, K. S. and BERGERON, C. J. (1959). Ibid. **30**, 1058.
BARRINGTON, A. E. (1963). *High vacuum engineering.* Prentice-Hall, New Jersey.
BELSER, R. B. (1954). Ibid. **25**, 180.
COCHRAN, J. F., MAPOTHER, D. E., and MOULD, R. E. (1956). *Phys. Rev.* **103**, 1657.
COLLINS, R. L., MAYER, K., and TRAVIS, J. C. (1967). *Rev. scient. Instrum.* **38**, 446.
CORAK, W. S. and WEXLER, A. (1953). Ibid. **24**, 994.
CRAIG, P. P., STEYERT, W. A., and TAYLOR, R. D. (1962). Ibid. **33**, 869.
CROFT, A. J. (1963). *Cryogenics* **3**, 65.
DUSHMAN, S. (1949). *Scientific foundations of vacuum technique.* Wiley, New York.
FRASER, D. B. (1962). *Rev. scient. Instrum.* **33**, 762.
GARFUNKEL, M. P. and WEXLER, A. (1954). Ibid. **25**, 170.
HICKMAN, R. S., KENNEY, R. W., MATHEWSON, R. C., and PERKINS, R. A. (1959). Ibid. **30**, 983.
HORWITZ, N. H. and BOHM, H. V. (1961). Ibid. **32**, 857.
LANE, C. T. (1949). Ibid. **20**, 140.
MEHR, D. L. and MCLAUGHLIN, E. F. (1963). Ibid. **34**, 104.
MOSS, T. H., KELLERS, C. F., and BEARDEN, A. J. (1963). Ibid. **34**, 1267.
PARTRIDGE, J. H. (1949). *Glass-to-metal seals.* Society of Glass Technology, Sheffield.
PIRANI, M. and YARWOOD, J. (1961). *Principles of vacuum engineering.* Chapman & Hall, London.
REEBER, M. D. (1961). *Rev. scient. Instrum.* **32**, 1150.
REIMANN, A. L. (1952). *Vacuum technique.* Chapman & Hall, London.
SEKI, H. (1959). *Rev. scient. Instrum.* **30**, 943.
STRONG, J. (1938). *Modern physical laboratory practice.* Blackie, London.
WEXLER, A., CORAK, W. S., and CUNNINGHAM, G. T. (1950). *Rev. scient. Instrum.* **21**, 259.
WHEATLEY, J. C. (1964). Ibid. **35**, 765.
YARWOOD, J. (1955). *High vacuum technique*, 3rd rev. edn. Chapman & Hall, London.

PART III
PHYSICAL DATA
CHAPTER XI
HEAT CAPACITY AND EXPANSION COEFFICIENT

1. Heat capacity of solids

It is well known (see, for example, texts by Mott and Jones 1936, Kittel 1953) that the specific heat at constant volume of most solids may be represented tolerably well by the Debye function

$$C_v = 9Nk(T/\theta)^3 \int_0^{\theta/T} \frac{x^4}{(e^x-1)(1-e^{-x})}\,dx$$
$$= 9R(T/\theta)^3 J_4(\theta/T).$$

The derivation of such a formula is based on the elastic continuum model of a solid for which the number of modes of vibration in a frequency interval ν, $\nu+d\nu$ is given by

$$f(\nu)\,d\nu = bV\nu^2\,d\nu = 4\pi\left(\frac{1}{v_l^3}+\frac{2}{v_t^3}\right)V\nu^2\,d\nu;$$

V is the volume of the solid and v_l, v_t are the wave propagation velocities for longitudinal and transverse waves respectively. The maximum frequency ν_m is given by the normalizing condition that in a volume V containing N atoms, there are only $3N$ modes:

$$3N = \int_0^{\nu_m} f(\nu)\,d\nu = \tfrac{1}{3}bV\nu_m^3.$$

Introducing the Debye characteristic temperature

$$\theta_D = h\nu_m/k,$$

and writing $x = h\nu/kT$, we find that the total internal energy

$$U = 9NkT(T/\theta)^3 \int_0^{\theta/T} \frac{x^3\,dx}{(e^x-1)}$$

and $$C_v = \left(\frac{\partial U}{\partial T}\right)_v = 9R(T/\theta)^3 \int_0^{\theta/T} \frac{x^4}{(e^x-1)(1-e^{-x})}\,dx.$$

In Table D (in the Appendix) are given values of C_v computed from Debye's expression for various values of θ/T.† Note that for $T > \theta_D$, $C_v \to 3R = 24{\cdot}94$ J/g-atom deg, and for $T \leqslant \theta_D/20$, $C_v = 1943{\cdot}7(T/\theta_D)^3$ J/g-atom deg.

As might be expected, experimental values for C_v often depart quite considerably from those given by the Debye model, a large part of the discrepancy arising from the fact that the real frequency spectrum in the atomic lattice is very different from the $f(\nu) \propto \nu^2$ spectrum assumed by Debye. These questions have been discussed in detail in reviews by Blackman (1955), Keesom and Pearlman (1956), de Launay (1956), Parkinson (1958), Gopal (1966), etc.

What is perhaps surprising is that the discrepancies in most solids are so small. Usually they are represented by deriving an appropriate value of θ_D from each experimental value of C_v (using the tabulated Debye function) and drawing a graph of θ_D versus T; in such a graph the Debye theory would demand complete constancy of θ from $T = 0°$ K upwards. Fig. 133 shows experimental θ_D plots for a number of solids.

At high temperatures, that is at temperatures comparable with θ_D, the value of θ_D calculated from experimental data appears to be fairly constant for any particular solid, and the major variations occur at temperatures below about $\theta_D/5$. For purposes of cryogenic design it is usually sufficient to assume the validity of the Debye approximation and calculate for any solid in question the specific heat on the basis of the values of θ_D listed in Table E (in the Appendix) and the tabulated Debye function in Table D. The values of θ_D shown for the various elements are calculated from specific heat data obtained in the region between $\theta_D/2$ and θ_D. They are not the limiting values obtained from low-temperature specific heats; the latter are usually signified by θ_0 and may be found in the various review articles on specific heats which were mentioned above, in compilations such as Corruccini and Gniewek (1960), or may be

† These values are taken from the Landolt–Bornstein physical tables. A more complete tabulation of the Debye energy and specific heat functions to six places has been given by Beattie (1926).

calculated from the elastic constants (Huntingdon 1958, Hearmon 1966) with the aid of the tables of de Launay (1956). Comprehensive tables or graphs of heat capacities can be found in Shiffman (1952), Corruccini and Gniewek (1960), *Landolt–Bornstein* (Vol. 2, Pt. 4, 1961, Springer-Verlag, Berlin) and a

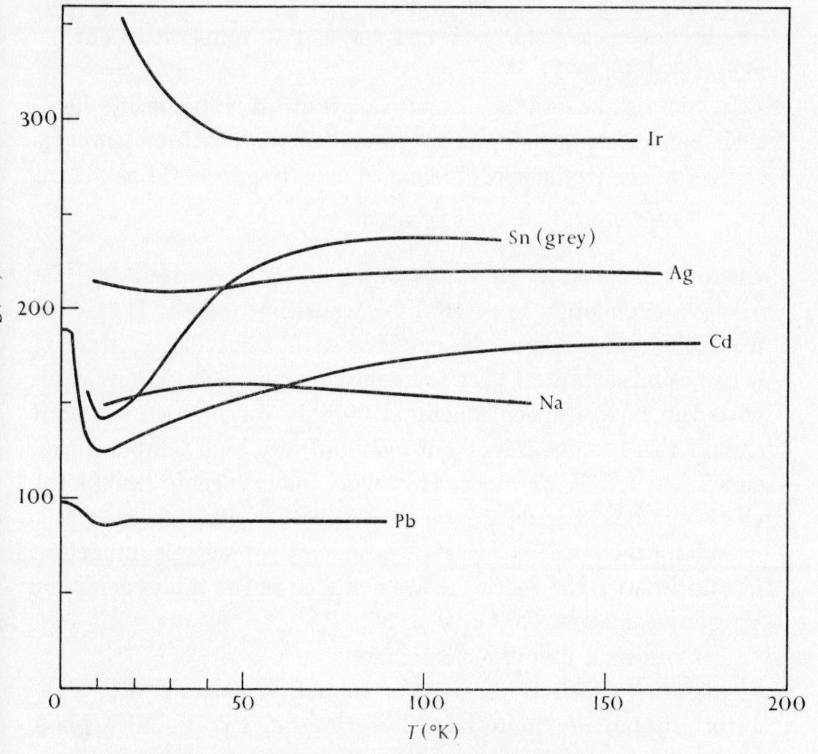

FIG. 133. Variation of θ_D with temperature.

Compendium of the properties of materials at low temperatures prepared by V. J. Johnson of N.B.S. Boulder Laboratories as *WADD Tech. Report* 60/56 (reprinted by Pergamon Press, 1961).

The values of specific heat discussed above are values, C_v, of the specific heat at constant volume. C_v is related to the quantity which is usually measured, namely the specific heat at constant pressure C_p, by the thermodynamic equation

$$C_p - C_v = VT\beta^2/\chi;$$

here β is the volume expansion coefficient and χ is the compressibility. It follows to a first approximation that

$$C_p - C_v = AC_v^2 T,$$

where the constant A may be calculated at one temperature. The difference $C_p - C_v$ is negligible at low temperatures but usually becomes of the order of 1 per cent at temperatures in the vicinity of $\frac{1}{2}\theta_D$.

In metals the free electrons contribute to the specific heat, their contribution being proportional to the absolute temperature. The electronic specific heat is usually expressed as

$$C_e = \gamma T,$$

where the constant γ varies from $\sim 10^{-3}$ J/g-atom'deg^2 for monovalent metals to $\sim 10^{-3}$ for transition metals. It is clear from the temperature dependence of C_e and C_{lattice} that C_e becomes important at very low temperatures, being comparable with the lattice specific heat at $\sim 5°$ K for many metals; it should also become important again at very high temperatures, e.g. $T = 1000°$ K or more. However, for cryogenic design the electronic heat capacity is not usually of major importance.

Among technical materials whose heat capacity is important in calorimetry, the following are included in the tables compiled by Corruccini and Gniewek (1–300° K):

(i) common metallic elements;
(ii) alloys such as constantan, monel, Wood's metal;
(iii) supporting materials like graphite, Pyrex, silica glass, Teflon; and
(iv) cements including Glyptal, Araldite type I, Bakelite varnish.

Subsequently further data have been published for a number of materials including:

(a) vitreous silica (Flubacher *et al.* 1959);
(b) glycerol glass (Craig, Massena, and Mallya 1965);
(c) soft solders (lead–tin alloy) containing 40, 50, and 60 wt per cent lead (de Nobel and du Chatenier 1963, Zeigler and Mullins 1964);

(d) stainless steel containing 17·5% Cr, 10·1 Ni, 0·86 Mn, for which du Chatenier, Boerstoel and de Nobel (1965) find

$$c = 4640T + 3{\cdot}8T^3 \text{ ergs/g degK} \quad (T < 10^\circ \text{K});$$

(e) constantan and manganin which Ho, O'Neal, and Phillips (1963) observed to have a substantial hyperfine contribution proportional to T^{-2}, below 1° K:

$$C \approx 28T^{-2} + 2050T \text{ ergs/g degK}$$
$$\text{(constantan below } 0{\cdot}3^\circ \text{K)}$$

$$C \approx 115T^{-2} + 600T + 29{\cdot}4T^3 \text{ ergs/g degK}$$
$$\text{(manganin below } 2{\cdot}5^\circ \text{K)};$$

(f) nylon and cold-setting Araldite below 4° K (Brewer *et al.* 1966) each of which gave approximately

$$C \approx 180T^3 \text{ ergs/g degK};$$

(g) Teflon, Kel-F, and nylon by Reese and Tucker (1965);

(h) GE 7031 cement (Phillips 1959) has a heat capacity

$$C \approx 400T^3 \text{ ergs/g degK};$$

(i) an Allen–Bradley carbon resistor, $\frac{1}{10}$ W and 10 Ω weighing 0·05 g had a heat capacity between 1 and 4° K of

$$C \approx 64T + 13T^3 \text{ ergs/degK}$$

(Keesom and Seidel 1959).

It is important to note that the heat capacities per gram of such things as epoxy resin, GE7031, or Glyptal varnish are far greater than that of copper at temperatures between 2° and 20° K. Thus even small quantities of these cements used in the assembly of a calorimeter can contribute appreciably to its heat capacity and must not be neglected. Also, below 1° K wires of manganin and constantan have relatively large heat capacities due to a hyperfine (T^{-2}) term.

Very frequently in low-temperature design we need to know approximate values for the heat capacity of a given weight of a metal, in order to calculate the approximate volume of liquid refrigerant required for cooling it. Obviously, a knowledge of the total heat, or enthalpy, H enables us to calculate this more readily. The enthalpy may be derived by integrating the C_p–T curve:

$$H(T) = U+PV = \int_0^T C_p\,dT.$$

Since $C_p - C_v$ is not a unique function of T/θ but depends on the expansion coefficient and compressibility of a substance, a general table of H as a function of T/θ cannot be compiled. However, for our practical purposes it may be assumed that $U = \int C_v\,dT$ will be sufficient approximation to $H = \int C_p\,dT$. Table XIX below, abbreviated from the Landolt–Bornstein physical tables, gives values of $(U-U_0/T$ as a function of θ/T.

TABLE XIX

Some values of $\dfrac{U-U_0}{T} = \dfrac{1}{T}\displaystyle\int_0^T C_v\,dT$ *in* cal/g-atom °K

θ/T	$\frac{U-U_0}{T}$	θ/T	$\frac{U-U_0}{T}$	θ/T	$\frac{U-U_0}{T}$
0·2	5·522	2·2	2·409	6·0	0·461
0·4	5·111	2·4	2·205	6·5	0·379
0·6	4·723	2·6	2·018	7·0	0·313
0·8	4·359	2·8	1·847	7·5	0·260
1·0	4·018	3·0	1·689	8·0	0·218
1·2	3·698	3·5	1·352	9·0	0·156
1·4	3·400	4·0	1·082	10·0	0·115
1·6	3·123	4·5	0·869	11·0	0·087
1·8	2·866	5·0	0·701	12·0	0·067
2·0	2·628	5·5	0·568	15·0	0·034

It should be remembered that this tabulation in the form $U-U_0$ neglects the presence of any specific heat anomalies due to order–disorder transitions or other excitation processes which will contribute to the internal energy and the enthalpy of a substance in which they occur. However, these are relatively uncommon in the range below 300° K.

As an example consider the cooling from 90° to 4° K of 1 g-atom of copper (63·6 g), for which $\theta_D \simeq 310$.

At 90° K, $T/\theta = 0{\cdot}290$ ($\theta/T = 3{\cdot}44$), therefore

$$U-U_0 = 90\times 1{\cdot}37 = 123 \text{ cal/g-atom.}$$

At 4° K, $T/\theta = 0{\cdot}012_9$ ($\theta/T = 77{\cdot}5$), therefore

$$U - U_0 \simeq 0{\cdot}01 \text{ cal/g-atom.}$$

Hence $U_{90} - U_4 \simeq 123$ cal/g-atom.

This indicates that about 180 cm³ of liquid helium would be required to cool 63·6 g of copper from 90° to 4° K, using only the latent heat of vaporization of the helium.

2. Coefficient of thermal expansion

When making low-temperature equipment it is usual to select the same material for the different pumping tubes. Not only is this simpler for reasons of availability but it lessens the risk of strains arising from differential expansion or contraction between different parts of the system. As an example, it is ill-advised to use a thin-walled inconel tube as the material for a metal sleeve running through the centre of a brass chamber, if the tube is constrained by being soldered to the chamber at each end. The difference in contraction of the materials imposes considerable strain on the light inconel tube when both are cooled from room temperature to say 4° K.

Frequently in cryogenic design, some knowledge is required of the expansion coefficients of the commonly used materials. Direct measurements of the expansion or contraction of many solids have been made in the past few years from room temperature down to about 4° K. The expansion coefficient of a solid has a temperature dependence very similar to that of the specific heat, so that at temperatures well below the Debye temperature the linear expansion coefficient (which we shall denote by α) becomes very small and difficult to measure.

The expansion coefficient α and the specific heat at constant volume may be related by the Grüneisen expression (see, for example, Mott and Jones 1936, or Kittel 1953)

$$3\alpha = \gamma\chi C_V/V,$$

where χ is the compressibility and C_V/V is the specific heat per unit volume; the proportionality factor γ is known as Grüneisen's constant, and may be alternatively defined by

$$\gamma = -\frac{d \ln V}{d \ln \theta}.$$

This parameter γ is not constant at all temperatures any more than is θ. They both are averages over the frequency spectrum (Barron 1957). However, in many solids, particularly metals, the limiting values of γ at low temperatures (γ_0) and at high temperatures (γ_∞) do not differ very much, just as θ_0 and θ_∞ are not very different. Certainly at normal temperatures (say $T > \theta/5$), γ remains reasonably constant, the values for most solids being between 1·5 and 2·5 (e.g. review of Collins and White 1964). From the aspects of technical design, the values of γ or α at the lowest temperatures are usually of less interest because the major part of the thermal contraction on cooling has already taken place in the range from 300° K to 40° or 50° K. Hence a change in the Grüneisen γ from say 2·1 to 1·7 which only occurs for $T < \theta/5$ would have negligible practical effect on the total change in dimensions of a solid which is being cooled from 300° K to 10° or 20° K.

In Table E in the Appendix are given experimental values for the linear expansion coefficient of a number of elements at room temperature. These are taken chiefly from the *Smithsonian physical tables* (9th edn., 1954) and W. B. Pearson's *Handbook of lattice spacings of metals and alloys* (Pergamon Press, London, 1958).

Table F in the Appendix gives experimental values of the expansivity, i.e. the change in length at a temperature T° K from the length at room temperature divided by the length at room temperature, i.e. values of $(L_{293} - L_T)/L_{293}$ for a number of commonly used materials. More complete data and references may be found in the very useful compilation by Corruccini and Gniewek (1961) or the N.B.S. *Compendium of the properties of materials at low temperatures* (editor V. J. Johnson, *WADD Technical Report* 60/56, reprinted by Pergamon Press, 1961). Subsequent data for alloys based on aluminium, cobalt, copper, iron, titanium, and nickel have been given by Arp *et al.* (1962)

and Rhodes *et al.* (1963). Other materials which have been measured and may be of technical interest are titanium carbide, sapphire, graphite (see Arp *et al.*, loc. cit.), and magnesium oxide (White and Anderson 1966).

If experimental values for $\alpha(T)$ are not available, a useful estimate can be made by assuming that α varies with T in the same way as the specific heat, e.g.

$$\frac{\alpha(T)}{\alpha(295)} = \frac{C_\mathrm{v}(T)}{C_\mathrm{v}(295)}.$$

A value for $\alpha(295)$ is obtained from Table E and $C_\mathrm{v}(T)/C_\mathrm{v}(295)$ is calculated with the aid of a θ_D value from this table and the Debye function (Table D). Of course, if experimental values of C_v at the required temperatures are available, these should be preferred.

For example: In the case of copper, α at 295° K is $0{\cdot}167 \times 10^{-4}$ per degC; since $\theta_\mathrm{D} = 310°$ K, $\theta/T = 1{\cdot}05$ at 295° K from which

$$C_\mathrm{v} \simeq 5{\cdot}64 \text{ cal/g-atom deg.}$$

$$\text{At a temperature T, } \alpha(T) = \frac{C_\mathrm{v}(T)}{5{\cdot}64} \times 0{\cdot}167 \times 10^{-4};$$

at 50° K, $\theta/T = 6{\cdot}2$ and hence $C_\mathrm{v} = 1{\cdot}48$, therefore

$$\alpha = 0{\cdot}044 \times 10^{-4};$$

and at 100° K, $\theta/T = 3{\cdot}1$ and hence $C_\mathrm{v} = 3{\cdot}85$, therefore

$$\alpha = 0{\cdot}115 \times 10^{-4} \text{ per degC};$$

cf. experimental values of $0{\cdot}039 \times 10^{-4}$ at 50° K and $0{\cdot}106 \times 10^{-4}$ at 100° K.

3. References

Arp, V., Wilson, J. H., Winrich, L., and Sikora, P. (1962). *Cryogenics* **2**, 230.
Barron, T. H. K. (1957). *Ann. Phys.* **1**, 77.
Beattie, J. A. (1926). *J. Math. Phys.* **6**, 1.
Blackman, M. (1955). *Handb. Phys.* **7**, 325.
Brewer, D. F., Edwards, D. O., Howe, D. R., and Whall, T. E. (1966). *Cryogenics* **6**, 49.
du Chatenier, F. J., Boerstoel, B. M., and de Nobel, J. (1965). *Physica* **31**, 1061.

Collins, J. G. and White, G. K. (1964). *Prog. low Temp. Phys.* **4**, 450.

Corruccini, R. J. and Gniewek, J. J. (1960). *Natn. Bur. Stand. Monograph* 21: *Specific heat and enthalpies of technical solids at low temperatures.* U.S. Govt. Printing Office, Washington, D.C.

—— and Gniewek, J. J. (1961). *Natn. Bur. Stand. Monograph* 29: *Thermal expansion of technical solids at low temperatures.* U.S. Govt. Printing Office, Washington, D.C.

Craig, R. S., Massena, C. W., and Mallya, R. M. (1965). *J. appl. Phys.* **36**, 108.

Flubacher, P., Leadbetter, A. J., Morrison, J. A., and Stoicheff, B. P. (1954). *Physics Chem. Solids* **12**, 53.

Gopal, E. S. R. (1966). *Specific heats at low temperatures.* Plenum Press, New York.

Hearmon, R. F. S. (1966). *Landolt–Bornstein*, New Series, Group III, Vol. 1, p. 1. Springer-Verlag, Berlin.

Ho, J. C., O'Neal, H. R., and Phillips, N. E. (1963). *Rev. scient. Instrum.* **34**, 782.

Huntingdon, H. B. (1958). *Solid St. Phys.* **1**, 213.

Keesom, P. H. and Pearlman, N. (1956). *Handb. Phys.* **14**, 282.

—— and Seidel, G. (1959). *Phys. Rev.* **113**, 33.

Kittel, C. (1953). *Introduction to solid state physics.* Wiley, New York.

de Launay, J. (1956). *Solid St. Phys.* **2**, 220.

Mott, N. F. and Jones, H. (1936). *The theory of the properties of metals and alloys.* Clarendon Press, Oxford.

de Nobel, J. and du Chatenier, F. J. (1963). *Physica* **29**, 1231.

Parkinson, D. H. (1958). *Rep. Prog. Phys.* **21**, 226.

Phillips, N. E. (1959). *Phys. Rev.* **114**, 676.

Reese, W. and Tucker, J. E. (1965). *J. chem. Phys.* **43**, 105.

Rhodes, B. L., Moeller, C. E., Hopkins, V., and Marx, T. I. (1963). *Adv. cryogen. Engng* **8**, 278.

Shiffman, C. A. (1952). *Heat capacities of the elements below room temperature.* General Electric Research Publication Services, Schenectady.

White, G. K. and Anderson, O. L. (1966). *J. appl. Phys.* **37**, 430.

Zeigler, W. W. T. and Mullins, J. C. (1964). *Cryogenics* **4**, 39.

CHAPTER XII

ELECTRICAL AND THERMAL RESISTIVITY

1. Electrical resistivity

Introduction

ALTHOUGH this book does not pretend to deal with the theory of transport properties of metals, a brief discussion of electron transport processes and the mechanisms which limit them, may help our appreciation of the value of certain physical data and of the extent to which these data may be extrapolated.

In an element which is a metallic conductor, some electrons are free or quasi-free so that they can be accelerated by the action of an applied electric field. The conductivity remains finite due to the influence of various electron scattering processes, chiefly thermal vibrations of the metallic ions in the crystal lattice and the chemical or physical impurities present in the lattice. If these processes are such as to produce a finite mean free time τ or mean free path l for the electrons, we should expect classically that the electrical conductivity

$$\sigma = \frac{Ne^2\tau}{m} = \frac{Ne^2l}{mv},$$

where N is the number of conduction electrons per cubic centimetre, e the electronic charge, m the electronic mass, and v their average velocity in the absence of an applied field.

On the quantum mechanical band model, the position is rather more restrictive. There must be vacant energy states into which an electron can make a transition after exchanging energy with the applied field or scattering centres; when scattered by a vibrating ion the energy exchanged is only of the order of kT so that only those electrons with energies near the top of the conduction band can take part in the conduction process. This restricts electronic conduction to elements in which the highest energy band is not completely filled, or some equivalent situation, e.g. a filled band which just touches or is overlapped by an empty band. In the monovalent metals (Cu, Ag, Au, and the alkalis) the s-band is half filled; in the divalent metals Mg, Zn, Cd, etc., conductivity is assured by overlapping of the s- and p-bands. Another important class are the transition elements in which an unfilled d-band overlaps with a succeeding s-band and produces a complex situation in which about 0·5–1 free electron per atom in the s-band appears to account for the electrical conductivity.

As in the classical picture, we can define an effective mean free path on the band model and deduce that

$$\sigma = \frac{Ne^2 l}{mv}.$$

Now v is the average velocity of the electrons at the top of the band, i.e. near the Fermi surface. The electrons involved are only those within a region of the order of the thermal energy kT from the Fermi surface; their energy at the surface is 1–7 eV (equivalent to thermal energy at $\geqslant 20\,000°$ K), therefore the velocity v at ordinary temperatures is practically temperature independent; also N, e, and m are constants. Calculation of the electron-free path is difficult although the temperature dependence can be estimated at high temperatures ($T > \theta$) or at low temperatures ($T \ll \theta$); when $T > \theta$, the wave number $\mathbf{k}$ of the electron and wave number $\mathbf{q}$ of the lattice wave or phonon with which it interacts are comparable in magnitude. Therefore the change in direction or momentum of an electron at a collision with a vibrating ion is considerable and l is simply the average distance an electron travels between collisions. When $T \ll \theta$, $\mathbf{q}$ is very small and collisions with phonons become rare and less important since static imperfections (chemical impurity atom, dislocation, grain boundary) are still present and cause elastic scattering.

So that:

(i) For $T > \theta$, resistivity is due chiefly to scattering by phonons, which is denoted by ρ_i.

Now $\rho_i \propto l^{-1}$

$\propto$ mean square amplitude of vibration or ions

$\propto T/Mk\theta^2$ (M = atomic weight).

(ii) For $T \ll \theta$, $\rho_i \to 0$

and $\rho \simeq \rho_r$

therefore $\rho \simeq$ constant, since $l =$ constant.

In practice we may expect that at moderately high temperatures the electrical resistance of a fairly pure metallic element is due to thermal vibrations and that the resistivity ρ_i should be proportional to T. For elements in a non-magnetic state this holds to a first approximation. Change in the Fermi surface with temperature and thermal expansion do affect the linearity of $\rho_i(T)$, but for many cryogenic calculations it is sufficient to assume $\rho_i \propto T$ over a range from $T \sim \frac{1}{2}\theta$ up to a temperature approaching a phase change or the melting temperature. At temperatures sufficiently low that $\rho_r \gg \rho_i$, the resistance becomes sensibly constant unless a superconducting transition occurs; in this context the phrase 'sufficiently low' depends on the purity of the element. Elements such as Mg, Sn, and In may be readily obtained with a chemical purity exceeding 99·999 per cent and elements may have a residual resistivity ρ_r, 100 000 times smaller than their resistivity at room temperature; in such instances ρ_r only becomes dominant for $T < \frac{1}{50}\theta$.

We have not discussed the behaviour of ρ_i at temperatures below the region ($T \geqslant \theta$) where scattering is elastic. The direction of motion and the energy of an electron before and after a collision with a phonon are governed by the conservation laws:

$$\mathbf{k}' = \mathbf{k} + \mathbf{q} \text{ (conservation of momentum or wave number)}$$

and $$E' = E \pm h\nu \text{ (conservation of energy).}$$

As $h\nu \sim kT$ and $E' \simeq E$ is the Fermi energy, the change in energy at a collision is small. However, at temperatures $T \sim \theta_D$, $|\mathbf{q}| \sim |\mathbf{k}|$ and the change in direction or momentum is large. As the temperature decreases, the wave number $\mathbf{q}$ or the dominant lattice waves decreases, unlike the electron wave number $\mathbf{k}$ which is sensibly constant. Since $h\nu \sim kT$, therefore $\mathbf{q} \propto T$ and it follows that to a first approximation the angle of deflexion of an electron by interaction with a lattice wave is given by

$$\phi \sim |\mathbf{q}|/|\mathbf{k}| \sim T/\theta.$$

At temperatures quite low in comparison with the Debye temperature we expect the lattice specific heat to vary as T^3 and the density of phonons to vary as T^3. This suggests that the probability of scattering of an electron should be proportional to T^3 but the angle ϕ may be so small as to make a single scattering a very inefficient resistive process. If, indeed, a series of scatterings of an electron is required to deflect it from its 'conduction path', then the probability of its being scattered through a large angle in a series of random scattering processes will vary as $(T/\theta)^2$.

Hence
$$\rho_i \propto T^3(T/\theta)^2$$
$$\propto T^5.$$

A more general expression for the resistivity due to thermal scattering, which reduces to the T^5-relation at low temperatures and to a T-dependence at normal temperatures, is that of Bloch and Grüneisen:

$$\rho \propto \frac{T^5}{\theta^6}\int_0^{\theta/T} \frac{x^5\,dx}{(e^x-1)(1-e^{-x})}. \qquad (53)$$

This should be valid for a simple free-electron model of a Debye solid in the absence of Umklapp processes. With suitable choice of characteristic temperature θ it can express the resistance–temperature relation tolerably well for some elements, without specifying the magnitude. Generally, however, θ (or θ_R) must be allowed to vary at least 20 or 30 per cent at lower temperatures to achieve agreement. This is not unlike the specific heat wherein departures from the Debye model may be represented by variations in θ_D of 10 to 30 per cent. Note that θ_R and θ_D are not necessarily the same: they represent different averages over the frequency spectrum, though naturally they are similar in magnitude.

Detailed accounts of the theoretical background may be found in Ziman (1960, 1964) or other texts on solids by Mott and Jones, Kittel, Dekker, etc.

2. Data on electrical resistivity

Metallic elements

As we have a rather better knowledge of the Debye characteristic temperature (θ_D), from specific heat data, than of any other characteristic temperature, it is instructive to compare experimental values for the variation of reduced resistivity ρ_i/ρ_θ as a function of reduced temperature T/θ_D with some theoretical functions. Fig. 134 shows on a logarithmic graph the reduced resistivity as a function of reduced temperature together with two theoretical curves of which curve A represents the Grüneisen–Bloch equation (53):

$$\frac{\rho_i}{\rho_\theta} = \left(\frac{T}{\theta}\right)^5 \int_0^{\theta/T} \frac{x^5\,dx}{(e^x-1)(1-e^{-x})} \Big/ \int_0^1 \frac{x^5\,dx}{(e^x-1)(1-e^{-x})}$$

$$= \left(\frac{T}{\theta}\right)^5 J_5(\theta/T) \div J_5(1)$$

$$= 4{\cdot}226(T/\theta)^5 J_5(\theta/T).$$

The second theoretical curve (B in Fig. 134) is based on a $J_3(\theta/T)$ function and is the type of functional relation between electrical resistance and temperature which Wilson (1938) has suggested might be valid for some transition elements, at least over a restricted temperature range. Briefly, the reason for this is as follows: the transition elements all show a high electrical resistivity at room temperature in comparison with monovalent metals. Mott (1935, 1936) suggested that this results from the additional high probability of s-electrons being scattered into the d-band by interaction with thermal vibrations rather than merely being scattered to other available states in the s-band. Wilson deduced a resistive contribution due to s–d scattering to be given by

$$\rho_{sd} = b(T/\theta)^3 \int_{\theta_E/T}^{\theta/T} \frac{x^3\,dx}{(e^x-1)(1-e^{-x})}.$$

We may derive a relation $\rho_i/\rho_\theta = f(\theta/T)$ based on the dubious assumptions that

$$\rho_i \simeq \rho_{sd}$$

and

$$\rho_{sd} = b(T/\theta)^3 \int_0^{\theta/T} \frac{x^3\,dx}{(e^x-1)(1-e^{-x})}.$$

Hence

$$\frac{\rho_i}{\rho_\theta} = (T/\theta)^3 J_3(\theta/T)/J_3(1)$$

$$= 2{\cdot}084\left(\frac{T}{\theta}\right)^3 J_3(\theta/T). \tag{54}$$

The experimental values of ρ_i/ρ_θ generally lie in the region between curves A and B (Fig. 134). Also, at low temperatures ($T \leqslant \frac{1}{10}\theta$) eqns (53) and (54) lead to T^5 and T^3 relations: for monovalent metals and some divalent metals we find $\rho_i \propto T^5$ at

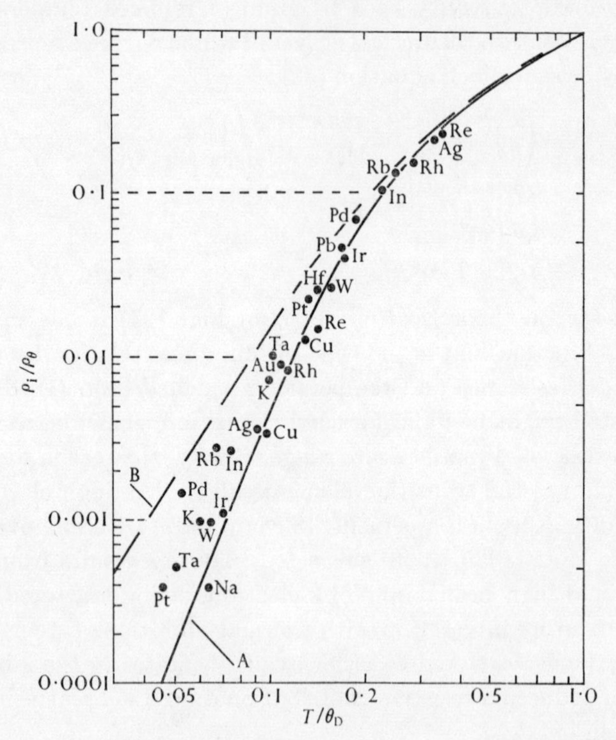

Fig. 134. Reduced electrical resistivity (ρ_i/ρ_θ) as a function of reduced temperature (T/θ_D).

——— Curve $\rho_i/\rho_\theta = 4{\cdot}226(T/\theta_D)^5 \int_0^{\theta/T} \frac{x^5\,dx}{(e^x-1)(1-e^{-x})}$.

– – – Curve $\rho_i/\rho_\theta = 2{\cdot}084(T/\theta_D)^3 \int_0^{\theta/T} \frac{x^3\,dx}{(e^x-1)(1-e^{-x})}$.

sufficiently low temperatures but for the transition elements the index of T in this region varies from 3 to 5, and in some transition elements there is evidence that $\rho_i \propto T^2$ for $T \sim \frac{1}{50}\theta$, which suggests that electron–electron interactions become important at these very low temperatures. The experimental points in

Fig. 134 have been plotted with the values of $\theta = \theta_D$ used in Table E (Appendix) and electrical resistivities given in Gerritsen (1956), Meissner and Voigt (1930), and White and Woods (1959); Meaden (1965) has also listed values of ρ_i for many elements in his book, together with an extensive bibliography. The values given in Table G for Ag, Al, Au, Cu, etc. are from the same sources. Numerical tabulations of the J_3 and J_5 integrals may be found in Rogers and Powell (1958) (see also Table D in the Appendix).

In cases where experimental values of ρ are not available for an element, then they can be estimated, assuming

$$\rho = \rho_i(T) + \rho_r \quad \text{(Matthiessens' rule)}$$

and that the temperature dependence of ρ_i is given approximately by the Bloch–Grüneisen relation. It is necessary to know ρ_r from a measurement at liquid-helium temperatures and ρ at room temperature. Obviously, for rather impure samples, ρ_r will be the major component at temperatures below $\theta/10$ (see Fig. 134) and the detailed behaviour of ρ_i will be a less important factor in determining the total resistivity than for a very high-purity specimen.

Procedure for calculating values of ρ

Given a bar or wire of a metallic element—purity unknown—we may require an estimate of its electrical resistance at temperatures from 4° to 300° K. The simplest procedure is then to measure the electrical resistance R_{295} at room temperature, and measure R_4 by dipping it into a dewar of liquid helium. If the specimen is a fine wire having a resistance of $10^{-2}\,\Omega$ or more, this may be measured with potentiometer or resistance bridge. If it is a solid bar of copper of resistance 10^{-5} or $10^{-6}\,\Omega$ then a very sensitive measuring device such as a galvanometer-amplifier or chopper-amplifier is needed; or an induction eddy-current method can be used on short fat samples (e.g. Meaden 1965, has details of these methods). Then, knowing R_{295} and $R_4 = R_r$, we can deduce R_i at room temperature ($= R_{295} - R_r$); if desired the shape factor $l/A = R/\rho$ may also be calculated, using a value of

$\rho(295^\circ\,\mathrm{K})$ from Table E. Then, at temperatures between 4° and 300° K,

$$\rho = \rho_{\mathrm{i}}(T) + \rho_{\mathrm{r}}$$

or

$$R = R_{\mathrm{i}}(T) + R_{\mathrm{r}},$$

where the constant R_{r} (or ρ_{r}) is known. $R_{\mathrm{i}}(T)$ or $\rho_{\mathrm{i}}(T)$ may be found, using a value of $\theta = \theta_{\mathrm{D}}$ from Table E, from (i) the Grüneisen–Bloch equation, (ii) Fig. 134, or (iii) experimental values of ρ_{i} in Table G. It follows from our earlier remarks that method (i) may be in serious error at temperatures $T \leqslant \frac{1}{10}\theta$ unless we apply a prior knowledge of the departures from the Grüneisen–Bloch equation for the metal in question.

Example: Given a rod of copper, it is found that

$$R_{295} = 0{\cdot}016\ \Omega, \quad R_4 = 0{\cdot}00027\ \Omega,$$

and it is required to know R at 110° K. First,

$$R_{\mathrm{i}}(295) = 0{\cdot}0160_0 - 0{\cdot}00027 = 0{\cdot}0157_3\ \Omega.$$

Assuming

$$\rho_{\mathrm{i}}/\rho_\theta = (T/\theta)^5 J_5(\theta/T) \div J_5(1)$$

we have

$$\frac{\rho_{\mathrm{i}}(110)}{\rho_{\mathrm{i}}(295)} = \frac{\rho_{\mathrm{i}}(110)}{\rho_{\mathrm{i}}(\theta)} \frac{\rho_i(\theta)}{\rho_i(295)}.$$

For $T/\theta = 110/310$ Table D gives $\rho_{\mathrm{i}}(110)/\rho_{\mathrm{i}}(\theta) = 0{\cdot}249$ and for $T/\theta = 295/310$ gives $\rho_{\mathrm{i}}(295)/\rho_{\mathrm{i}}(310) = 0{\cdot}945$; hence

$$\frac{\rho_{\mathrm{i}}(110)}{\rho_{\mathrm{i}}(295)} = 0{\cdot}263$$

and

$$R_{\mathrm{i}}(110) = 0{\cdot}263 \times 0{\cdot}016 = 0{\cdot}00414\ \Omega.$$

Now

$$R = R_{\mathrm{i}}(110) + R_4$$
$$\simeq 0{\cdot}0044\ \Omega \text{ at } 100^\circ\,\mathrm{K}.$$

Data on metallic alloys

In most alloys, excepting those very dilute alloys which may be regarded as slightly impure metallic elements, the residual or impurity resistance is relatively high. In many important alloys such as the stainless steels, monel, cupro-nickel, etc., ρ_{r} is much greater than the resistivity due to thermal vibrations and hence $\rho \simeq \rho_{\mathrm{r}}$ is practically independent of temperature. As we mentioned earlier (§ 4.7) some alloys (manganin and constantan) show an anomalous decrease in resistance as the temperature falls below 100° K, but the total change in resistance in cooling from 300° to 4° K is still less than 20 per cent of the total.

For some of the alloys commonly used in low-temperature equipment, Table H (Appendix) gives values of the electrical resistivity at room temperature, 90° or 77° K, and at 4° K. These values are taken chiefly from Berman (1951) and Estermann and Zimmerman (1952); the NbZr data are given by Evans and Erickson (1965). Other values and an extensive bibliography are listed by Meaden (1965).

It is useful to note the effect of different impurity atoms present in solid solution in a parent metal (see discussions in Gerritsen 1956, Ziman 1960, 1964, etc.). In a monovalent solute, like copper, the increase in residual resistivity due to the addition of small amounts of a neighbouring element from Group IIA, IIIA, IVA, etc. varies approximately as the square of the valency difference (Z) between the constituents, i.e. the matrix and impurity; the addition to copper of 1 atomic per cent of Zn ($Z = 1$), Ga ($Z = 2$), Ge ($Z = 3$), As ($Z = 4$) increases the resistivity by approximately 0·3, 1·4, 3·8, 6·8 $\mu\Omega$ cm respectively. Generally for metals such as Al, Cu, Ag, Au the increase, $\Delta\rho$, in $\mu\Omega$ cm per atomic per cent is 0·1–0·5 for a homovalent impurity, and increases to 5 or more for a valence of 3 or more. Vacancies generally increase the resistivity by 1–2 $\mu\Omega$ cm per atomic per cent while dislocation densities of N lines cm^{-2} give rise to increases of 10^{-19}–10^{-18} N Ω cm.

Up to a fairly large concentration of impurity—provided it is within the range of solid solubility—the increase in resistance varies nearly linearly with concentration in many binary systems. This is to be expected from Nordheim's rule which states that for a disordered solid solution $\Delta\rho$ varies with concentration c as

$$\Delta\rho = ac(1-c)+bc.$$

Nordheim's rule breaks down when long-range ordering occurs. For example, well-ordered specimens of the alloys CuAu or Cu_3Au may have a room temperature resistivity of only *ca.* 3–4 $\mu\Omega$ cm.

Note that severe cold working of a pure metal will generally affect its residual resistance quite markedly, in silver, for example, to the extent of about $0{\cdot}2\,\mu\Omega$ cm; however, the effect of such working may be quite lost in an alloy for which the residual resistance is already $10\,\mu\Omega$ cm or more. Incidentally, this is not necessarily the case when considering the heat conduction in an alloy of high electrical resistance, e.g. stainless steel, because an appreciable part of the heat is carried by lattice waves at low temperatures; the lattice waves are much more sensitive at low temperatures to the presence of line or plane imperfections such as dislocations or boundaries than they are to point imperfections. The heat conductivity at 10° K of a stainless-steel rod may be decreased appreciably by cold work while its electrical resistance is not noticeably altered.

3. Discussion of thermal resistivity

Electronic thermal conduction

In those pure metallic elements for which there is approximately one charge carrier per atom, i.e. excepting elements such as bismuth, the electronic thermal conductivity λ_e is much larger than the lattice conductivity λ_g. The same scattering processes limit λ_e as limit the electrical conductivity. We may adopt a simple kinetic model for which $\lambda_e = C_e vl/3$, where the electronic specific heat C_e is proportional to T and v is practically constant.

Then at temperatures for which the mean free path l is the same for electrical or thermal transport we expect theoretically that

$$\rho\lambda_e = \frac{\lambda_e}{\sigma} = \frac{\pi^2}{3}\left(\frac{k}{e}\right)^2 T = LT. \tag{55}$$

The Lorenz constant L has the value $2{\cdot}45\times 10^{-8}$ W Ω deg^{-2}.

At high temperatures ($T \geqslant \theta_D$), large angle elastic scattering of the electrons by thermal vibrations occurs, and we expect the effective mean free path l to vary as T^{-1} and to have the same magnitude for thermal or electrical transport; hence just as we expect $\rho \propto T$, so should λ_e be constant and

$$\frac{\rho\lambda_e}{T} = L \quad (T \geqslant \theta_D).$$

Experimental values for a number of metals at or near room temperature indicate that this Wiedemann–Franz–Lorenz law is moderately well obeyed.

At sufficiently low temperatures where impurity scattering is dominant, the mean free path should be constant so that $\rho \simeq$ constant, and $\lambda \propto T$. Measurement at liquid-helium temperatures on a large number of metallic elements have confirmed the validity of the eqn (55), namely that

$$\frac{\rho_r \lambda_r}{T} = \frac{\rho_r}{W_r T} = 2{\cdot}45 \times 10^{-8}\ \mathrm{W}\,\Omega\,\mathrm{deg}^{-2}$$

within the limits of experimental accuracy which are about ± 1 per cent; here λ_r and $W_r(=1/\lambda_r)$ are respectively the electronic thermal conductivity and electronic thermal resistivity limited by static impurities.

At all temperatures we expect the thermal resistivity due to impurities W_r, and that due to thermal vibrations W_i, to be additive, at least to a first approximation; that is, the thermal equivalent of Matthiessen's rule should be approximately correct:

$$1/\lambda = W = W_r + W_i,$$

where

$$W_r = A/T \quad (A \text{ is a constant})$$

and

$$W_i = W_i(T)$$

$$= W_\infty, \text{ a constant, for } T \geqslant \theta_D.$$

The theoretical solutions of the transport equation which gives $W_i(T)$ have been discussed in detail by Wilson (1953) and Klemens (1956). Suffice it to say that the agreement between experiment and theory is not good, which is not altogether surprising in view of the neglect in the theory of Umklapp processes and dispersion at high temperatures. However, it has been shown (MacDonald, White, and Woods 1956) that a function of the form $(T/\theta)J_3(\theta/T)$ appears to fit the observations of $W_i(T)/W_\infty$. The graph in Fig. 135 illustrates the degree of correlation between observations and this empirical function. The curve represents the variation of

$$2(T/\theta)^2 \int_0^{\theta/T} \frac{x^3\,dx}{(e^x-1)(1-e^{-x})}$$

with the reduced temperature T/θ. The experimental values of W_i/W_∞ are calculated from data on W_i given in Klemens's review article, in Rosenberg (1955), Powell and Blanpied (1954), and White and Woods (1959), etc. The values of W_∞, the fairly constant values of thermal resistivity at $T \geqslant \theta_D$, and values of θ are from Table E, W_∞ being the approximate reciprocal of λ_{295}.

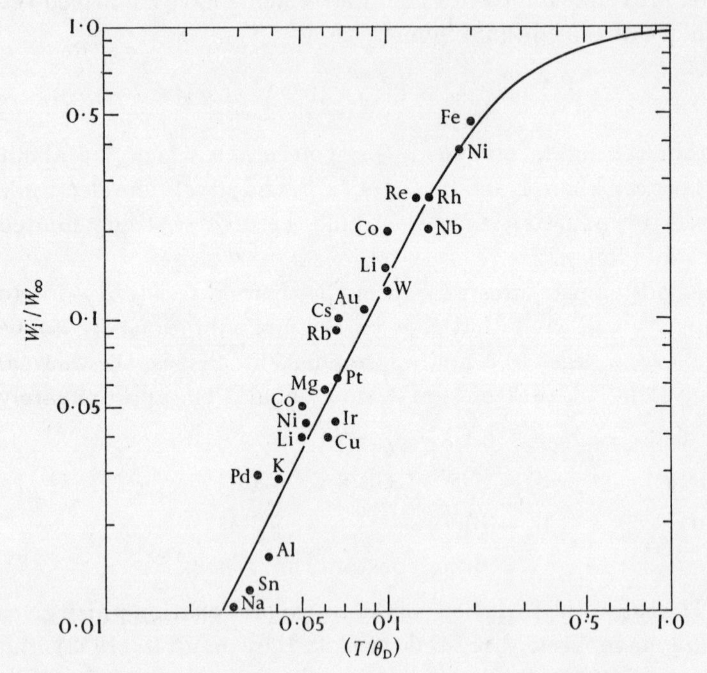

FIG. 135. Reduced thermal resistivity (W_i/W_∞) as a function of reduced temperature (T/θ_D)

$$\text{——— } W_i/W_\infty = 2(T/\theta)^2 \int_0^{\theta/T} \frac{x^3\,dx}{(e^x-1)(1-e^{-x})}.$$

The J_3 function tabulated by Macdonald and Towle (1956; see also Rogers and Powell 1958, for tables of $J_n(x)$ integrals where $n = 2, 3, \ldots, 16, 17$) has been used to obtain the values of

$$2(T/\theta)^2 \int_0^{\theta/T} \frac{x^3\,dx}{(e^x-1)(1-e^{-x})} \simeq W_i/W_\infty$$

used in Fig. 135 and given in Table D.

This partial agreement suggests that with a knowledge of the thermal resistivity of a metallic element at room temperature and at liquid-helium temperatures, both $W_i(T)$ and $W_r = A/T$ can be calculated approximately at any intermediate temperature; hence the electronic thermal conductivity, $\lambda_e = (W_i + W_r)^{-1}$, can be obtained. More simply, the Wiedemann–Franz law allows us to calculate an approximate value for W_∞ and a good value for $A = W_r T$ directly from electrical resistance measurements made at room temperature and at the temperature of liquid helium. This procedure is outlined in more detail in § 4 below.

Lattice thermal conductivity

In insulators all the heat current is carried by lattice waves; in semiconductors and in semi-metals like bismuth, graphite, or antimony, lattice conductivity is usually dominant at low temperatures; while in many disordered alloys the lattice heat conductivity may be at least comparable with the electronic heat conductivity at low temperature so that $\lambda = \lambda_e + \lambda_g$. The temperature dependence of the lattice conductivity is rather more complex and therefore less predictable than that of the electronic conductivity; this arises from the variation in frequency or wavelength of the dominant lattice waves with temperature which results in a very different temperature dependence for scattering by the different classes of lattice imperfections.

At high temperatures the lattice conductivity λ_g, is chiefly limited by thermal vibrations, i.e. interaction between the waves (or phonons) themselves due to the anharmonic nature of the coupling between the vibrating atoms. If we denote this thermal resistance by W_u, then it is dominant at temperatures $T \geqslant \theta_D$ except in highly disordered solids.

The effect of physical and chemical impurities on lattice waves is complicated because the important frequencies are temperature sensitive. As we should expect from physical intuition it is found that at very low temperatures, where only long waves are excited in the solid, planar irregularities scatter the waves much more effectively than do small point defects. As discussed by Ziman (1960) and Klemens (1956, 1967) grain boundaries (resistivity W_B), dislocations (W_D), and vacancies or impurity atoms (W_P) provide three distinct scattering processes for which the proportionality relations between effective mean free path and frequency ν are as follows:

(i) grain boundaries: l_B is frequency independent, generally

$$\lambda \simeq \tfrac{1}{3}Cvl;$$

therefore, since $C \propto T^3$ for $T \ll \theta_D$,

$$\lambda_B \propto W_B^{-1} \propto T^3 \quad \text{for } T \ll \theta_D$$

as the phonon velocity v is constant;

(ii) dislocations: $$l_D \propto \nu^{-1}$$

and, since $$\nu \propto T,$$

therefore $$\lambda_D \propto W_D^{-1} \propto T^2;$$

(iii) point defects:

$$l_P \propto \nu^{-4} \propto T^{-4} \quad \text{(Rayleigh scattering)};$$

therefore $$\lambda_p \propto W_p^{-1} \propto T^{-1}.$$

It is only in the case of boundary scattering that we can make any reliable *a priori* estimate of the mean free path of the lattice waves. Casimir (1938) has deduced that in an extended crystalline solid

$$\lambda_B = 2{\cdot}31 \times 10^3 RA^{\frac{2}{3}} pT^3 \text{ W/cm deg},$$

where R is the radius of the crystal, assumed to be of circular cross-section; p is approximately a constant which depends on the ratio of sound velocities in different directions, the usual value being $p \simeq 1{\cdot}4$. A is the constant in $C_v = AT^3$ J/cm³ deg.

The contribution from phonon-interaction, W_u, varies approximately as T^{-1} at higher temperature ($T \gg \theta_D$). The approximate magnitude has been calculated by Leibfried and Schlömann (1954) on the basis of a simplified model of a cubic crystal:

$$\lambda_u \approx 12k^3 M a \theta^3 / 5h^3 \gamma^2 T$$
$$\approx 3{\cdot}6 a A \theta_D^3 / \gamma^2 T \text{ W cm}^{-1} \text{degK}^{-1}.$$

Here M is the atomic mass, a is the lattice constant, γ is the anharmonicity or Grüneisen parameter (about 2 for many solids); h and k are Planck's and Boltzmann's constants. This formula leads to a slight overestimate of the lattice conductivity in some cubic solids; it overestimates λ_u badly (by a factor of 5 or more) in diamond-structure solids such as germanium and silicon where the γ determined from expansion measurements is relatively small (*ca.* 0·7) and is probably an inappropriate average to use in this equation.

At lower temperatures, λ_u changes exponentially with temperature as the Umklapp processes, responsible for the resistance, become more improbable.

In metals a most important source of scattering of lattice waves are the free electrons. Makinson (1938) first showed that this thermal resistance (W_E) is proportional to T^{-2} at low temperatures and (see also Klemens 1956) is related to the electron–phonon coupling which is responsible for ρ_i and W_i in the respective electron transport processes.

An earlier figure (Chapter VII, p. 214) illustrates the differing temperature dependences of the total thermal conductivity in a pure metallic element, a crystalline dielectric solid, an alloy, and a glass.

More detailed discussions of heat conduction in solids are given in reviews by Klemens (1956, 1958), Mendelssohn and Rosenberg (1961), Ziman (1960), and in *Thermal conductivity* (ed. R. P. Tye, Academic Press, New York, 1967).

4. Thermal-resistivity data

Pure metallic elements

A practical example of the folly of assuming particular values of thermal conductivity for a material merely because it is labelled by the name of a certain element is illustrated by Fig. 136. These curves are experimental values for various samples of copper and show the tremendous variation of conductivity, particularly from 10° to 30° K, that arises with slight change in chemical or physical purity.

Data for a large number of metallic elements, as well as alloys, non-metals, glasses, etc., are given in the following compendia:

(i) *N.B.S. Circular* 556, *Thermal conductivity of metals and alloys at low temperatures*, by R. L. Powell and

W. A. Blanpied (U.S. Govt. Printing Office, Washington, D.C., 1954);

(ii) *WADD Tech. Report* 60/56, *Properties of materials at low temperatures*, by V. J. Johnson of the N.B.S. Boulder Laboratories (also Pergamon Press, 1961);

(iii) *Thermal conductivity of selected materials* NSRDS–NBS8, by R. W. Powell, C. Y. Ho and P. E. Liley (U.S. Govt. Printing Office, Washington, D.C., 1966).

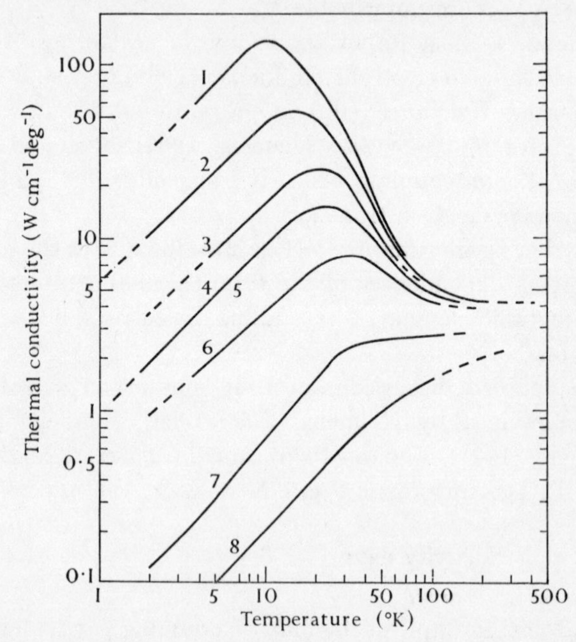

FIG. 136. Thermal conductivity of copper samples. 1, pure (99·999%, annealed, American Smelting and Refining); 2, pure (99·999%, annealed, Johnson Matthey); 3, coalesced (99·98%, oxygen-free annealed, Phelps Dodge); 4, pure (99·999%, cold worked, Johnson Matthey); 5, electrolytic tough pitch (99·9+%, representing some tubes, much sheet and plate); 6, free-cutting tellurium (99%+0·6% Te, representing machining rods and bar); 7, pure Cu+0·056% Fe (annealed); 8, phosphorus deoxidized (99·8%+0·1% P, representing some tubes, pipe, sheet, and plate). Curves 1, 3, 5, 6, and 8 from Powell, Rogers, and Roder (1957); curves 2 and 4 from White (1953), and curve 7 from White and Woods (1955).

Bearing in mind that the data for pure materials at low temperatures ($T < \theta/10$) are only representative of the actual samples used, these compendia are a very useful guide to the behaviour pattern of many solids. These are particularly useful for alloys, glasses, and plastics whose conductivity is not too sensitive to composition even at low temperatures; also for elements or dielectric crystals at temperatures greater than $\theta/5$ or $\theta/10$, where a limited number of defects are less important.

At higher temperatures $\lambda \simeq \lambda_\infty$ is relatively insensitive to purity or temperature; the values of λ_{295} which are listed in Table E are intended to represent the best experimental estimates of the conductivity of the metallic elements in states of high purity and at a temperature sufficiently high that $\lambda \to \lambda_\infty$.

When the practical problem arises of calculating approximate values of $\lambda(T)$ for a particular sample of a metallic element the suggested procedure is:

(i) Measure its electrical resistance at room temperature and in liquid helium to obtain R_{295} and R_4; by comparing $R_{295}-R_4$ with ρ_i at 295° K given in Table E, the shape factor l/A may be calculated.

(ii) Then calculate $W_\infty = \rho_i(295°\,\text{K})/2{\cdot}45\times10^{-8}\,.\,295$ cm deg W^{-1} or obtain W_∞ direct from Table E, assuming $W_\infty = 1/\lambda_{295}$. Also obtain $W_0 T = \rho_4/2{\cdot}45\times10^{-8}$ cm deg^2 W^{-1}.

(iii) At each temperature T, calculate $W_0 = \rho_4/2{\cdot}45\times10^{-8}T$; and using Table D or Fig. 135 calculate

$$W_i = 2W_\infty(T/\theta_D)^2\,.\,J_3(\theta_D/T).$$

(iv) Then at each temperature $W \simeq W_0 + W_i$.

For example: Consider a bar of iron; measurements with a galvanometer amplifier give

$$R_{295} \simeq 0{\cdot}00141\ \Omega$$

and
$$R_4 \simeq 0{\cdot}00011\ \Omega.$$

Step (i) gives $R_i(295) = 0{\cdot}00130\ \Omega$ and since

$$\rho_i(295) = 9{\cdot}8\times10^{-6}\ \Omega,$$

$$l/A = 133.$$

Hence $\rho_4 = 0{\cdot}83\times10^{-6}\ \Omega\,\text{cm}$.

Step (ii). We may calculate

$$W_\infty = 9{\cdot}80 \times 10^{-6}/2{\cdot}45 \times 10^{-8}\,.\,295 = 1{\cdot}36,$$

or obtain an experimental value from Table E

$$W_\infty = 1/\lambda_{295} = 1{\cdot}25$$

and $$W_0 T = \rho_4/L = 34.$$

Step (iii). Using $\theta_D = 400°$ K we may proceed to calculate W or λ at, say, 40° K.

For $T/\theta = 0{\cdot}1$, Fig. 135 (Table D) gives $W_i/W_\infty = 0{\cdot}144$, therefore

$$W_i = 0{\cdot}18 \quad \text{and} \quad W_0 = 34/40 = 0{\cdot}85.$$

Step (iv). Thus at 40° K, $W = 0{\cdot}85+0{\cdot}18 = 1{\cdot}03$,

hence $$\lambda \simeq 0{\cdot}97 \text{ W cm}^{-1}\,\text{deg}^{-1}.$$

A source of error in this method of predicting λ lies in the possible departure of W_i from the assumed $T^2\,.\,J_3(\theta/T)$ relation. This semi-empirical relation leads to $W_i/W_\infty = C(T/\theta)^2$ for $T \leqslant \frac{1}{10}\theta$ where $C = 14{\cdot}4$, whereas experiment shows that C may vary from 10 to about 25; in general a range of 12–20 covers most metallic elements. Again at temperatures $T < \frac{1}{10}\theta$, W_i appears to fall more rapidly than T^2 in many elements, so that an index of 2·3–2·5 (rather than 2·0) is not unusual. At worst such discrepancies may result in our estimate of λ being wrong by 30 or 40 per cent in the vicinity of the conductivity maximum. A much better estimate may be obtained if we merely use our electrical resistance data to estimate $W_0 T$ and use existing experimental data on λ (for other samples of the element) to give values of W_i.

Either procedure is usually more reliable than to attempt to guess the physical or chemical purity of a specimen, since residual electrical resistance and low temperature heat conductivity are very sensitive to specific impurities and their state of segregation in the metal.

Metallic alloys

As we discussed above for pure metallic elements, a relatively easy measurement, that of $\rho_4 = \rho_r$, allows a rough estimate of the electronic thermal conductivity λ_e, to be made. In cases where the impurity scattering is not excessive, for example wherever $\rho_r/\rho_{295} < 0{\cdot}5$, it may be assumed that electronic conductivity λ_e, is the dominant mechanism, i.e.

$$\lambda = \lambda_e + \lambda_g$$
$$\simeq \lambda_e.$$

It is in the technically important alloys such as monel, stainless steel, and cupro-nickel that the residual resistance is very high and λ_e is reduced to become comparable with λ_g. Estimation or prediction of λ_g from other physical data appears extremely difficult at the present time. By making certain assumptions about the process of phonon–electron interaction, a relation between W_E (or λ_E) and W_i can be deduced which enables us to estimate crudely the lattice conductivity in annealed alloys at temperatures below about $\frac{1}{20}\theta$, where λ_g is chiefly limited by the presence of free electrons (see White and Woods 1957):

$$\lambda_E = \frac{0{\cdot}53T_1 \times 10^{-6}}{\rho_i(T_1)}(T/\theta)^2 \quad \text{for } T \leqslant \tfrac{1}{20}\theta,$$

where temperature T_1 is comparable with θ_D and may be conveniently 295° K or thereabouts.

In most instances, the ideal electrical resistivity ρ_i at 295° K for the major constituent of an alloy is known, then

$$\lambda_E \simeq 157 . 10^{-6}(T/\theta)^2/\rho_i(295).$$

As an example we may choose annealed inconel and assume, since nickel is the major constituent, that

$$\rho_i(295) = 7{\cdot}1 \times 10^{-6}\,\Omega\,\text{cm} \quad \text{and} \quad \theta_D \simeq 390°\,\text{K};$$

then λ_E, the thermal conductivity due to phonons when scattering by free electrons is dominant, is

$$\lambda_E \simeq \frac{157 \times 10^{-6}}{7{\cdot}1 \times 10^{-6}}(T/390)^2 = 1{\cdot}4 \times 10^{-4}T^2\ W/\text{cm deg}.$$

This result appears to approximate to the experimentally determined values for the lattice conductivity of annealed inconel at temperatures below 20° K (see, for example, Estermann and Zimmerman 1952).

However, even if this estimate of W_E were always correct, it could only yield a value of λ_g over a narrow temperature range and be correct for annealed alloys in which dislocation scattering is unimportant.

Because of these difficulties, data are given in Table I for measured values of total heat conductivity in a number of common alloys, some of them in both annealed and cold-worked conditions. The data have been largely taken from Powell and Blanpied (1954) with the exception of data on brass, soft solder, and Wood's metal which were published by Berman, Foster, and Rosenberg (1955), the non-magnetic alloy silicon bronze (Powell, Rogers, and Roder 1957), and the aluminium alloys (Powell, Hall, and Roder 1960).

Dielectric crystals

These are not of great interest in cryogenic design with the exception of paramagnetic crystals used in cooling (see § 9.5) and some ceramics used for heat-sinks. The latter are generally buttons of alumina (or sapphire), beryllia (see § 7.7), which are metallized on each side so that an electrical lead can be thermally anchored with solder to one side and the other side is soldered to a brass or metal surface in the cryostat.

The conductivity of some dielectric crystals is illustrated in Fig. 137. As discussed above, the conductivity is limited by phonon–phonon interactions at higher temperatures and by grain boundaries at the lowest temperatures. At intermediate temperatures dislocations, point defects, etc., reduce the conductivity. Therefore data in Fig. 137 or in Powell and Blanpied's compilation may be considered to be representative at higher temperatures ($T > \theta/5$); but at low temperatures the conductivity may differ markedly from sample to sample.

Pure crystals of beryllia and alumina both have conductivities which can reach 10 W cm^{-1} degK^{-1} or more. At temperatures below 10° K, the conductivity is controlled by the grain size and by the Casimir equation discussed in § 3.

Glasses, nylon, Teflon, etc.

Glass is a highly disordered substance, and the conductivity, which is due to lattice waves, might be expected to be given by the product $Cvl/3$ where the mean free path l is of the order of the

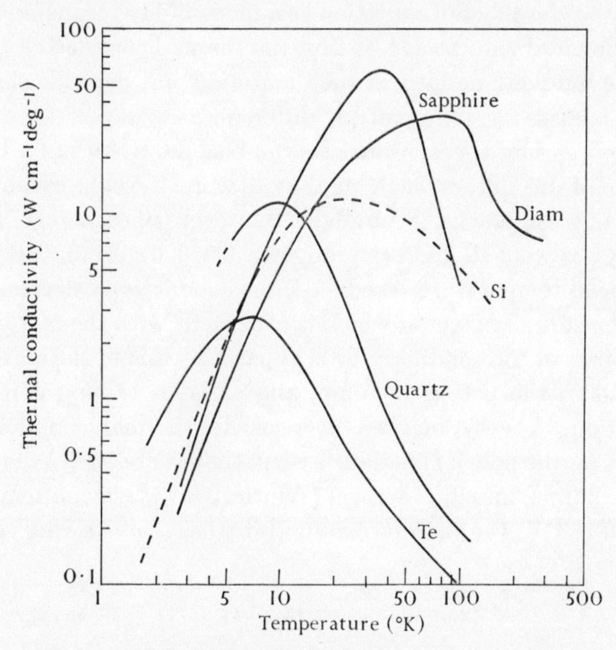

FIG. 137. Thermal conductivity of some crystalline dielectric solids.

interatomic distance. This seems to be approximately true at least down to temperatures of 20°–30° K, below which longitudinal lattice waves of long wavelength make a significant contribution to λ_g (see Klemens, 1951, 1956). Table I includes conductivity values for some glasses as well as for plastic materials used in cryostats. The data listed at 4° K and above are taken from Powell and Blanpied (1954), Berman *et al.* (1955), Powell and Rogers (1955); those below 4° K are from Anderson, Reese, and Wheatley (1963) or extrapolated.

Powder insulators

The importance of some powdered materials in providing thermal insulation for liquid-air vessels has already been mentioned (§ 2.1).

The effective thermal conductivities of many powder insulators and cellular glass or cellular plastic insulators are very similar at room temperature in the presence of an atmosphere of air. Due partly to the effect of radiation as a mode of heat transfer, some experimental values suggest that the thermal conductivity may not be uniquely defined in such materials but depends slightly on thickness or temperature difference. However, to a first approximation, the apparent thermal conductivities of a broad group of insulators such as Zerolite, rock-wool, corkboard, cellular glass blocks, Styrofoam, Santocel (silica aerogel), and fibre-glass wool all lie in the range of 300–600 μW cm^{-1} deg^{-1}. If the mean temperature is reduced, the conductivity decreases in an almost linear fashion which is consistent with the fact that a large part of the conductivity is apparently due to the air or gas present in the porous medium, and a lesser fraction is due to radiation. A very marked decrease in thermal conductivity occurs for the powder insulators when the gas pressure is reduced. For example, for silica aerogel (White 1948) at a mean temperature of 183° K, the thermal conductivity has the following values.

Pressure	100	1	10^{-2}	10^{-3}	10^{-4}	10^{-5}	mm Hg
λ	125	53	17	13	11	11	μW/cm deg.

If the pressure is kept below 10^{-3} mm Hg, λ is relatively insensitive to pressure (see, for example, Johnston, Hood, and Bigeleisen 1955). Fig. 138 below, taken from the work of Fulk, Devereux, and Schrodt (1957) at the National Bureau of Standards, illustrates the dependence of λ on pressure. This latter group of workers has also shown that the conductivity decreases when finely divided aluminium is added to a powder insulator. Fig. 139 shows the extent to which the addition of aluminium powder reduces the apparent thermal conductivity in Perlite or Santocel, presumably by reducing the radiation transfer.

Fulk (1959) has written an interesting review of the role of powder insulation in low-temperature equipment.

Of the cellular or foamed plastic insulators, perhaps the most commonly used in recent years has been Styrofoam, an exploded polystyrene which has a conductivity of about 430 μW cm^{-1} deg^{-1} at room temperature. From data given for Styrofoam by Waite

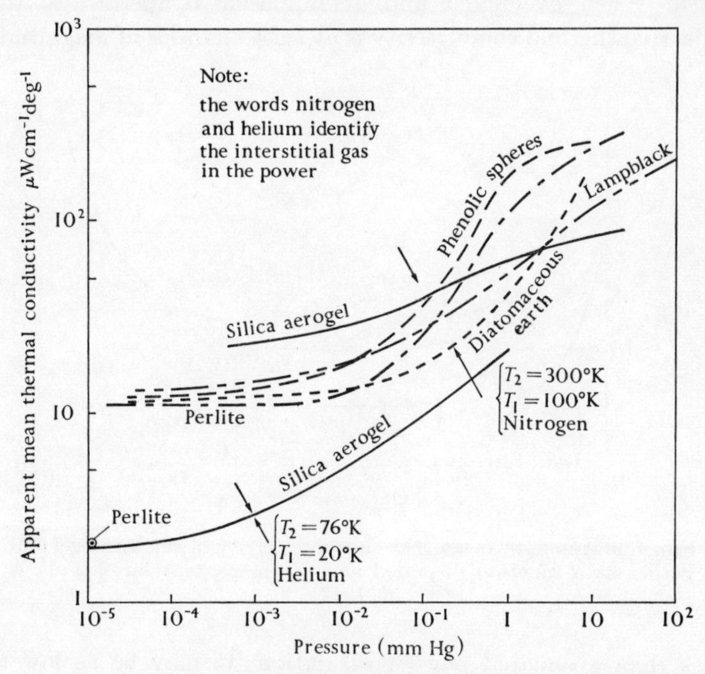

FIG. 138. Apparent mean thermal conductivity of some power insulators as a function of gas pressure (after Fulk, Devereux, and Schrodt 1957).

(1955) we may obtain: λ is 430 at a mean temperature of 300° K, 290 at 227° K, 220 at 173° K, and 160 at 116° K. Such figures indicate that Styrofoam insulation is comparable with Santocel insulation when air at or near atmospheric pressure is present in the Santocel. If, however, the interstitial air is removed from the Santocel its conductivity becomes much lower.

The addition of a reflecting powder to any of the insulators reduces the conductivity substantially (Fig. 139). Insulation can be made even more effective if it consists of multilayer sandwiches of reflecting sheets with insulating layers between. These are so-called 'superinsulations', which normally use aluminium foil or aluminized Mylar for the reflecting layers and fibreglass mat or nylon net between them. Commercial insulation of this type uses foil of *ca.* 1 mil thickness and has 50 or more layers per inch. When evacuated and at liquid-air temperatures, the effective thermal conductivity is at least an order of magnitude

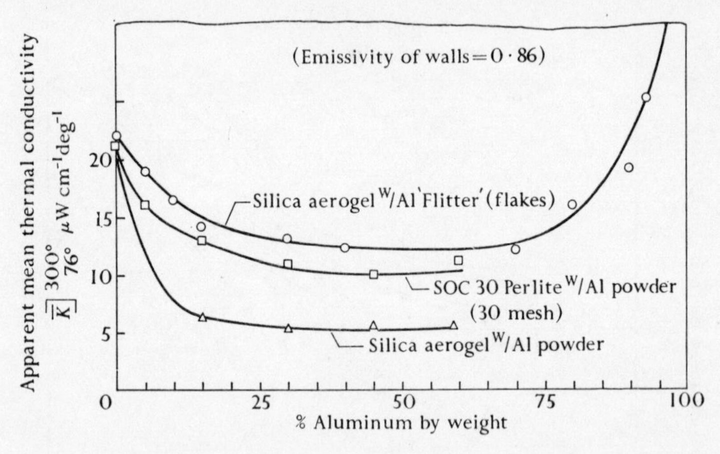

Fig. 139. Apparent mean thermal conductivity of silica aerogel and Perlite as a function of added aluminium content (after Fulk *et al.* 1957).

less than evacuated powder insulation. It may be as low at 10^{-7} W/cm degK. Naturally such insulation has become widely used in storage dewars for liquid air, hydrogen, or helium. It has made it possible for large liquid-air containers to have evaporation rates of a few tenths of a per cent in 24 hours; also liquid-helium storage dewars can be made with superinsulation instead of a liquid-air shield and still preserve a satisfactory evaporation rate of 2–3 per cent per 24 hours.

Details of the first use of superinsulation may be found in Section D of the Proceedings of the 1959 Cryogenic Engineering Conference (*Advances in cryogenic engineering*, Vol. 8, 1960, Plenum Press, New York); see also Cockett and Molnar (1960).

5. References

ANDERSON, A. C., REESE, W., and WHEATLEY, J. C. (1963). *Rev. scient. Instrum.* **34**, 1386.

BERMAN, R. (1951). *Phil Mag.* **42**, 642.

—— FOSTER, E. L., and ROSENBERG, H. M. (1955). *Br. J. appl. Phys.* **6**, 181.

CASIMIR, H. B. G. (1938). *Physica* **5**, 495.

COCKETT, A. H. and MOLNAR, W. (1960). *Cryogenics* **1**, 21.

ESTERMANN, I. and ZIMMERMAN, J. E. (1952). *J. appl. Phys.* **23**, 578.

EVANS, D. J. and ERICKSON, R. A. (1965). Ibid. **36**, 3517.

FULK, M. M. (1959). *Prog. Cryogen.* **1**, 65.

—— DEVEREUX, R. J., and SCHRODT, J. E. (1957). *Proc.* 1956 *Cryogenic Engng. Conf., N.B.S., Boulder, Colorado*, p. 163.

GERRITSEN, A. N. (1956). *Handb. Phys.* **19**, 137.

JOHNSTON, H. L., HOOD, C. B., and BIGELEISEN, J. (1955). *Proc.* 1954 *cryogen. engng conf., N.B.S. Report No.* 3517, p. 139.

KLEMENS, P. G. (1951). *Proc. R. Soc.* A**208**, 108.

—— (1956). *Handb. Phys.* **14**, 198.

—— (1958) *Solid St. Phys.* **7**, 1.

—— (1967) in *Thermal conductivity* (ed. R. P. Tye). Academic Press, New York.

LEIBFRIED, G. and SCHLÖMANN, E. (1954). *Nachr. Ges. Wiss. Göttingen* IIa, 71.

MACDONALD, D. K. C. and TOWLE, L. (1956). *Can. J. Phys.* **34**, 418.

—— WHITE, G. K., and WOODS, S. B. (1956). *Proc. R. Soc.* A**235**, 358.

MAKINSON, R. E. B. (1938). *Proc. Camb. phil. Soc. math. phys. Sci.* **34**, 474.

MEADEN, G. T. (1965). *Electrical resistance of metals.* Plenum Press, New York.

MEISSNER, W. and VOIGT, B. (1930). *Annln Phys.* **7**, 761, 892.

MENDELSSOHN, K and ROSENBERG, H. M. (1961). *Solid St. Phys.* **12**, 223.

MOTT, N. F. (1935). *Proc. phys. Soc.* **47**, 571.

—— (1936). *Proc. R. Soc.* A**153**, 699.

POWELL, R. L. and BLANPIED, W. A. (1954). *Natn. Bur. Stand. Circular* 556. U.S. Govt. Printing Office, Washington, D.C.

—— HALL, W. J., and RODER, H. M. (1960). *J. Appl. Phys.* **31**, 496.

—— and ROGERS, W. M. (1955). *Nat. science foundation conf. low temp. phys. and chem.* 1955, Baton Rouge, Louisiana.

—— —— and RODER, H. M. (1957). *Proc.* 1956 *cryogen. engng conf., N.B.S., Boulder, Colorado*, p. 166; *J. appl. Phys.* **28**, 1282.

ROGERS, W. M., and POWELL, R. L. (1958). *N.B.S. Circular* 595, *Tables of transport integrals.* U.S. Govt. Printing Office, Washington, D.C.

ROSENBERG, H. M. (1955). *Phil. Trans. R. Soc.* A**247**, 441.

WAITE, H. J. (1955). *Proc.* 1954 *cryogen. engng conf., N.B.S. Report No.* 3527, p. 158.

White, G. K. (1953). *Aust. J. Phys.* **6**, 397.
—— and Woods, S. B. (1955). *Can. J. Phys.* **33**, 58.
—— —— (1957). *Proc.* 1956 *cryogen. engng conf. N.B.S., Boulder, Colorado*, p. 120.
—— —— (1959). *Phil. Trans. R. Soc.* **251**A, 273.
White, J. F. (1948). *Chem. Engng Prog.* **44**, 647.
Wilson, A. H. (1938). *Proc. R. Soc.* A**167**, 580.
—— (1953). *The theory of metals*. Cambridge University Press.
Ziman, J. (1960). *Electrons and phonons*. Clarendon Press, Oxford.
—— (1964). *Principles of the theory of solids*. Cambridge University Press.

APPENDIX

TABLES

TABLE A

Vapour pressure–temperature

(See Chapter IV, § 5, for references and comments)

p	^{3}He	^{4}He	*Equilibrium* —H_2	*Normal* —H_2	N_2	O_2
mm Hg *at* 0° C, *standard gravity*	1962 *scale*	1958 *scale*	*Woolley, Scott, and Brickwedde* (1948)	*Woolley, Scott, and Brickwedde* (1948)	*Armstrong* (1964) ($T > 63{\cdot}2°$ K) *Keesom and Bijl* (1937) ($T < 63{\cdot}2$ K)	*Muijlwijk, Moussa, and Van Dijk* (1966)
	(° K)	(° K)	(° K)	(° K)	(° K)	(° K)
800	3·2395	4·2700	20·44	20·56	—	90·671
780	3·2152	4·2427	20·37	20·48	77·58	90·427
760	3·1905	4·2150	$20{\cdot}27_3$	$20{\cdot}39_0$	$77{\cdot}36_4$	90·178
740	3·1653	4·1868	20·18	20·30	77·14	89·924
720	3·1396	4·1580	20·09	20·21	76·91	89·664
700	3·1134	4·1287	20·00	20·11	76·67	89·399
680	3·0866	4·0989	19·90	20·02	76·43	89·128
660	3·0594	4·0684	19·80	19·92	76·18	88·851
640	3·0315	4·0373	19·70	$19{\cdot}82_5$	76·93	88·567
620	3·0030	4·0056	19·60	19·72	75·68	88·276
600	2·9739	3·9731	19·50	19·62	75·41	87·978
580	2·9441	3·9399	19·40	19·52	75·13	87·671
560	2·9136	3·9059	19·30	19·41	74·85	87·357
540	2·8824	3·8711	$19{\cdot}18_5$	19·30	74·56	87·034
520	2·8504	3·8354	$19{\cdot}06_5$	$19{\cdot}18_5$	74·27	86·701
500	2·8175	3·7987	18·94	19·06	73·97	86·358
480	2·7838	3·7611	18·82	$18{\cdot}94_5$	73·66	86·004
460	2·7491	3·7224	18·69	18·82	73·34	85·639
440	2·7133	3·6825	18·57	18·69	73·00	85·260
420	2·6765	3·6414	$18{\cdot}43_5$	18·55	72·66	84·869
400	2·6385	3·5990	18·29	$18{\cdot}40_5$	72·30	84·462
380	2·5992	3·5551	18·15	18·26	71·92	84·038
360	2·5586	3·5097	18·00	18·11	71·54	83·597
340	2·5164	3·4625	$17{\cdot}83_5$	17·96	71·13	83·136
320	2·4726	3·4134	17·67	17·79	70·70	82·653
300	2·4269	3·3622	17·50	17·62	70·25	82·146
290	2·4033	3·3357	$17{\cdot}41_5$	17·53	70·01	81·882
280	2·3792	3·3086	$17{\cdot}32_5$	$17{\cdot}43_5$	69·77	81·611
270	2·3545	3·2808	17·22	17·34	69·53	81·332
260	2·3292	3·2524	17·13	17·25	69·28	81·045
250	2·3033	3·2231	17·03	17·15	69·02	80·749

TABLE A (*cont.*)

p	^3He	^4He	*Equilibrium* —H_2	*Normal* —H_2	N_2	O_2
mm Hg *at* 0° C, *standard gravity*	1962 *scale* (° K)	1958 *scale* (° K)	*Woolley, Scott, and Brickwedde* (1948) (° K)	*Woolley, Scott, and Brickwedde* (1948) (° K)	*Armstrong* (1964) ($T > 63{\cdot}2°$ K) *Keesom and Bijl* (1937) ($T < 63{\cdot}2$ K) (° K)	*Muijlwijk, Moussa, and Van Dijk* (1966) (° K)
240	2·2767	3·1931	16·93	17·05	68·75	80·443
230	2·2493	3·1622	$16{\cdot}82_5$	16·94	68·48	80·127
220	2·2212	3·1304	$16{\cdot}71_5$	$16{\cdot}83_5$	68·18	79·800
210	2·1922	3·0976	16·61	$16{\cdot}71_5$	67·88	79·460
200	2·1623	3·0637	16·49	16·60	67·57	79·108
190	2·1314	3·0287	16·36	$16{\cdot}48_5$	67·25	78·740
180	2·0994	2·9924	$16{\cdot}23_5$	$16{\cdot}36_5$	66·91	78·358
170	2·0663	2·9546	16·10	16·23	66·56	77·957
160	2·0318	2·9153	15·97	16·10	66·20	77·537
150	1·9959	2·8744	15·83	15·95	65·81	77·096
140	1·9585	2·8315	15·68	15·80	65·42	76·630
130	1·9192	2·7865	15·52	15·64	64·98	76·137
120	1·8779	2·7390	15·35	15·47	64·52	75·611
110	1·8343	2·6888	15·18	15·29	64·03	75·069
100	1·7881	2·6354	15·00	15·10	63·50†	74·444
90	1·7387	2·5781	14·77	14·89	62·94	73·787
80	1·6856	2·5163	14·56	14·68	62·39	73·068
70	1·6279	2·4489	14·30	14·42	61·77	72·270
60	1·5646	2·3745	14·04	14·15	61·06	71·373
50	1·4939	2·2911	13·73‡	13·83§	60·24	70·344
45	1·4550	2·2450	$13{\cdot}57_5$	13·67	—	69·763
40	1·4131	2·1952‖	13·40	13·51	59·28	69·127
35	1·3676	2·1413	13·21	13·30	—	68·420
30	1·3176	2·0827	12·99	13·09	58·08	67·624
25	1·2617	2·0174	12·74	12·85	—	66·709
20	1·1978	1·9427	12·46	12·57	56·48	65·624
18	1·1692	1·9092	12·32	12·43	—	65·125
16	1·1384	1·8729	$12{\cdot}17_5$	12·28	55·65	64·577
14	1·1049	1·8333	12·02	$12{\cdot}11_5$	—	63·968
12	1·0679	1·7893	11·83	11·92	—	63·281
10	1·0266	1·7396	11·62	11·72	—	62·488
9	1·0038	1·7120	11·50	11·59	—	62·040
8	0·9791	1·6820	11·36	11·47	—	61·547
7	0·9523	1·6490	11·23	11·34	—	60·998

† Triple point at 63·15° K (94·0 mm Hg) from Keesom and Bijl (1937).
‡ Triple point at 13·81° K (52·8 mm Hg).
§ Triple point at $13{\cdot}95_7$° K (54·0 mm Hg).
‖ λ-point at 2·1720° K (37·80 mm Hg).

TABLE A (*cont.*)

p	^3He	^4He	*Equilibrium* —H_2	*Normal* —H_2	N_2	O_2
mm Hg *at* 0° C, *standard gravity*	1962 *scale*	1958 *scale*	*Woolley, Scott, and Brickwedde* (1948)	*Woollley, Scott, and Brickwedde* (1948)	*Armstrong* (1964) ($T > 63\cdot2^\circ$ K) *Keesom and Bijl* (1937) ($T < 63\cdot2$ K)	*Muijlwijk, Moussa, and Van Dijk* (1966)
	(° K)	(° K)	(° K)	(° K)	(° K)	(° K)
6	0·9227	1·6123	11·08	11·17	—	60·378
5	0·8894	1·5707	10·88	10·98	—	59·662
4	0·8512	1·5221	10·65	10·76	—	58·810
3·5	0·8295	1·4943	10·53	10·64	—	58·813
3·0	0·8055	1·4632	10·38	10·50	—	57·750
2·5	0·7785	1·4277	10·23	10·32	—	57·100
2·0	0·7474	1·3863	10·02	10·12	—	56·325
1·5	0·7101	1·3359	—	—	—	55·359¶
1·0	0·6624	1·2699				
0·9	0·6508	1·2536				
0·8	0·6383	1·2359				
0·7	0·6246	1·2162				
0·6	0·6093	1·1942				
0·5	0·5920	1·1691				
0·4	0·5720	1·1395				
0·3	0·5477	1·1032				
0·2	0·5163	1·0554				
0·1	0·4693	0·9814				
0·08	0·4556	0·9595				
0·06	0·4391	0·9325				
0·04	0·4174	0·8967				
0·02	0·3844	0·8407				
0·01	0·3557	0·7907				
0·005	0·3305					
0·001	0·2829					

¶ Triple point at 54·36° K (1·14 mm Hg).

TABLE B

$Z = (R_T - R_4)/(R_{273} - R_4)$ and average $\Delta Z/\Delta T$ for some platinum thermometers (see Chapter IV, § 6). Values of $W_{4\cdot 2} = R_{4\cdot 2}/R_{273}$ indicate the electrical 'purity'

T	$10^6 Z$					$\Delta Z/\Delta T \times 10^6$
	'*Ideal*' (*Berry*)	*Meyers* 459 (*NSL*)	*Meyers* 373 (*NSL*)	*T*4 (*NRC*)	*CCT*64	
($10^6 W_{4\cdot 2}$	≈ 0	372	425	453	536)	—
10·5	—	$287\cdot_5$	288	—	308	103
11	$356\cdot_2$	345	344_5	—	368	115
11·5	—	411	408_5	434	434·3	130
12	489·7	485	481_5	510	508·9	147
12·5	—	$569\cdot_5$	565	597	594·1	168
13	$663\cdot_3$	665	659	693	690·3	190
13·5	—	772	766	801	798·7	213
14	$884\cdot_5$	893	886	922	920·4	242
14·5	—	1 028	1 020	1 058	1 056·3	270
15	$1\,161\cdot_4$	1 179	1 169	1 209	1 207·1	300
15·5	—	1 345	1 335	1 375	1 374	332
16	$1\,502\cdot_5$	1 529	1 517	1 558	$1\,558\cdot_2$	366
17	$1\,915\cdot_5$	1 951	1 936	1 980	$1\,980\cdot_3$	416
18	$2\,408\cdot_7$	2 453	2 435	2 480	$2\,481\cdot_0$	496
19	$2\,987\cdot_3$	3 040	3 021	3 067	3 067·4	582
20	$3\,660\cdot_1$	3 721	3 699	3 746	3 746·8	675
21	4 431	4 501	4 476	4 529	$4\,525\cdot_5$	775
22	5 307	5 385	5 359	5 411	$5\,408\cdot_9$	880
23	6 292	6 378	6 351	6 406	6 402·0	990
24	7 394	7 485	7 456	7 513	$7\,508\cdot_4$	1105
25	8 613	8 708	8 678	8 736	8 731·5	1220
26	9 950	10 050	10 019	10 080	10 073	1340
27	11 409	11 513	11 481	11 544	11 536	1460
28	12 989	13 097	13 065	13 128	13 120	1580
29	14 692	14 802	14 770	14 838	14 826	1705
30	16 516	16 630	16 596	16 667	16 653	1825
32	20 528	20 646	20 612	20 689	20 670	2007
34	25 015	25 137	25 103	25 182	25 158	2245
36	29 956	30 069	30 049	30 133	30 096	2470
38	35 326	35 455	35 421	35 508	35 460	2690
40	41 091	41 222	41 188	41 280	41 224	2884
42	47 220	47 352	47 319	47 414	47 351	3065
44	53 683	53 814	53 782	53 881	53 816	3231
46	60 449	60 580	60 547	60 651	60 584	3383
48	67 489	67 620	67 588	67 694	67 625	3520
50	74 773	74 904	74 872	74 982	74 912	3642
55	93 868	93 999	93 967	94 083	94 021	3819
60	113 934	114 065	114 035	114 147	114 109	4013

TABLE B (*cont.*)

T	10^6Z 'Ideal' (Berry)	*Meyers* 459 (*NSL*)	*Meyers* 373 (*NSL*)	*T*4 (*NRC*)	*CCT*64	$\Delta Z/\Delta T \times 10^6$
($10^6W_{4\cdot2}$	≈ 0	372	425	453	536)	—
65	134 706	134 837	134 806	134 928	134 874	4154
70	155 916	156 048	156 027	156 140	156 085	4242
75	177 370	177 503	177 473	177 599	177 572	4291
80	198 990	199 122	199 092	199 215	199 214	4324
85	220 721	220 853	220 824	220 947	220 928	4346
90	242 495	242 625	242 597	242 719	242 655	4354
95			264 311	264 491	264 359	
100			285 944	286 063	286 011	
110			328 975	329 090	329 101	
120			371 707	371 819	371 867	
130			414 161	414 269	414 313	
140			456 354	456 456	456 466	
150			498 304	498 400	498 360	

TABLE C

Thermoelectric potential difference E with respect to 0° K *and thermopower dE/dT* (*after Powell, Bunch, and Corruccini* 1961, *in Chapter* 4)

T	Au+2·1% Co *versus* Cu		*Constantan versus* Cu		*Normal silver versus* Cu
(° K)	E (μV)	dE/dT (μV/degK)	E (μV)	dE/dT (μV/degK)	E (μV)
1	0·53	1·05	0·17	0·33	
2	2·09	2·07	0·66	0·66	
3	4·66	3·07	1·48	0·98	
4	8·22	4·04	2·62	1·30	
5	12·74	4·99	4·07	1·61	0·00
6	18·20	5·92	5·83	1·92	
7	24·57	6·82	7·90	2·22	
8	31·83	7·70	10·26	2·52	
9	39·96	8·55	12·92	2·81	
10	48·93	9·38	15·88	3·10	0·01
12	69·30	10·97	22·64	3·66	
14	92·75	12·46	30·50	4·20	

TABLE C (*cont.*)

T (° K)	Au+2·1% Co *versus* Cu E (μV)	Au+2·1% Co *versus* Cu dE/dT (μV/degK)	*Constantan versus* Cu E (μV)	*Constantan versus* Cu dE/dT (μV/degK)	*Normal silver versus* Cu E (μV)
16	119·1	13·85	39·43	4·73	
18	148·1	15·17	49·40	5·25	
20	179·6	16·43	60·40	5·77	0·2
22	213·7	17·64	72·42	6·27	
24	250·1	18·79	85·43	6·76	
26	288·8	19·89	99·43	7·24	
28	329·6	20·94	114·4	7·71	
30	372·5	21·94	130·3	8·17	1·8
32	417·3	22·88	147·1	8·62	
34	464·0	23·79	164·7	9·05	
36	512·4	24·64	183·3	9·48	
38	562·5	25·45	202·7	9·90	
40	614·2	26·23	222·9	10·31	7·2
45	749·9	28·00	276·8	11·28	
50	893·9	29·56	335·6	12·20	15·2
55	1045·2	30·93	398·8	13·07	19·5
60	1202·9	32·13	466·2	13·89	24·0
65	1366·2	33·19	537·5	14·66	28·3
70	1534·5	34·11	612·7	15·38	32·7
75	1707·1	34·91	691·2	16·03	37·2
80	1883·5	35·64	773·0	16·69	41·6
85	2063·4	36·35	858·1	17·37	45·8
90	2246·8	37·00	946·7	18·04	49·8
100	2622·6	38·11	1133·7	19·36	57·2
120	3402·8	39·80	1546·4	21·90	71·2
140	4211·2	40·97	2009·5	24·41	84·1
160	5039·1	41·77	2522·7	26·86	96·5
180	5880·4	42·32	3083·1	29·16	109·2
200	6730.6	42·67	3688·6	31·37	121·8
250	8875·6	43·03	5388·0	36·51	155·2
300	11025·5	42·92	7330·2	41·09	192·3

Table D

Some important physical functions (see Chapters XI and XII)

T/θ	θ/T	C_V cal/mol deg	ρ_i/ρ_θ $(\propto T^5J_5)$†	ρ_i/ρ_θ $(\propto T^3J_3)$‡	W_i/W_∞ $(\propto T^2J_3)$§
(∞)	0	5·957	∞	∞	1·00
(10)	0·1	5·954	10·55	10·42	1·00
(5)	0·2	5·945	5·268	5·201	0·998
(2·5)	0·4	5·909	2·617	2·588	0·993
(2·0	0·5	5·833	2·083	2·062	0·990
(1·667)	0·6	5·851	1·725	1·711	0·985
(1·25)	0·8	5·770	1·274	1·268	0·974
(1·0)	1·0	5·669	1·000	1·000	0·960
(0·833)	1·2	5·549	0·813	0·8186	0·943
(0·714)	1·4	5·412	0·678	0·6873	0·923
(0·667)	1·5	5·337	0·623	0·6341	0·913
(0·625)	1·6	5·259	0·574	0·5873	0·902
(0·556)	1·8	5·094	0·493	0·5084	0·878
(0·500)	2·0	4·918	0·426	0·4444	0·853
(0·400)	2·5	4·444	0·3043	0·3273	0·785
(0·333)	3·0	3·947	0·2220	0·2478	0·713
(0·286)	3·5	3·459	0·1639	0·1910	0·642
(0·250)	4·0	2·996	0·1216	0·1491	0·572
(0·222)	4·5	2·573	0·0906	0·1176	0·508
(0·200)	5·0	2·197	0·0679	0·09360	0·449
(0·1667)	6·0	1·582	0·03849	0·06082	0·350
(0·143)	7·0	1·137	0·02220	0·04084	0·274
(0·125)	8·0	0·8233	0·01308	0·02832	0·217
(0·111)	9·0	0·6041	0·00791	0·02025	0·175
(0·100)	10·0	0·4518	0·00492	0·01490	0·143
(0·0833)	12	0·2667	0·002071	0·00868	0·100
(0·0769)	13	0·2109	0·001401	0·00684	0·0853
(0·0714)	14	0·1688	0·000972	0·00548	0·0736
(0·0625	16	0·1133	0·000500	0·00367	0·0563
(0·0556)	18	0·0796	0·000278	0·00258	0·0445
(0·0500)	20	0·0580	0·000164	0·00188	0·0360
(0·0400)	25	0·0298	0·0000526	0·00096	0·0231
(0·0333)	30	0·0172	0·0000220	—	—

† ρ_i/ρ_θ tabulated as $(T/\theta)^5J_5(\theta/T)\div J_5(1)$.
‡ ρ_i/ρ_θ tabulated as $(T/\theta)^3J_3(\theta/T)\div J_3(1)$.
§ W_i/W_∞ tabulated as $2(T/\theta)^2J_3(\theta/T)$.

Table E

Physical properties of some elements at room temperature (295° K). For anisotropic elements the values listed are generally appropriate to the polycrystalline form

Element	*Atomic weight*	*Structure*	*Density* (g cm^{-3})	θ_D (° K)	*Coefficient of linear thermal expansion* × 10^4 *per* deg K	*Electrical resistivity* ($\mu\Omega$ cm)	*Thermal conductivity* (W cm^{-1} deg^{-1})	*Superconducting transition temperature* (° K)
Aluminium	26·97	f.c.c.	2·70	380	0·24	2·76	2·4	1·20
Antimony	121·76	rhombohedral	6·68	210	0·11	41	0·18	—
Arsenic	74·91	rhombohedral	5·73	290	0·05	29	—	—
Barium	137·36	b.c.c.	3·5	110	0·18–0·26	39	—	—
Beryllium	9·013	h.c.p.	1·84	920	0·12	3·25	2·2	0·03
Bismuth	209·0	rhombohedral	9·8	120	0·13	115	0·085	—
Boron	10·82	hexagonal (?)	2·34	$\simeq$ 1300	0·08	$\simeq 10^{12}$	(0·2)	—
Cadmium	112·41	h.c.p.	8·65	175	0·30	7·3	0·92	0·56
Calcium	40·08	f.c.c.	1·55	210	0·22	3·6	0·98	—
Carbon:								
(graphite)	12·01	hexagonal	2·22	$\simeq$ 400	—	$\sim 10^2$ } $\sim 10^6$ }	$\sim$2	—
(diamond)	—	diamond	—	$\simeq$ 2000	—	$> 10^{12}$	6·5	—
Caesium	132·91	b.c.c.	1·9	45	0·97	20·0	(0·4)	—
Chromium	52·01	b.c.c.	7·2	480	0·044–0·075	12·9	$0{\cdot}8_6$	—
Cobalt	58·94	h.c.p.	8·92	380	0·13–0·16	5·80	(1·2)	—
Copper	63·54	f.c.c.	8·96	310	0·167	1·68	4·0	—
Gallium	69·72	orthorhombic	5·97	240	0·18	14·9	—	1·09
Germanium	72·60	diamond	5·32	400	0·06	$\sim 5\times10^7$	$\simeq 180/T$	—
Gold	197·2	f.c.c.	19·3	185	0·14	2·21	3·1	—
Hafnium	178·6	h.c.p.	13·09	210	0·06	30·6	0·22	0·08
Indium	114·76	tetragonal	7·31	110	0·30	8·8	(0·8)	3·40
Iridium	193·1	f.c.c.	22·5	290	0·065	5·05	1·45	—
Iron	55·85	b.c.c.	7·87	400	0·12	9·80	0·80	—

TABLE E (*cont.*)

Element	*Atomic weight*	*Structure*	*Density* (g cm^{-3})	θ_D (° K)	*Coefficient of linear thermal expansion* × 10^4 *per* deg K	*Electrical resistivity* ($\mu\Omega$ cm)	*Thermal conductivity* (W cm^{-1} deg^{-1})	*Superconducting transition temperature* (° K)
Lead	207·21	f.c.c.	11·34	88	0·29	21·0	0·35	7·19
Lithium	6·94	b.c.c.	0·53	360	0·45	9·3	0·72	—
Magnesium	24·32	h.c.p.	1·74	330	0·25	4·30	1·6	—
Manganese	54·93	cubic (complex)	7·44	410	0·37	140	—	—
Mercury (near 220° K)	200·61	rhombohedral	14·2	110	0·61	21	$\simeq$ 0·3 (200° K)	4·15
Molybdenum	95·95	b.c.c.	10·2	380	0·052	5·33	1·4	0·92
Nickel	58·69	f.c.c.	8·9	390	0·13	7·05	$0{\cdot}9_0$	—
Niobium	92·91	b.c.c.	8·57	250	0·071	14·4	0·53	9·25
Osmium	190·2	h.c.p.	22·48	$\simeq$ 350	0·048	9·1	(0·9)	0·66
Palladium	106·7	f.c.c.	12·02	290	0·12	10·55	$0{\cdot}7_2$	—
Platinum	195·23	f.c.c.	21·45	225	0·089	10·42	$0{\cdot}7_0$	—
Potassium	39·10	b.c.c.	0·85	98	0·83	7·2	0·98	—
Rhenium	186·31	h.c.p.	21·0	275	0·066	18·7	(0·5)	1·70
Rhodium	102·91	f.c.c.	12·44	350	0·084	4·78	1·50	—
Rubidium	85·48	b.c.c.	1·53	61	0·66	12·5	(0·6)	—
Ruthenium	101·7	h.c.p.	$12{\cdot}2_5$	450	$\simeq$ 0·08	7·4	(1·0)	0·47
Selenium	78·96	vitreous	4·30	—	0·49	$> 10^{12}$	0·002–0·003	—
	—	hexagonal	4·80	250	0·4	$\sim 10^7$	$5/T$	—
Silicon	28·06	diamond	2·32	$\simeq$ 700	0·024	$> 10^{10}$	$\simeq 400/T$	—
Silver	107·88	f.c.c.	10·49	220	0·19	1·64	4·2	—
Sodium	22·99	b.c.c.	0·97	160	0·71	4·75	1·4	—
Strontium	87·63	f.c.c.	2·6	140	—	21·5	—	—
Tantalum	180·88	b.c.c.	16·6	230	0·065	13·0	0·55	4·48
Tellurium	127·61	hexagonal	6·24	180	$\simeq$ 0·17	$\sim 0{\cdot}4 \times 10^6$	$10/T$	—
Thallium	204·39	h.c.p.	11·85	94	0·29	16·4	0·4	2·39

TABLE E (*cont.*)

Element	*Atomic weight*	*Structure*	*Density* (g cm^{-3})	θ_D (° K)	*Coefficient of linear thermal expansion* × 10^4 *per deg* K	*Electrical resistivity* ($\mu\Omega$ cm)	*Thermal conductivity* (W cm^{-1} deg^{-1})	*Superconducting transition temperature* (° K)
Thorium	232·12	f.c.c.	11·7	140	0·11	15·6	0·4	1·37
Tin	118·7	tetragonal	7·3	160	$\simeq$ 0·25	11·0	0·7	3·722
		(diam. cubic)						
	—	(grey tin)	5·76	260	—	—	—	—
Titanium	47·90	h.c.p.	4·5	360	0·085	43·0	0·20	0·39
Tungsten	183·92	b.c.c.	19·3	315	0·046	5·32	1·7	0·012
Uranium (α)	238·07	orthorhombic	19·05	160	0·13	$\leqslant$ 25	0·28	0·8
Vanadium	50·95	b.c.c.	6·11	380	0·08	20·0	0·36	5·3
Zinc	65·38	h.c.p.	7·14	240	0·27	$5{\cdot}9_5$	1·2	0·87
Zirconium	91·22	h.c.p.	6·5	250	0·059	42·4	0·21	0·546

(i) Values of atomic weight, density, coefficient of linear thermal expansion, and structures have been taken from standard tables, e.g. *Smithsonian physical tables* (9th revised edn, Smithsonian Institution, 1954), *Manual of lattice spacings and structures of metals and alloys* (W. B. Pearson, Pergamon Press, 1958), and *American Institute of Physics handbook* (2nd edn, McGraw-Hill, 1963).

(ii) Values of the Debye temperatures θ_D have been obtained from specific heat data in the range $\theta/2-\theta$ which are given in reviews or compilations listed in Chapter XI, e.g. Blackman 1955, Keesom and Pearlman, 1956, Gopal 1966, Corruccini and Gniewek 1960.

(iii) Values for electrical resistivity are from sources listed in Chapter XII: Meaden 1965, Gerritsen 1956, White and Woods 1959.

(iv) Thermal conductivity data are also from sources listed in Chapter XII, e.g. in the compilation by Powell and Blanpied (1954), compendium by V. Johnson (1960–1), and the *American Institute of Physics handbook* (2nd edn, McGraw-Hill 1963).

(v) Experimental values of T_c, the superconducting transition temperature, are generally those listed by B. W. Roberts (*Prog. Cryogen.* **4**, 159, 1964).

TABLE F

Linear thermal contractions, relative to 293° K. *Units are* $10^4 \times (L_{293} - L_T)/L_{293}$. *Sources of data are discussed in* § 11.2, *e.g. compilation by Corruccini and Gniewek* 1961

Substance T° K	0	20	40	60	80	100	150	200	250
Aluminium	41·5	41·5	41·3	40·5	39·1	37·0	29·6	20·1	9·7
Chromium	9·8	9·8	9·8	9·7	9·5	9·0	7·4	5·05	2·4
Copper	32·6	32·6	32·4	31·6	30·2	28·3	22·2	14·9	7·1
Germanium	9·2	9·2	9·2	9·2	$9{\cdot}0_5$	8·7	7·1	4·9	2·4
Iron	19·8	19·8	19·7	19·5	18·9	18·1	14·8	10·2	4·9
Lead	70·8	70·0	66·7	62·4	57·7	52·8	39·9	26·3	12·4
Nickel	22·4	22·4	22·3	21·9	21·1	20·1	16·2	11·1	5·0
Silver	41·3	41·2	40·5	38·9	36·6	33·9	25·9	17·3	8·2
Titanium	15·1	15·1	15·0	14·8	14·2	13·4	10·7	7·3	3·5
Tungsten	8·6	8·6	8·5	8·3	8·0	7·5	5·9	4·0	1·9
Brass (65 Cu, 35 Zn)	38·4	38·3	38·0	36·8	35·0	32·6	25·3	16·9	8·0
Constantan	—	—	26·4	25·8	24·7	23·2	18·3	12·4	$5{\cdot}8_5$
German silver	37·6	37·6	37·3	36·2	34·5	32·3	25·4	17·0	8·1
Invar*	4·5	4·6	4·8	4·9	4·8	4·5	3·0	2·0	1·0
304, 316 Stainless steel	—	29·7	29·6	29·0	27·8	26·0	20·3	13·8	6·6
Pyrex	5·6	5·6	5·7	5·6	5·4	5·0	$3{\cdot}9_5$	2·7	0·8
Vitreous silica	−0·7	−0·65	−0·5	−0·3	$-0{\cdot}1_5$	0·0	$0{\cdot}1_8$	$0{\cdot}2_5$	$0{\cdot}1_5$
Araldite	106	105	102	98	94	88	71	$50{\cdot}_5$	25·0
Nylon	139	138	135	131	125	117	95	67	34·0
Polystyrene	155	152	147	139	131	121	93	63	30·0
Teflon	214	211	206	200	193	185	160	124	75·0

* The expansion of Invar or NiFe alloys containing *ca.* 36 per cent Ni is very sensitive to composition and heat treatment.

TABLE G

Electrical resistivities of ideally pure elements in $\mu\Omega$ cm (*see* § 12.2)

T° K	Ag	Al	Au	Cu	Fe	Na	Ni	Pb	Pt	W
295	1·61	2·74	2·20	1·70	9·8	4·75	$7{\cdot}0_4$	21·0	10·42	$5{\cdot}3_3$
273	1·47	2·50	2·01	1·55	8·7	4·29	6·2	19·3	9·59	$4{\cdot}8_2$
250	1·34	2·24	1·83	1·40	$7{\cdot}5_5$	3·82	5·4	17·6	8·70	$4{\cdot}3_2$
200	1·04	1·65	1·44	1·06	5·3	2·88	3·7	13·9	6·76	$3{\cdot}2_2$
150	$0{\cdot}73_5$	$1{\cdot}0_6$	1·04	$0{\cdot}70_5$	$3{\cdot}1_5$	2·00	$2{\cdot}2_4$	10·2	4·78	$2{\cdot}1_1$
100	0·42	0·47	0·63	0·35	$1{\cdot}2_4$	1·15	$1{\cdot}0_0$	6·5	2·74	1·02
80	0·29	0·25	0·46	$0{\cdot}21_5$	0·64	0·81	0·55	5·0	1·91	0·60
50	0·11	0·05	0·20	0·050	$0{\cdot}13_5$	0·32	0·15	$2{\cdot}7_5$	0·72	0·15
25	0·010	—	0·027	0·0025	$0{\cdot}012_5$	0·039	0·017	1·0	0·084	$0{\cdot}011_5$
15	0·0011	—	0·0037	$0{\cdot}0001_7$	$0{\cdot}003_4$	0·005	$0{\cdot}004_5$	0·25	0·0116	$0{\cdot}002_4$

TABLE H

Electrical resistivity of alloys in $\mu\Omega$ cm (*see* § 12.2)

Alloy	*Physical*	295° K	90° K	77° K	4·2° K
Brass (70 Cu, 30 Zn)	strained	7·2	5·0	—	4·3
	annealed	6·6	4·2	—	3·6
Constantan	as received	52·5	45	—	44
Cupro-nickel (80 Cu, 20 Ni)	as received	26	—	24	23
Evanohm	as received	134	133	—	133
German silver	as received	30	27·5	—	26
Gold + 2·1 Co	as received	*ca.* 13	12·3	—	12·0
Manganin	as received	48	46	45	43
Monel	annealed	50	—	32	30
Niobium + 25 Zr	as received	40	28	—	25 (11° K)

Table I

Thermal conductivity in mW cm^{-1} deg^{-1} *for some technical alloys, glasses, and plastics* (*see* § 12.4 *for sources of data*)

	0·1° K	0·4	1	4	10	40	80	150	300° K
Al 5083 (*ca.* 4·5 Mg, 0·7 Mn)	—	—	7	30	82	340	560	800	1200
Al 6063 (*ca.* 0·6 Mg, 0·4 Si)	—	—	85	350	870	2700	2300	2000	2000
Brass (70 Cu, 30 Zn)	0·6	2·5	7	30	100	375	650	850	1200
Brass (70 Cu, 27 Zn, 3 Pb)	—	—	—	30	90	350	550	900	1100
Cu + 30 Ni	0·06	0·3	0·9	5	20	~120	~200	~250	300
Constantan	0·06	0·23	1	8	35	140	180	200	230
German silver	—	—	—	7	28	130	170	180	220
Gold + 2·1 Co	—	—	2	10	40	135	200	—	—
Inconel	—	—	0·5	4	15	70	100	125	140
Manganin	—	—	0·6	5	21	75	130	160	220
Silicon bronze (96 Cu, 3 Si, 1 Mn)	—	—	—	—	15	69	140	—	250
Soft solder (60 Sn, 40 Pb)	—	—	—	160	425	525	525	—	~500
Stainless steels (18/8)	—	—	—	2·5	7	46	80	110	150
Wood's metal	—	—	—	4·0	120	200	230	—	—
Nylon	—	0·006	$0·02_5$	$0·12_5$	$0·3_9$	—	—	—	—
Perspex	—	—	0·21	0·56	0·62	—	—	—	—
Plexiglass	—	—	0·21	0·50	—	—	—	—	—
Polystyrene	—	—	0·12	0·26	—	—	—	—	—
Pyrex	0·003	0·03	0·12	—	—	—	4·8	7·6	11
Soft glass	—	—	0·15	1·15	1·9	2·6	4·6	—	—
Teflon	0·0002	0·004	0·04	$0·4_5$	$0·9_5$	$1·9_6$	2·3	—	—
Vitreous silica	—	—	—	1	1·2	2·5	4·8	8·0	14

Cryogenic equipment

The list of suppliers below is arranged under chapter headings so that, for example, firms manufacturing level indicators or transfer siphons for liquefied gases are listed under Chapter II: Storage and Transfer of Liquefied Gases. The list is not intended to be exhaustive but should include many of the major producers (in 1967) of *cryogenic* equipment in North America and Western Europe. Inevitably such a list soon becomes incomplete or in error as new firms arise and others are taken over by the larger concerns.

In Britain the Science Research Council has compiled a list of U.K. suppliers with some European and American firms included. This list in booklet form is called *Cryogenic equipment* and new editions have been issued periodically from the Science Research Council (State House, High Holborn, London, W.C.1). Another valuable source of information will be *Cryogenic laboratory equipment* by A. J. Croft and P. V. E. McClintock (Plenum Press, London, to be published). Finally, the *Cryogenic information report*, published monthly by Technical Economics Associates, Estes Park, Colo., lists many new cryogenic items that are produced industrially.

Chapter I. *Production of Low Temperatures*

Liquefiers and/or refrigerators

Air Products and Chemicals Inc., P.O. Box 538, Allentown, Pa. 18105, U.S.A.

British Oxygen Cryoproducts, Deer Park Rd., London, S.W.19, U.K.

Cryosystems Ltd., 40 Broadway, London, S.W.1, U.K.

Leybold's Nachfolger, 5 Köln-Bayental, West Germany.

Linde's Eismaschinen A.G., Hoellriegelskreuth, near Munich, West Germany.

A. D. Little Inc., 30 Memorial Drive, Cambridge, Mass. 02142, U.S.A.

Philips, Scientific Equipment Dept., Eindhoven, Netherlands.

T.B.T., Boîte Postale 10, Sassenage, Isère, France.

Chapter II. *Storage and Transfer of Liquefied Gases*

Dewars, level indicators, level controllers

Andonian Associates, 26 Thayer Rd., Waltham, Mass. 02154, U.S.A.
British Oxygen Cryoproducts, Deer Park Rd., London, S.W.19, U.K.
Cryenco Engineering Co., 200 West 48th Ave., Denver, Colo. 89216, U.S.A.
Cryosystems Ltd., 40 Broadway, London, S.W.1, U.K.
Cryotronics Inc., High Bridge, N.J. 98829, U.S.A.
Gardner Cryogenics Corp., 150 William St., Highstown, N.J. 08520, U.S.A.
Guest and Chrimes Ltd., P.O. Box 9, Rotherham, Yorks, U.K.
Hofman Labs. Inc., 5 Evans Terminal, Hillside, N.J. 07205, U.S.A.
Janis Research Company, 21 Spence St., Stoneham, Mass. 02180, U.S.A.
Kelvin Vessels Ltd., 23 Broadwall, London, S.E.1, U.K.
Linde Company, Div. of Union Carbide Corp., 270 Park Ave., New York, N.Y. 10013, U.S.A.
Oxford Cryogenics Ltd., Osney Mead, Oxford, U.K.
Spembly Technical Products, New Road Ave., Chatham, Kent, U.K.
Standard Air Co., 78 Gardner Ave., Brooklyn, N.Y. 11237, U.S.A.
Sulfrian Cryogenics Inc., 1290 Central Ave., Hillside, N.J. 07205, U.S.A.
Superior Air Products Co., 132 Malvern St., Newark, N.J. 07105, U.S.A.
T.B.T., Boîte Postale 10, Sassenage, Isère, France.

Chapter IV. *Temperature Measurement*

Thermometer wire

Engelhard Industries, Baker Platinum Div., 113 Astor St., Newark, N.J. 07114, U.S.A.
Johnson Matthey & Co. Ltd., 73 Hatton Garden, London, E.C.1, U.K.
Leeds & Northrup Co., 4907 Stenton Ave., Philadelphia, Pa. 19144, U.S.A.
Sigmund Cohn Corp., 125 So. Columbus Ave., Mount Vernon, N.Y. 10553, U.S.A.
ThermoElectric Co. Inc., Saddle Brook, N.J. 07663, U.S.A.

Germanium thermometers

CryoCal, Inc., 1371 Avenue E., Riviera Beach, Florida 33404, U.S.A.
Honeywell Inc., Philadelphia Div., 1100 Virginia Drive, Fort Washington, Pa. 19034, U.S.A.

Radiation Research Corp., 312 Florida Ave., West Palm Beach, Florida 33401, U.S.A.

Solitron Devices, Inc., 1177 Blue Heron Blvd., Riviera Beach, Florida 33404, U.S.A.

Texas Instrument Inc., P.O. Box 66027, Houston, Texas 77006, U.S.A.

Platinum thermometers

Degussa, Weissfrauenstr. 9, Frankfurt/Main, West Germany.

Hartman and Braun A.G., Falkstr. 5, Frankfurt/Main, West Germany.

Leeds & Northrup Co., 4908 Stenton Ave., Philadelphia, Pa. 19144, U.S.A.

Minco Products Inc., 740 Washington Ave. North, Minneapolis, Minn. 55401, U.S.A.

Research & Engineering Controls Ltd. (Rosemount Engineering Co.), Durban Rd., Bognor Regis, Sussex, U.K.

Rosemount Engineering Co., 4896 W. 78th St., Minneapolis, Minn. 55424, U.S.A.

Sensing Devices Inc., 333 Culver Boulevarde, Playa del Reg. Calif. 90291, U.S.A.

Tinsley & Co. Ltd., Werndee Hall, South Norwood, London, S.E. 25, U.K.

Chapter VI. *Introduction to Cryostat Design*

Cryostat dewars

See following firms listed for Chapter II: Andonian, B.O.C., Hofman, Janis, Oxford Cryogenics, Spembly, Sulfrian, T.B.T.

Also:

Texas Instruments Ltd., P.O. Box 66027, Houston, Texas 77006, U.S.A.

H. S. Martin, 1916 Greenleaf St., Evanston, Ill. 60202, U.S.A. (glass dewars).

Chapter IX. *Adiabatic Demagnetization*

Magnets

Avco-Everett Research Laboratory, 2385 Revere Beach Parkway, Everett, Mass. 02149, U.S.A.

Cryosystems Ltd., 40 Broadway, London, S.W.1, U.K.

International Research & Development Co. Ltd., Fossway, Newcastle-upon-Tyne 6, U.K.

Magnion Incorporated, 195 Albany St., Cambridge, Mass. 02139, U.S.A.

Mullard Ltd., Mullard House, Torrington Place, London, W.C.1, U.K.

Newport Instruments Ltd., Newport Pagnell, Bucks, U.K.

Oxford Instrument Co., Osney Mead, Oxford, U.K.

Varian Associates Inc., 611 Hansen Way, Palo Alto, Calif. 94304, U.S.A.

Westinghouse Electric Corporation, P.O. Box 868, Pittsburgh, Pa. 15230, U.S.A.

Wires and/or tape for supermagnets

Radio Corporation of America, Harrison, N.J. 07029, U.S.A.

Supercon Div. of National Research Corp., 9 Erie Drive, Natick, Mass. 01762, U.S.A.

Westinghouse Electric Corporation, P.O. Box 868, Pittsburgh, Pa. 15230, U.S.A.

AUTHOR INDEX

SUBJECT INDEX